ANIMAL SCIENCE, ISSUES AND PROFESSIONS

BEES

BIOLOGY, THREATS AND COLONIES

Animal Science, Issues and Professions

Additional books in this series can be found on Nova's website under the Series tab.

Additional E-books in this series can be found on Nova's website under the E-books tab.

Insects and Other Terrestrial Arthropods: Biology, Chemistry and Behavior

Additional books in this series can be found on Nova's website under the Series tab.

Additional E-books in this series can be found on Nova's website under the E-books tab.

ANIMAL SCIENCE, ISSUES AND PROFESSIONS

BEES

BIOLOGY, THREATS AND COLONIES

RICHARD M. FLORIO
EDITOR

Nova Science Publishers, Inc.
New York

For permission to use material from this book please contact us:
Telephone 631-231-7269; Fax 631-231-8175
Web Site: http://www.novapublishers.com

Library of Congress Cataloging-in-Publication Data

Bees : biology, threats, and colonies / editor, Richard M. Florio.
p. cm.
Includes bibliographical references and index.
ISBN 978-1-61324-825-6 (hardcover : alk. paper) 1. Honeybee. 2. Honeybee--Diseases. I. Florio, Richard M.
QL568.A6B384 2011
595.79'9--dc23
2011017612

Published by Nova Science Publishers, Inc. † New York

CONTENTS

PREFACE

For students of animal behavior, honey bees are an intriguing organism, interacting in a complex eusocial colony setting as well as with the environment as they forage over wide areas. Much of that behavior is moderated by odors, which honey bees can detect at extremely low concentrations. This book presents current research from across the globe in the study of bees, including the importance of odor in learning and behavior of the honeybee; the role of honeybees in pollination ecology; threats to the stingless bee in the Brazilian Amazon; honeybee viruses and age-related associative and non-associative learning performance in honeybees

Chapter 1 - Honeybees play an important ecological role as pollinators of many plant species, and their products are the basis for a multi-million dollar commercial industry in the US and more than a thousand million Bath in Thailand. This chapter provides a summary of the natural history of Thai honeybees. The authors focus on the role of Thai honeybees in pollination ecology, potential threats to honeybees, and commercial applications of products derived from honeybees. This chapter covers how honeybees reproduce, variation in caste development among Thai species, and how sex determines the division of labor in these populations. The discovery of the oldest bee species and the evolution of honeybees also will be discussed. In addition to behavior related to nesting and colony defense. The authors will also examine honeybee pheromones: how honeybees produce pheromones, how they detect pheromones and other odorants, and how they respond after exposure to pheromones. Finally, the authors will focus on the role of parasites (i.e., wax moth, mites), predators and pathogens on the ecology of Thai honeybees, and explore the consequences for honeybee commercial product and pollination.

There are seven sections in this chapter on Thai honeybees, 1) natural history, evolution and taxonomy; 2) castes, development, and age polyethism; 3) anatomy, including the structure of the pheromone glands and exocrine glands; 4) olfaction, odor production, odor perception and pheromones, 5) pollination, 6) beekeeping, and 7) honeybee pathogens, parasites and predators. The authors goal is not to exhaustively discuss each topic, but provide relevant information about native species of Thai honeybees in regional context. Much more information is known about the European honeybee (*Apis mellifera*) than other species of *Apis*. In some of these sections, the authors will therefore be necessarily brief.

Chapter 2 - Asian continent is blessed with eight indigenous and one exotic honeybee species. Among these, five species are nesting in cavities and construct multiple parallel combs. In contrast, remaining four are open-nesting species which build single exposed combs. Asian honeybees exhibit a rich biodiversity in respect of distribution, nesting behavior, sex pheromone communication, defensive behavior etc. Honeybees have got an immense value for their various products and also for meticulous services they render in cross pollination of varieties of crops. Thai sac brood, a major viral disease is primarily responsible for mortality of millions of *Apis cerana* colonies. Frequent incidence and infestation of brood mites, *Varroa destructor* and *Tropilaelaps clareae,* predation by various species of ants, wasps and birds render a greater hindrance in establishment of *Apis mellifera* in most parts of Asia. Similarly, traditional methods of honey harvesting, habitat destruction, deforestation and pesticide poisoning are apparently threatening open-nesting and hive honeybee species. There is a greater scope to safeguard honeybees from pests, predators, parasites and diseases through economically viable and environmentally sustainable bee-friendly scientific methods. Obviously, creating awareness on anthropogenic impact on these bee fauna would play a tremendous role in conservation of honeybees in Asia.

Chapter 3 - This chapter presents an overview of the literature on the importance of odor in learning and behavior of the honey bee. The authors begin by illustrating the importance of odor in complex odor environments, and move on to describing the behavior of crop-attached and naïve recruit foragers. Information is also provided on the neural mechanisms of odor detection, limitations of cognitive interpretations of odor learning, and the role of odor perception in repellents and pesticides.

Chapter 4 - Deforestation and nest destruction by "professional" honey gatherers has drastically reduced the population of stingless bees in the Amazon, especially species of the *Melipona* genera, which are very good honey producers among the native stingless bees group. Here, the authors present some of the peculiar aspects of *Melipona* nesting biology, its ecology and its puzzling sex and caste determining system and show how these aspects are correlated with the rapidly growing rates of nest extinction by human activity. Conservation efforts are needed to promote more knowledge of the *Melipona* species and its biology as well as to enhance the beekeeping of stingless bees (Meliponiculture) in many communities and villages throughout the Brazilian Amazon in time to slow down nest destruction. The authors discuss issues and results from cryptic species as well as the logistical challenges of doing research in the Brazilian Amazon. Finally, the authors present suggestions for the development of appropriate and advanced technical training for beekeepers that can be used to strengthen and expand a rational way of exploiting the honey from stingless bees in the region.

Chapter 5 - The Chronic bee paralysis virus (CBPV), Acute bee paralysis virus (ABPV), Israeli acute paralysis virus (IAPV), Deformed wing virus (DWV), Kashmir bee virus (KBV), Black queen cell virus (BQCV) and Sacbrood virus (SBV) are considered to be able to cause severe disease in honeybees and hence they have been objects of active research currently.

In this chapter, information about the geographical spread of the virus, its genome (briefly) and close relatives, the transmission pathways in nature and especially the significant relationships of the virus with other bee enemies are presented. Tissue tropism along with the symptoms are also noted. Old and modern methods of bee viruses diagnosis are quoted. The recent case of colony depopulation in Greece and the first detection of viruses in Greek adult bee populations with the use of molecular techniques is presented separately.

Chapter 6 - In recent years massive honeybee colony losses have been reported mainly in the north hemisphere, including North America and several European countries. Scientists and beekeepers from all over the world are working to elucidate the causes of this phenomenon. Among different factors potential involved, the presence of different pathogens such as the mite Varroa destructor, the microsporidium Nosema ceranae and the presence of different RNA viruses may have a role in honeybee losses. It has been proposed that these events cannot be due to a single factor and a combination of causes should be considered.

In Uruguay (South America) there are about 4,000 beekeepers and more than 400,000 beehives. The most important pathogens of honeybees are present and widely distributed in the country.

V. destructor is found along the country, but their virulence varies from region to region. Almost every beekeeper applies acaricides on the end of the summer or the beginning of the fall, owing to avoid colony loses. The acaricides fluvalinate, flumetrin, amitraz and coumaphos are very efficient, except for some regions where resistant mites to some of these molecules have been detected.

N. ceranae is present in Uruguay before 1990 and nowadays is the main species detected of *Nosema* spp.. However, there have not been reports of colony loses whose cause was without doubts nosemosis. This is confirmed by thousands of colonies that are moved to Eucalyptus grandis plantations in the north of the country at the end of the summer, where every colony suffers severe infections with Nosema spp. and even in that way, reach the following spring without difficulties.

On the other side, the presence of acute bee paralysis virus (ABPV), chronic bee paralysis virus (CBPV), black queen cell virus (BQCV), sacbrood virus (SBV) and deformed wing virus (DWV) have been detected in Uruguay since 2005. Their detection in different geographic regions, the simultaneous co-infection of colonies by several viruses, the high prevalence found in the country and a high incidence in different seasons indicates that they are widely spread. Interestingly, the seasonal incidence of DWV around the year was similar to the incidence of V. destructor, according with previous reports. Kashmir virus and Israeli Acute paralysis virus, other honeybee viruses potentially related to honeybee looses have not been detected so far.

The presence and co-existence of these pathogens are maybe involved in the reduction of colonies production detected during the last years. However, massive honeybee looses have not been reported. In the present chapter the authors present in detail the sanitary situation of Uruguay focussed on the presence of *V. destructor*, *N. ceranae* and RNA viruses and their prevalence and coexistence along the year, and discuss possible reasons why honeybee looses episodes have not been detected

Chapter 7 - The biogenic amines include the three catecholamines (dopamine, norepinephrine, and epinephrine), and the tyrosine metabolites (octopamine and tyramine). An ester of acetic acid and choline (acetylcholine), indoleamine (serotonin), and imidazoleamine (histamine) are also considered as biogenic amines that play important roles in brain functions. Biogenic amines regulate many functions in the brain, including endocrine secretion, cognitive function, aggression, emotional states, motivation, reward circuitry, and learning and memory. Any disruption in the biogenic amines-mediated signaling alters the levels of second messengers, therefore impairs various normal physiological functions. This chapter focuses on biogenic amines-based pesticides (such as acetylcholine mimics and octopamine mimics) that disrupt biogenic amines-mediated signaling in honeybees, impairing their olfactory learning and memory. European honeybees *Apis mellifera* are responsible for the pollination of fruits, vegetables and nuts that accounts ~1/3 of the United States crop. In addition to pollination, honeybee products (honey, propolis, royal jelly and bee venom) are known to have many therapeutic uses. A continuous decline in western honeybee colonies "Colony Collapse Disorder" (CCD) in Europe and in North America is lately drawing a lot of attention due to pollination crisis. If the problem of CCD is not resolved soon enough then this could have a major impact on food industry, affecting United State's economy a big time. The main objective of this review is to discuss the significance of biogenic amines-mediated signaling in relation to olfactory learning and memory in honeybees and shed the light on a potential molecular mechanism underlying CCD.

Chapter 8 - Honeybee learning studies often utilize the well known proboscis extension reflex(PER) technique, a method that is known to be influenced by a honeybee's age. In this study, the authors investigated whether the PER technique may be linked to some of the discrepancies that have been previously noted between associative and non-associate learning performance in honeybees. Whilst associative learning has been shown to improve with age, non associative learning, such as habituation, has shown an apparent decline in performance with age. Here the authors investigated changes in the sugar elicited PER threshold using 6 different sugars to evaluate the relative value of a given sugar solution as a bee ages. The authors results revealed an interesting switch in sugar response between the ages of 7 and 8 days when sucrose or fructose was presented. The authors found that the sucrose threshold decreased suddenly between 7 and 8 days of age and response to fructose ceased at day 8. The same pattern of results was found when comparing the fructose response's of pollen and nectar foragers. Further results obtained with nectar and pollen foragers suggest that this was not due to a cessation of the perception of fructose.

Our sucrose results explain the apparent discrepancies that have been observed between associative and non associative learning performance in honeybees. The authors suggest that in associative learning the perceived value of the sugar reward increases with age and so learning only appears to improve. Conversely, in non associative learning the stimulus value increases with age, and subsequently habituation of the response is harder to achieve. The authors discuss the importance of perspective in interpreting behavioural data, especially when conducting age related experiments in honeybees.

Chapter 9 - Formation and functioning of aggregations of humans, animals, birds, and also development of many geophysical processes is attended with generation of a quasistationary acoustic noise, described by different mathematical methods in accordance with its temporal, spectral and spatial structure. In the majority of cases the noise is regarded as interference, but not at all seldom it is used as a source of information. For example, by the structure of acoustic noise it is possible in general features to determine the mood of human crowds , the state of families or clusters of social insects, while by the radio emission of stars and other cosmic formations a study is made of the physical processes ongoing there.

For analysis of nonstationary acoustic processes, use is traditionally made of spectral-correlation analysis. However, it is ineffective if the temporal scale of the evolving nonstationarity is smaller than the duration of the analyzed noise process, because in this case there is averaging of the spectrum of fluctuation power over the entire time interval of observation of the possible detected signal. As a rule, spectral analysis is conducted on a data sample of large extension. This in a significant measure excludes the contribution of separate local regions of nonstationarity into the resulting Fourier spectrum of the signal. But in many analyzed time series it is the local changes in the properties of investigated noises on small time scale that can contain useful information. In addition upon a statistical approach to analysis of acoustic processes the dynamic entity that gives birth to them is, as a rule, not traced.

In recent time still broader application is found by special spectral methods oriented onto analysis of stationary processes. Of such methods the greatest distribution has been gained by window Fourier analysis, wavelet analysis and polyspectral analysis. Application of algorithms with sliding windows permits substantially increasing the resolving power of analysis in the time domain with retention of a high enough resolution in the frequency domain. But this is coupled with a significant increase of the computation volume and accordingly with an increase of time expenditures in analysis. Besides, three-dimensional (frequency-amplitude-time) surface images are too complex for formal recognition.

The goal of the present work included development of a noninvasive method of investigation of nonstationary acoustic noises, not introducing uncontrolled error and ensuring simplification of interpretation on the basis of visualization and statistical analysis. The method has been approved in studying the connections between the changes in the physiological state of bees and the components of sounds generated by them, giving rise to the sound noise (background) of a bee family. The presence of these connections was established earlier by methods of spectral analysis.

Chapter 10 - Certain cyclicity in thermoregulatory activity under constant environmental conditions is characteristic, to different degrees, of both poikilothermal and homoio-thermal organisms. In poikilotherms, a high variability of body temperature is related to changes in their loco-motor activity. When this activity decreases, the body temperature of a poikilotherm approaches the ambient temperature. The influence of loco-motor activity on body temperature in homoiothermal animals is lower than in poikilotherms, but variation in their body temperature obviously tends to increase with a decrease in body weight.

When at rest, endogenous rhythmic changes of body temperature in homoiothermal animals are maintained by a number of hierarchically interrelated oscillators differing in their ability to generate autonomous oscillations. In vertebrates, the role of the leading oscillator may be played by the suprachiasmatic nuclei of specialized cells in the central nervous system (Minnis and Pit-tendrigh, 1968). Unlike a single organism, consolidated aggregations of insects (families or isolated groups) have no coordinating center that could play the role of this oscillator (Es'kov, 1992, 1995, 2003). Every individual in a family of social insects has all the properties of an integral organism, but none of the individuals in social insect families can survive for long being out of touch with the family. Consolidation, interrelation, and interdependence of family members are especially high in bees of the genus *Apis* (Es'kov, 1995), which makes them a convenient model for studying the nature of periodic oscillations in complex systems.

The purpose of this study was to analyze the nature of rhythms in the thermoregulatory activity of honeybees *Apis mellifera* L., which stand out among social insects due to the perfect mechanisms of temperature regulation within the nest.

Chapter 11 - The body temperature of an individual honeybee, as in other poikilotherms, is highly dependent on the ambient temperature. However, bee clusters and families actively regulate their temperature; the most precise control is maintained over the brood. Whatever the physiological state of the family, its size, and the ecological situation, the averaged brood temperature varies by fractions of a degree, and only in some extreme cases may deviate more than 1 C. Lacking any centralized mechanism of thermal control, bee aggregations successfully adjust heat production and heat loss in the brood zone through activities of adult worker bees.

Clearly, such regulatory activity requires some inborn program, but it is unknown how it is launched, what signals from the brood may urge the worker bees to provide heat or stop doing it. This work is an attempt to shed some light on these questions.

Chapter 12 - The amplitude and temporal structure of electrical oscillations related to the functioning of the heart (ECG) in insects varies over a wide range depending on their physiological state and ecological situation. In wasps and bumblebees, specific changes of ECG were observed under hypo- and hyperthermia. This study deals with the effects of temperature on ECG in the honeybee, which differs from bumblebees and wasps in taxonomic position, level of social organization, and adaptations to the thermal factor.

Worker bees, drones, and queens of Apis mellifera L. were used in the study. The ECG was recorded with an N-338 analog recorder. The input resistance of the preamplifier for a range of up to 5 kHz was not less than 3 gΩ, and the noise level at the input was not less than 50 μV. The active electrolytically sharpened tungsten electrode was introduced into the cervical articulation, and the indifferent electrode was inserted into the abdominal region. The bees were preliminary immobilized by injecting 10% urethane solution into the abdominal cavity.

The experimental insects were placed in a thermo-stated shielded chamber. For controlling insect body temperature, the microsensor of an electronic thermometer (ST11-19 microthermistor) was placed 1-2 cm away from the insect. Hypothermia was usually limited to cooling up to a state of profound cold torpidity. In some experiments, exposure to hyper- and hypothermia reached lethal values.

Chapter 13 - Honeybee as an insect is classed with poikilotherms. However, poikilothermy is typical only of an individual bee outside the family. Bee colonies in moderate and cold climate can winter at ambient temperatures down to -35…-45°C. This is far beyond the individual cold tolerance, which is determined by the freezing temperature of body liquids, -7…-16°C in wintering bees. The ability of the species to survive the long winter is based on a complex of hereditary programmed eto-physiological reactions stimulated by cold. Protection against life-endangering chilling is attained through formation of ordered clusters that ensure accumulation and rational use of heat.

The present work undertakes mathematical modeling of thermoregulation mechanisms and temperature field distribution in bee clusters, which is topical for understanding the principles of self-organization in complex biological systems.

Chapter 14 - The colony of honey bees, as well as other species of the genus *Apis,* is a consolidated group of related and mutually dependent organisms which cannot survive and reproduce outside it. In the case of eusocial species, natural selection and other factors of evolution act on colonies rather than on individuals. Reproduction and dispersal of bee colonies proceeds by part (approximately half) of worker bees leaving the parental colony together with a fertilized or unfertilized female (queen). Beekeepers usually call this process "swarming" but this name does not properly reflect the phenomenon and is inconsistent with the strict entomological use of this term. Therefore, the process of fission of a bee colony will be referred to as sociotomy.

Preparation for sociotomy may take from several days to several weeks. During this period many bees keep from replenishing the food reserves, building combs, feeding the brood, etc., which leads to decreasing production of the colony. Since the bees leaving the parental colony tend to settle as far from it as possible, the beekeeper should be able to recognize the colonies about to undergo fission, and to estimate the timing of this process.

Chapter 15 - Although the honey bee, as a representative of the class of insects, belongs to the poikilotherm animals, it can actively regulate the temperature inside the nest, which is usually the most stable in the zones occupied by the brood. However, outside the nest the bee responds to changes in the external temperature as a typical poikilotherm.

The overcoming by a honey bee long wintering in conditions of moderate and cold climate is based on using a large complex of individual and social adaptations. Their acquisition is related to the evolution of sociality. In this direction in the species phylogenesis adaptations providing forage supplies accumulation, their economical utilization during wintering and inner-nest temperature regulation have appeared to be of dominating importance.

The honey bee colonies overwinter in the state of low activity and maintain a relatively high temperature in their nests. Data on the thermal regime in the nests of wintering bees were mostly obtained by various sensors (thermoelectric couples, thermal resistors) installed in the hives. Such methods, however, do not allow one to analyze the entire system of thermal processes taking place in the nest, much less to measure the temperature of individual bees, follow their responses to temperature dynamics, and determine their contribution to regulation of thermal processes.

In this work the role of fat accumulations gained by bees in the period of preparation to wintering and changing of the maximum overcooling temperature (MOT) of the bee's liquid fractions during their wintering are considered. The task of the research was also to study temperature fluctuations dynamics in the nests of wintering bees. However, the influence of external temperature on the thermal processes in the nests of wintering bee colonies was studied by analysis of infrared radiation. In addition, the authors analyzed the response to cooling in individual bees located in different inter-comb spaces.

In: Bees
Editor: Richard M. Florio

ISBN: 978-1-61324-825-6

Chapter 1

BIOLOGY OF THAI HONEYBEES: NATURAL HISTORY AND THREATS

Guntima Suwannapong[1], Mark Eric Benbow[2] and James C. Nieh[3]

[1]Department of Biology, Faculty of Science, Burapha University Chon Buri, Thailand
[2]Department of Biology, University of Dayton, College Park, OH, US
[3]Section of Ecology, Behavior and Evolution, Division of Biological Sciences, University of California, San Diego, US

Abstract

Honeybees play an important ecological role as pollinators of many plant species, and their products are the basis for a multi-million dollar commercial industry in the US and more than a thousand million Bath in Thailand. This chapter provides a summary of the natural history of Thai honeybees. We focus on the role of Thai honeybees in pollination ecology, potential threats to honeybees, and commercial applications of products derived from honeybees. This chapter covers how honeybees reproduce, variation in caste development among Thai species, and how sex determines the division of labor in these populations. The discovery of the oldest bee species and the evolution of honeybees also will be discussed. In addition to behavior related to nesting and colony defense. We will also examine honeybee pheromones: how honeybees produce pheromones, how they detect pheromones and other odorants, and how they respond after exposure to pheromones. Finally, we will focus on the role of parasites (i.e., wax moth, mites), predators and pathogens on the ecology of Thai honeybees, and explore the consequences for honeybee commercial product and pollination.

There are seven sections in this chapter on Thai honeybees, 1) natural history, evolution and taxonomy; 2) castes, development, and age polyethism; 3) anatomy, including the structure of the pheromone glands and exocrine glands; 4) olfaction, odor production, odor perception and pheromones, 5) pollination, 6) beekeeping, and 7) honeybee pathogens, parasites and predators. Our goal is not to exhaustively discuss each topic, but provide relevant information about native species of Thai honeybees in regional context. Much more information is known about the European honeybee (*Apis mellifera*) than other species of *Apis*. In some of these sections, we will therefore be necessarily brief.

1. Natural History, Evolution and Taxonomy

Life History of Thai Honeybees

In recent years, interest in tropical bees has increased. This is appropriate because honeybees like originated in Tropical Africa and spread from South Africa to Northern Europe and East into India and China (Otis, 1990). The first bees appear in the fossil record in deposits dating about 40 million years ago in the Eocene. The oldest bee fossil is preserved in a piece of amber found from a mine in northern Burma. It is believed to date back as far as 100 million years to the time when bees and wasps split into two different lineages. The fossilized insect appears to share features both common to the bee and wasp, but is more similar to bees than wasps (Danforth et al., 2006). The earliest known honeybee fossil (genus *Apis*) was found in Europe dating back 35 million years. About 30 million years ago, honeybees appear morphologically very similar to modern honeybees (Koning, 1994). The genus *Apis* is evidently tropical in origin. It is native to Asia, Africa and Europe including such continental islands as Japan, Taiwan and the Philippines (Seeley, 1985). Honeybees did not appear in the Americas, Australia or New Zealand until European settlers introduced them in the 17th century (Zander and Weiss, 1964). Thailand is a tropical country that has a wide variety of flowering plants and animals. Perhaps due to high temperature, the native tropical bees generally build their nest as a single air open nest and rely upon aggressive behavior to defend these exposed nests (Collins and Kubasek, 1982).

Honeybees of the genus *Apis* are the most studied because of their fascinating and complex lifestyle, communication systems (Nieh, 1998; Nieh and Roubik, 1995), role as keystone pollinators of native plants, pollination of agricultural crops, and the valuable hive products that they produce, such as honey, royal jelly, bee wax, bee pollen, propolis and even bee venom. Honeybees belong to the order Hymenoptera, superorder Apocrita, infraorder Acuelata, superfamily Apoidea, family Apidae, subfamily Apinae, tribe Apini. There are more than 11 extant species of *Apis* worldwide (Michener, 2000). Four species are native to Thailand: *Apis andreniformis, A. cerana, A. dorsata and A. florea* (Oldroyd and Wongsiri, 2006). *A. mellifera* was introduced for beekeeping. Honeybees of Thailand are classified as follows:

Kingdom Animalia
Phylum Arthropoda
Class Insecta
Order Hymenoptera
Family Apidae
Genus *Apis*
Species *A. andreniformis*
A. cerana
A. *dorsata*
A. *florea*
A. mellifera (Ruttner, 1988)

Honeybees are hymenopterans, a group that generally feed on pollen and nectar and constitute about 20,000 species throughout the world, known taxonomically as the superfamily Apoidea (Michener, 2000). Although the question of how many honeybee species exist is still debated among taxonomists, at least four species are commonly recognized: the dwarf or midget bee (*A. florea*), the giant or rock bee (*A. dorsata*), the Asian bee (*A. cerana*), and the common European honeybee (*A. mellifera*). The existence of another giant bee (*A. laboriosa*), was recently confirmed in Nepal, but little is known about its biology (Seeley, 1985; Otis, 1990). *Apis* species are classified into two groups, based upon nesting. The first group builds single comb, open-air nests: *A. andreniformis, A. florea, A. dorsata, A. breviligula A. binghami* and *A. laboriosa*. These bees are restricted to the Asian tropics and subtropics. The second group consists of species that nest inside cavities where they build multiple combs: *A. cerana, A. koschevnikovi, A. nigrocincta, A. nuluensis,* and *A. mellifera* (Hepburn and Radloff, 2011; Michener, 2000).

Honeybees that Build Single-Comb, Open-Air Nests

The architectural design of the comb of all honeybee species is essentially similar. It consists of adjoining hexagonal cells made of wax secreted by the workers' wax glands. The bees use these cells to rear their brood and to store their food. The general utilization of comb space is also similar among the species. Honey is stored in the upper part of the comb. Beneath the honey storage area, there are commonly rows of pollen-storage cells, worker-brood cells, and drone-brood cells, respectively. The larger queen cells are normally built at the lower edge of the comb (Seeley, 1985; Wongsiri et al., 1991; Otis, 1990).

The Black Dwarf Honeybee, *Apis Andreniformis* Smith, 1858

Apis andreniformis was reported in Thailand by Wongsiri et al in 1990 from the coastal flats and near the foothill areas (1-100 meter above sea level) of Chantaburi province, Thailand to high mountainous and forest areas (approximately 1600 m altitude) in the northern parts of Thailand (Wongsiri et al., 1996a). This species was rediscovered in South China in the same habitat as a. It was the fifth honeybee species to be described of the eleven known *Apis* species, and its biology and geographic distribution remain relatively poorly understood (Maa, 1953; Tirgari, 1971; Wongsiri et al., 1996a; Wu and Kuang, 1987). Only recently, this species has been diagnostically separated from the closely related *A. florea* because there are sites where both species live conspecifically. Both species are distributed throughout tropical and subtropical Asia, including Southeast China, India, Burma, Laos, Vietnam, Malaysia, Indonesia (Java and Borneo), and the Philippines (Palawan) (Akratanakul, 1976; Maa, 1953; Otis, 1990; Tirgari, 1971; Wongsiri et al., 1996; Wu and Kuang, 1987). In body size, *A. andreniformis* is smaller than *A. florea*.

Morphologically, *A. andreniformis* has stripes of black hairs on the metathoracic tibia and on the dorsolateral (back and side) surface of the metathoracic basitarsus (Wongsiri et al., 1996a). Additionally, the pigmentation of *A. andreniformis* is blackish, while that of *A. florea* is yellowish. Other distinguishing characteristics include a difference in cubital indexes, which is the ratio of two of the wing vein segments of honeybees. The pattern of the veins of the fore wings is specific for each breed of bees. The cubital index is consistent for a given race of bee. It can be used to distinguish between similar populations of honeybees and to determine degrees of hybridization: *A. andreniformis* has an index of 6.37, and that of *A.*

Figure 1. The single open nesting of *Apis andreniformis*.

florea is 2.86. The proboscis of *A. andreniformis* has a length of 2.80 mm, while that of *A. florea* is 3.27 mm. This physical difference may contribute to differences in the types of flowers visited by different species. Finally, there are differences in the barbs of the stinger, and in the basitarsus of the drones (Rinderer et al., 1995; Wongsiri et al., 1996a). *Apis andreniformis* nests in quiet forests, generally in darker areas where there is 25 to 30% of ambient sunlight. The hive is built in the branches of trees or shrubs usually 1-15 m above ground, although the average height is 2.5 m. The nest typically ranges from 70-90 mm in size (Wongsiri et al., 1996a).

Almost nothing is known about the recruitment communication behavior of *A. andreniformis*. However, all studied *Apis* species can waggle dance to communicate food location (Nieh et al., 2003), and thus it is highly likely that *A. andreniformis* also uses this behavior (Dyer and Seeley, 1991; Dyer, 2000; 2002). Given, its similarity with *A. florea*, and the fact that it also builds single-comb, open nests it is reasonable to expect that *A. andreniformis* also waggle dances on exposed comb. Drone "dances" have been observed on the open comb surface of *A. andreniformis* nests (Wongsiri et al., 1996a), although the significance of this behavior is unclear. Further studies of this species' communication abilities are needed.

The Red Dwarf Honeybee, *Apis florea* Fabricius, 1787

As its name implies, the dwarf honeybee is small in body size. This species does well in very hot, arid climates. A nest of *A. florea* consists of a single comb, typically built in small trees. *Apis florea* nests in the open, but nests are camouflaged. Most nests are hung from slender branches of trees or shrubs covered with relatively dense foliage, from 0.3 to 8 m

Figure 2. The single open nesting of *Apis florea* showing ant barriers made of resin (propolis) at the edges of a branch.

above the ground (Hepburn, and Radloff, 2011; Wongsiri et al., 1996a). In Oman, where *A. florea* nests are frequently found in caves, such combs lack the crest that is the honey storage area and that surrounds the branch on which the comb is suspended (Akratanakul, 1976). *Apis florea* is generally distributed throughout Thailand and it is an economically important species as well as important to crop and wild plant pollination. The single comb nest contains cells of four sizes. The large storage cells for the honey are very deep and constructed in such a manner that the comb bulges out on either side and at the top. The small worker cells (2.7-3.1 mm) are located below the honey storage cells. The considerably larger drone cells (4.2-4.8 mm) are mostly found in the lower part of the comb. The pear shape queen cells, which are the largest of the cells, are located near the bottom. These can be observed when a colony loses its queen, and are emergency queen-cells (Ruttner, 1988).

This species applies a sticky resin (propolis) like substance to branches to support its comb and prevent ants and other insects from invading the nest (Akratanakul, 1976; Wongsiri et al., 1996a). The communication dance by scouts, announcing the discovery of a food source, also takes place on the platform of honey storage area and is a classic "figure eight" waggle dance However, unlike the cavity-nest species, *A. florea* foragers dance upon the relatively flat upper comb area above the branch supporting the nest. Because they dance upon this horizontal surface, foragers orient towards celestial cues rather than to gravity or towards landmarks if the celestial cues are unavailable (Dyer, 2002). As a result, these foragers dance with the waggle phase pointing directly at the indicated resource, with visual cues (celestial or landmark) used to correctly orient the waggle phase (Lindauer, 1956; 1961).

The Giant Honeybee, *Apis dorsata* Fabricius, 1798

This species has the largest individual body size of all honeybees (Michener, 2000). Interestingly, queen, workers, and drones of this species are all produced in cells similar in size and shape (Richards, 1953). The average cell diameter ranges between 5.42-6.35 mm (Dietz, 1992; Graham, 1992). The nest is made as a single comb about 1-2 meters long, approximately 0.5 m high on thick branches (20-40 cm diameter to support comb weight) on the upper parts of large trees that are 30-60 meters high. Other preferred nesting sites include overhanging rocks, cliffs, or cavities of large buildings (Lindauer, 1961). In general, *A. dorsata* may occur singly or several nests are formed aggregately, usually 20-50 nests in a single tree in the forests of Thailand.

Some populations of this species are very aggressive (Wongsiri et al., 1990; 1996b). About three-quarters of the worker population of a colony of giant honeybees is engaged in colony defense, forming a protective curtain that is three to four bees thick, similar to *A. florea*. While birds are common predators of *A. dorsata*, they are evidently well protected against ant invasion. Thus, the sticky bands of propolis around the nests of the dwarf honeybee are not found surrounding the nests of *A. dorsata*, nor are the nests hidden by dense foliage (Akaratanakul, 1976).

In Thailand, *A. dorsata* has relatively high levels of genetic diversity. There is a lack of genetic population differentiation between *A. dorsata* originating from geographically different regions when using microsatellite polymorphisms (Insuan et al., 2007). However, there are significant genetic differences between bees from the north-to-central region (north, northeast, and central regions), peninsular Thailand, and Samui Island (Insuan et al., 2007).

The range of the giant honeybee is similar to that of the dwarf honeybee. It occurs from Pakistan (and, perhaps, parts of southern Afghanistan) in the west, through the Indian subcontinent and Sri Lanka to Indonesia and parts of the Philippines in the east. Its north-south distribution spans southern China to Indonesia; it is not found in New Guinea or Australia. In Thailand this species is particularly found in forested areas with a large variety of wild plant species. The organization of the comb is similar to that in the other honeybee species: honey storage at the top, followed by pollen storage, worker brood and drone brood (Hepburn, and Radloff, 2011; Otis, 1990; Wongsiri et al., 1996b).

At the lower part of the nest is the colony's active area, known as the mouth, where workers take off and land and where communication dances by scouts announcing the discovery of food sources take place (Akaratanakul, 1976). This dance takes place on the vertical surface of the comb, and during its progress, the bees must have a clear view of the sky to observe the exact location of the sun. Unlike other species of honeybees, *A. dorsata* has the unusual habit of continuing foraging after sunset on bright moonlit nights (Wongsiri et al., 1991), a nocturnal ability that is not reported in other honeybee species (Suwannapong and Wongsiri, 1999, Suwannapong et al., 2010a). *Apis dorsata* foragers can fly at night in part because of large concentrations of visual pigment in the retinular cells of their ommatidia (Suwannapong and Wongsiri, 1999).

Figure 3. The aggregated single open nesting of *Apis dorsata*.

Honeybees that Build Multi-Comb Nests in Enclosed Cavities

The Asiatic Hive Honeybee, *Apis cerana* Fabricius, 1793

The Asiatic hive honeybee, *A. cerana* is widespread in temperate and tropical Asia (Smith et al., 2000). The range for this species is greater than that of *A. florea* and *A. dorsata*. It is found throughout Thailand. There are two subspecies of *A. cerana* in Thailand: *A. cerana cerana* and *A. cerana indica*. *Apis cerana* populations have high genetic diversity in the mainland (north, central, northeast and peninsular Thailand), but limited diversity in the Samui population, implying that genetic drift or founder effects may have occurred in this population (Hepburn et al., 2001; Hepburn, and Radloff, 2011; Otis, 1990; Wongsiri et al., 1996). There are five conspecific populations of *A.* cerana in Thailand. These are assigned to four different genetic groups: north and central region, peninsular Thailand, Samui island and northeast (Sittipraneed et al., 2001).

Apis cerana provides honey, beeswax, and the invaluable service of crop pollination. They tend to swarm, abscond and migrate quite frequently (Akratanakul, 1976; Maa, 1953; Morse and Moch, 1971; Otis, 1990; Richards, 2001; Smith et al., 2000; Wongsiri et al., 1996; Wu and Kuang, 1987). Among the native bees of Asia, *A. cerana* will likely become an increasingly important beekeeping resource. Its nest structure is similar to that of *A. mellifera* because it builds multiple combs inside a nest cavity (Figure 4). It also performs waggle dances inside the nest, usually in the dark, like *A. mellifera* (Lindauer 1956). Because *A.*

Figure 4. The cavity hive of *Apis cerana* in a clay jar.

cerana has not been domesticated to the same extent as *A. mellifera*, it poses some problems for apiculture such as high swarming and absconding rates, a different set of parasites, and more limited honey storage capabilities. However, they are strongly resistant to *Varroa jacobsoni* and predatory wasps (Kerr et al., 1974, Ruttner, 1988; Wongsiri et al., 1996a).

In the wild, these bees construct their nests in dark enclosures such as caves, rock cavities and hollow tree trunks. The normal nesting site is usually close to the ground, not more than 4-5 meters high. The bees' habit of nesting in the dark enables one to keep them in specially constructed vessels. For thousands of years *A. cerana* has been kept in various kinds of hives (clay pots, logs, boxes, wall openings, etc). Despite the relatively recent introduction of Langstroth-frame hives, colonies of *A. cerana* kept in traditional hives are still common in the villages of most Asian countries (Akratanakul, 1976; Hepburn, and Radloff, 2011). As a result, feral nests of *A. cerana* are less hunted by man than nests of dwarf and giant honeybees (Akratanakul, 1976; Maa, 1953; Otis, 1990; Tirgari, 1971; Wongsiri et al., 1996b; Wu and Kuang, 1987). The several combs in an *A. cerana* colony are built parallel to each other, and have a uniform distance known as the "bee space" between them. The body size of *A. cerana* workers is relatively small and there are two sizes of brood comb cells: smaller for worker and larger for drone brood. Queen cells are built on the lower edge of the comb. As in the other *Apis* species, honey is stored in the upper part of the combs, but also found in the outer combs, adjacent to the hive walls (Akratanakul, 1976; Maa, 1953; Otis, 1990; Wongsiri et al., 1996b).

Figure 5. The cavity hive of *Apis cerana* in the hole of a coconut tree trunk.

The European Honeybee, *Apis mellifera* Linnaeus, 1758

Apis mellifera was brought to Thailand for beekeeping about 60 years ago (Suppasat et al., 2007). Three common and five rare composite haplotypes exist among colonies in North, Central, Northeast and South Thailand. This species builds multiple-comb nests in dark cavities (like *A.* cerana), has an intermediate individual body size, and shares a similar social organization and division of labor with other honeybee species (Akratanakul, 1976; Maa, 1953; Otis, 1990; Tirgari, 1971). Beekeeping with *A. mellifera* in Thailand is quite successful. This species is used for honey production and is an integral part of Thai agriculture. It is used for pollination of longan, litchi, durian, rambutan and other crops. Although this species is one of the most studied, efforts over the past few decades to introduce *A. mellifera* into Asia have encountered a number of problems, such as the inter-species transmission of bee pests and diseases (Crane, 1990; Seeley, 1985). In addition, this species needs much more sugar feeding during dearth periods and is highly susceptible cold temperatures (Partap and Verma, 1994; 1998).

This species builds their multiple combs nest in the cavity of the hole that provides protection of the colony against predators, and from climatic changing including rain (Schmidt and Hurley, 1995). However, this species displays the ancestral characteristics of open nesting behavior since they can also nest in the open (Butler, 1975; Butler et al., 1970). This open nesting behavior could indicate a swarming nest that honeybees choose for a temporary place before determining the final cavity nest site (Raffiudin, 2002; Raffiudin and Crozier, 2007).

Figure 6. The multiple comb nest of *Apis mellifera*.

With respect to communication, *A. mellifera* is the best studied of all species of honeybees. All species of honeybees share the ability to "dance," performing repetitive cycling behaviors on the surface of the comb that indicate the presence of a resource outside the nest and, in the case of the waggle dance, the resource location (von Frisch 1967). Resources communicated include pollen, nectar, water, resin (propolis), and nest sites (Dyer 2002). Near the nest (the exact cutoff distance varies with the species and the bee but is generally less than 50 m for *A. mellifera*), bees perform a round dance which consists of the dancer moving in a circle and then periodically making a sharp turn to double back. Far away from the nest, bees perform a repeating waggle dance that consists of a figure-eight pattern in which bees waggle during the center portion. Generally, this occurs for resources greater than 100 m away from the nest (in *A. mellifera*, von Frisch 1967). The waggle portion communicates the distance to the resource and its direction.

There are differences in the coding of distance in different species of honeybees (Lindauer, 1956; Punchihewa et al., 1985) that are thought to be tuned to the foraging range of the species. Thus, species with different foraging ranges should have a different distance codings. Dyer and Seeley (1991) found no significant differences between the distance coding "dialects" of the different Asian species of honeybees, *A. florea, A. cerana,* and *A. dorsata* although, based upon reading dances for natural food sources, there were significant differences in the flight range. The much larger-bodied species, *A. dorsata*, had approximately twice the foraging range (95% of dances were for distances estimated to be less than 3.8 km) of *A. florea* (1.3 km for 95% point) or *A. cerana* (0.9 km for 95% point). The authors point out that observations over a variety of seasons is required for a more robust test of the dance dialect tuning hypothesis (Beekman et al., 2008; Dyer and Seeley, 1991; Lindauer, 1956; Otis, 1991; Punchihewa et al., 1985; Raffiudin and Crozier, 2007; Wongsiri et al., 1996b).

In general, much remains to be learned about the communication of Asian honeybees. For example, drone dances are reported in the dwarf honeybees (*A. andreniformis* and *A. florea*). Drones of *A. andreniformis* are reported to engage in runs on the surface of the colony's protective curtain, running in circular loops (similar to the round dance) with their wings somewhat spread out. A dancing drone can attract another drone to follow its dance and result in both drone followers and the drone flying off together. Drones of *A. florea* are not reported to dance (Rinderer et al., 1995; Wongsiri et al., 1996a). Unfortunately, there have been no subsequent studies of this interesting phenomenon.

2. Honeybee Development, Castes and Age Polytheism

Honeybee Caste and Development

There are three main forms or "castes" of honeybee in every honeybee colony: a single queen, a few hundred drones and several thousand workers (all female). The queen is a fertile, functional female that can produce males and females, the worker is an unfertilized female capable of only producing males (due to the haplodiploid sex determination system found in honeybees) and the drone is male (Tribe and Fletcher, 1977; Winston, 1979). Honeybees undergo complete metamorphosis, and all of the honeybee castes (worker, queen, and drone) pass through the same four stages during their development: egg, larva, pupa and adult. All three castes spend three days for their egg stages (Winston, 1979; 1987; 1992). The larval stages last for different amounts of time, depending caste, genetics, and the environment. The mean duration of the uncapped larval period is about 4.5 days for queen, 5.5 days for workers and 6.5 days for drones. The total developmental times average 16, 21 and 24 days for queen, workers and drones, respectively (Tribe and Fletcher, 1977; Winston, 1979; 1987; 1992).

The Queen: There is generally one queen in honeybee colony. The queen honeybee is effectively the "mother" of the colony. Her main role is to lay eggs. Through the production of queen pheromone, she influences the physiology and behaviour of the workers. Under normal circumstances, the queen is the only egg-laying bee in the colony. If there are enough cells available she will lay up to 2500 a day (Winston, 1992). A queen starts out as a simple fertilized egg that can be laid in a queen cell (for most species, a larger cell specially built for queen-rearing) or laid in a worker cell that is subsequently transformed (by the workers) into a queen cell. To create a queen cell, workers extend one or more existing worker cells in the honeycomb to form "queen cups". These will usually be towards the bottom or edge of the main brood area of the comb. The queen lays an egg in these queen cups and once hatched (3 -3½ days after being laid) the new larva is fed a rich diet of "royal jelly" by the "nurse" bees. Immediately after being laid, there is no difference between an egg that will become a queen or a worker. Whether or not an egg becomes a queen is largely due to worker actions. If the colony is ready to swarm, detects that it has a failing queen, or lost its queen, "house" worker bees will alter how they feed one or more larvae that hatch out. This rich food allows the larvae to grow larger and more quickly than a normal worker (Allen, 1955; Winston, 1979; 1992).

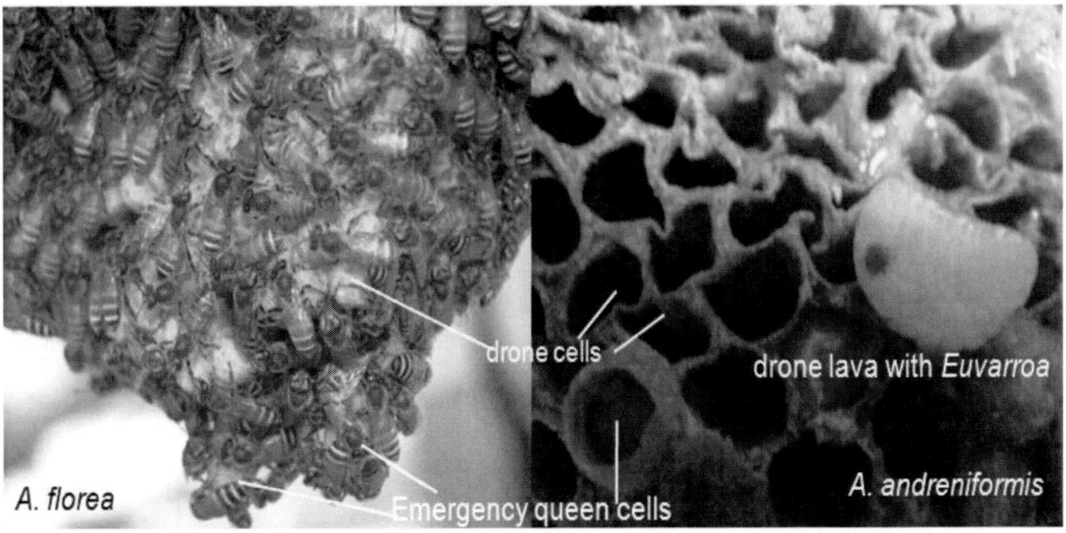

Figure 7. The emergency queen cell of *A. andreniformis* and *A. florea* surrounded by drone cells.

The new queen (known as the "virgin queen") emerges 16 days after the egg was laid. This compares to 21 days for a worker bee or 24 by the drone. If a queen is lost for whatever reason (e.g. is killed accidentally by predators), the colony will resort to making emergency queen cells (Seeley, 1985; Winston, 1987; 1992). Providing there are eggs or very young larvae (up to 3 days old) present, the colony will use these young eggs to raise several new queens. They will build out existing cells and start feeding what would have been ordinary worker larvae with royal jelly. The rich diet fed to the developing queen larva alters the anatomy of the larva (from that of the worker). The abdomen becomes notably longer than that of her sister workers. The larger abdomen accommodates the enlarged reproductive organs, namely the ovaries and spermatheca (Anderson, 1963; Mackensen, 1943; Winston, 1992). When the virgin queen is ready to emerge, one of four fates await her; 1) She will be attacked and possibly killed by another queen that emerged ahead of her, 2) She will seek out and find other emerging queens and kill them (sometimes by stinging through the cell before the victim has emerged), 3) She will be ushered out of the colony by workers to fly off with a number of other bees from the colony (a swarm), or 4) She will be accepted by the colony, the workers of which will immediately start to care for her(Anderson, 1963; Mackensen, 1943; Winston, 1992).

The virgin queen will typically stay in the colony for a few days in order to feed and gain strength and allow her reproductive organs to mature a little further (Mackensen, 1943; Winston, 1992; Woyke, 1963; 1969, 1973). Following this initial "rest period," the virgin queen emerges from the colony and makes a several reconnaissance flights. She may do this to ensure she knows where the hive is and also to find where drones congregate (Anderson, 1963; Mackensen, 1943; Winston, 1987; Woyke, 1969). Due to her large size, the queen is particularly vulnerable to being captured by predator like birds, and therefore avoids long flights (Anderson, 1963; Mackensen, 1943; Winston, 1992; Woyke, 1969, 1973).

When ready, usually any time up to around 21 days from emerging from her cell, the queen will make a "mating flight". She will head straight for the drone congregation site usually in the vicinity of the queen's colony and up to 30 metres from the ground. The queen flies up to the drones and mates repeatedly with several of them. Once her spermatheca is full, she heads back to the colony. She relies on the attendant workers for her every need. They feed her, groom her, and remove her wastes (Mackensen, 1943; Winston, 1987; 1992; Woyke, 1963; 1969). The queen will spend the remainder of her life laying eggs. The number she lays will largely depend on the activities of the colony. Workers control how many cells remain vacant for egg laying (increasing and decreasing the amount of stores of pollen and honey according to seasonal variations) and will control her egg production by varying the amount of food she is fed (Anderson, 1963; Brouwers, 1982; 1983; Mackensen, 1943; Winston, 1992; Woyke, 1973).

The queen produces a pheromone called queen mandibular pheromone. As the workers clean her, they distribute queen substance around the colony. This serves to maintain the colony in a "normal" state. However, as the queen gets older, typically coinciding with a steady reduction in the number of eggs laid, she produces less queen substance (Winston, 1992; Woyke, 1969; 1973). The colony senses this and begins preparation for her ultimate removal. This occurs by a process of supercedure where a new queen is raised and the old one is killed off either by the workers or the new queen. Supercedure is a natural process that will face nearly every queen. Supercedure may occur within the colony or once a colony has swarmed. A prime swarm (the first to leave a colony) will have the incumbent queen. Once the swarm establishes its new home, a new queen is raised and the old queen is killed. Queens occasionally lay fertilized eggs in worker cells, which develop into males called diploid drones (Mackenson, 1943). Male diploid drones are quickly eaten by workers (Woyke, 1963; 1969, 1973).

Workers: The worker bees, as the name implies, do almost all colony tasks. Workers are all sterile females. They carry out almost all the duties that go into building and maintaining a colony such as brood rearing, comb building, house cleansing, foraging, and colony defence (Winstons, 1992; Wongsiri et al., 1996a; 1996b). Workers generally have a smaller body size than either drones or the queen. There are about 8,000-25,000 workers in *A. andreniformis* and *A. florea* colonies, 40,000-50,000 workers in *A. mellifera* colony, 20,000-40.000 workers for *A. cerana* and, 50,000-80,000 for *A. dorsata* (Otis, 1990; Winston, 1992; Wongsiri et al., 1991, 1996). Worker bees, like the queen, possess a sting that is a modified ovipositor or egg laying tube. Despite the presence of reproductive organs, the workers are infertile and lack the reproductive capacity of the queen.

The worker bee starts out as a normal egg laid by the queen. The larva is fed "brood food" which is largely a combination of nectar, pollen and enzymes from the saliva of the "nurse" bees (Anderson, 1963; Mackenson, 1943; Winston, 1992). The pollen provides the necessary protein required for rapid growth. Sugar is the energy source. A worker larva is fed approximately half the quantity of brood food fed to a queen larva. However, there is little difference in the ingredients of the food that is fed to each. Instead, the difference is related to the relative proportions of those ingredients. Research suggests that the queen larva is fed a significantly higher proportion of sugar in its food (royal jelly) and that this extra energy boost given to a larva within its first 3 days promotes an increase in certain growth hormones that cause what would otherwise become a worker to become a queen. Recently, Kamakura (2011) made a major breakthrough and discovered a 57 kDa protein, royalactin, that induces honey bee larvae to become queens. Workers may lay eggs, under certain conditions, which develop into drones since workers never mate and they have no sperm to fertilize their eggs (Anderson, 1963; Mackenson, 1943). However, in a normal queenright colony, worker

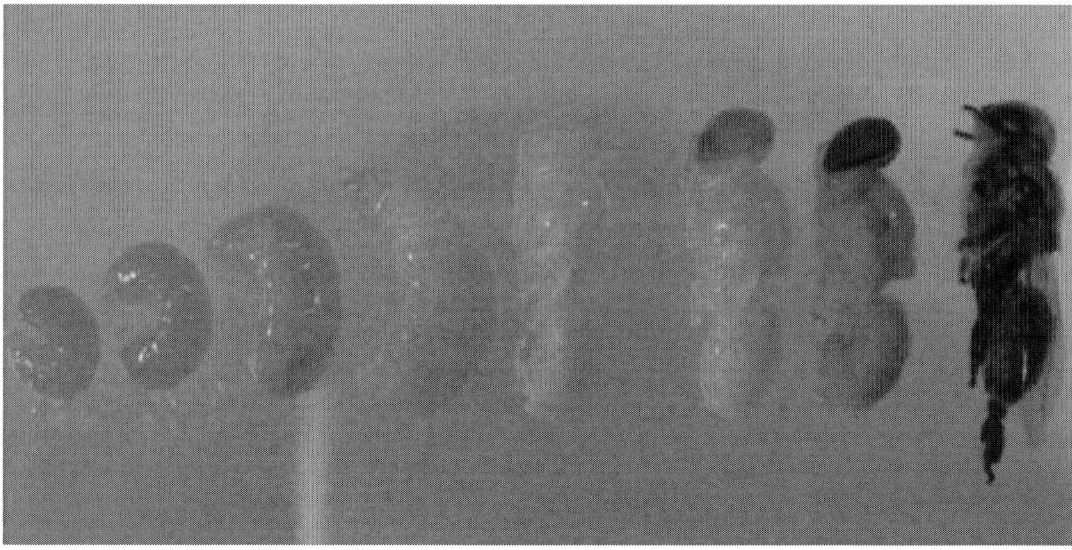

Figure 8. The different developmental stages of *A. florea* workers.

policing occurs and workers consume eggs produced by other workers (Ratnieks, 1993). In *A.* cerana, unlike *A. mellifera*, there can be a relatively large number of laying workers in a queenright colony (Partarp and Verma, 1998).

Drones: Drone bees are males. They do not work in the nest, and die after workers stop feeding them and then drive them from the nest. They seemingly have little or no purpose within the colony. However, drones perform one vital role. Their primary purpose is to mate with virgin queens, contributing their genes to different colonies. The body sizes of drones are larger than workers, but normally smaller than the queen. They are characterised by their large compound eyes, which appear to come together at the top, and rounded stocky bodies. They arise from unfertilised eggs laid by a queen or worker. Drones are "haploid", they only posses one half of the pairs of genes found in the "diploid" workers and queen (Anderson, 1963; Mackenson, 1943; Winston, 1992).

Usually, drone eggs are laid only by the queen as unfertilized eggs. The queen holds the eggs and sperm separately and has the choice as to whether or not an egg is fertilized as it is laid. When the queen lay an egg, she determines whether that egg is being laid in a worker or in a drone cell, likely by inspection of the cell within her forelegs prior to egg laying or by the angle of her abdomen during oviposition (Koeniger, 1979, 1983). The queen will lay drone eggs in drone cells that are larger in diameter in order to accommodate the larger body size of the drone. The queen can detect the difference in cell type and lays worker or drone eggs accordingly. If the egg is going into the drone cell, the queen does not release any sperm: the unfertilized egg is haploid (Winston, 1992). The drone larva is progressively fed by workers until it spins a cocoon within the cell and undergoes metamorphosis, turning into a pupa. The drone emerges as a young adult on day 24. The young drone, being unable to gather food for itself, is fed by the workers (Koeniger, 1969, 1970; Winston, 1992).

After a few days, the drone will begin to carry out reconnaissance flights. Their ability to navigate is probably the best of all three castes because they typically have around 75-80% more facets in their compound eyes than the workers or queen (Gary, 1963; Koeniger, 1969, 1970; Ruttner, 1966; Winston, 1992). This should give them better vision. Around 2 weeks after the drone emerges from the cell, it will be mature enough to mate. Drones typically come together in areas known as “drone congregation areas” or “mating yards”. These areas are often several metres from the ground (up to 30 m) and will typically be in the vicinity of the apiary or cluster of natural colonies. However, drones can fly several miles to find other established congregation areas (Gary, 1963; Koeniger, 1969, 1970; Ruttner, 1966; Winston, 1992).

How drones find these congregation areas, particularly from kilometers away, remains a major mystery. They may be able to sense the odours of other drones over some distance using their enlarged and highly sensitive antennae. However, this may not account for their long-distance orientation to drone congregation sites. It is possible that all drones possess similar innate preferences for congregation sites that lead to choose a few locations. Once they are close enough to a favored site, the odor of other drones may then enhance their orientation. However, this explanation is speculative and requires experimental verification.

A congregation area may have several hundred to several thousand drones (Gary, 1963; Koeniger, 1969, 1970; Ruttner, 1966). A virgin queen will have already found such a congregation area from early reconnaissance flights. On her mating flight the queen will head straight for the congregation area (Gary, 1963; Koeniger, 1969, 1970). The drones will sense her arrival (their notably large antennae have 10 times more sensory receptors than those of workers or queens) and will immediately pursue her. Several drones will mate with the queen. Each drone mounts the queen in turn and inserts his endophallus into the queen. The drone’s sex organs inflate (almost explosively) within the queen, pumping sperm into her. This rapid inflation and expulsion of sperm causes him to spring backwards and fall away from the queen. The endophallus is gripped by the opening of the queen’s oviduct and results in the drone’s endophallus and part of the drone’s gut contents being ripped out of his abdomen. Each additional drone will remove from the queen the endophallus of his predecessor before undergoing the same fate. After mating, the drone will often fall to the ground where he will die (Gary, 1963; Koeniger, 1969, 1970; Ruttner, 1966; Winston, 1992).

Honeybee Development and Differentiation

The first stage in the development of a worker bee occurs when the queen deposits a single fertilized egg at the bottom of a worker cell. The newly egg is shaped like an ellipsoid. It is elongated, rounded at the ends, and slightly convex in the center (Snodgrass and Ericson, 1992). After three days, the egg hatches into a first-instar larva and is then fed regularly by nurse bees. After successive stages of growth and molts, the larva completely covers the floor of its cell, and then changes position, stretching out along the depth of the cell. During its life, the larva goes through five larval instars (Benholf, 1925; Michener, 2000). When the larva is fully grown and no longer needs to be fed, house bees cap its cell with a thin layer of wax. At this stage, the larvae are, called "sealed brood." The sealed larva spins a cocoon around itself and begins to pupate. It sheds its last larval integument and differentiates into a pupa. The

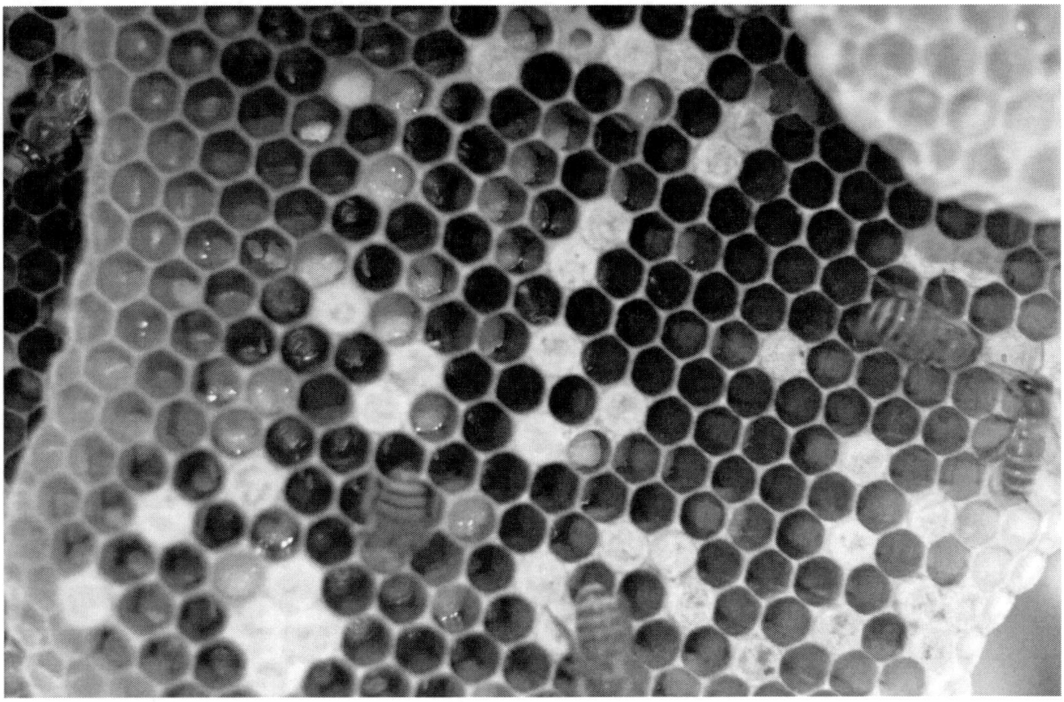

Figure 9. The different developmental stages of *A. cerana*.

pupa has all the adult bee's distinct body parts, but they all adhere tightly to the body, and some appendages are not yet fully expanded. Before emerging, the pupa becomes gradually darker in color. Finally, transformed into an adult, it slowly chews its way out of the cell. In *A. mellifera*, the complete metamorphosis from newly laid egg to emerging adult worker requires 21 days: three as an egg, six as a larva, and 12 as a pupa (Snodgrass and Ericson, 1992; Tribe and Fletcher, 1977; Winston, 1979).

For a drone, life begins when the queen deposits an unfertilized egg in drone brood cell at the bottom of the comb. Like the eggs of worker brood, drone-brood eggs require three days to hatch. When the larvae are full grown, the nurse bees cease feeding them, their cells are capped, they spin their cocoons, and pupation takes place. It requires 24 days for a drone to develop from newly laid egg to emerging adult. Emerging drones are fed on honey and royal jelly until they are about a week old. Their flight activity begins when they are from 6 to 8 days old, but they are sexually mature only after 12-14 days.

Whereas worker and drone brood are reared in hexagonal cells, queen development takes place in cells shaped somewhat like peanuts (Michener, 2000). There are three types of queen cells: swarm cells, supercedure cells, and emergency cells. Swarm queen cells are built along the lower edge of the comb, often in large numbers: as many as 20 cells of various ages may be seen in a colony. Supercedure queen cells, fewer in number, are generally about the same age and are perpendicular to the comb surface (Michener, 2000; Winston, 1987; 1992). They are usually formed from old, darker wax than swarm queen cells that, built at times of high food availability, consist of whiter, newly secreted wax. A distinctive feature of emergency queen cells is that they are expanded from ordinary worker cells already containing young larvae, and protrude directly from worker-brood cells on the surface of the comb. The development period of a queen is significantly faster than workers and drones: 16 days from egg to adult. The queen larva is well provided by nurse bees with royal jelly for her entire development. The food is deposited very frequently in the cell, and the queen larva simply lies on a bed of its food. The remains of uneaten royal jelly is often seen in the cell after the young queen emerges. Although larvae destined to become queens and workers are genetically similar (both are hatched from fertilized eggs), qualitative and quantitative differences in the diet they receive, particularly in the early stages of their larval lives, give rise to major differences in anatomical and physiological development (Michener, 2000; Winston, 1987; 1992).

Division of Labour

A colony of honeybees usually consists of a queen, several thousand workers, and (during the breeding season, a few hundred drones). Among the members of the colony, there is division of labor and specialization in the performance of biological functions (Winston, 1987). Workers can flexibly shift among different tasks, depending upon colony need (Ferguson and Winston, 1988; Smith et al., 2008). The tasks performed are primarily age related (Lindauer, 1961; Wang and Moeller, 1970). There is also a strong genetic component to division of labor with workers from different strains, races within the colony showing differences in task ontogeny (Winston and Katz, 1982). Both genetics and environment are important. Workers can perform a subset of multiple tasks at all ages (Lindauer, 1952; Winston, 1992).

Figure 10. Brood rearing by nurse bees in *A. cerana* comb.

Figure 11. *Apis florea* forager.

In general, young workers work inside the nest and older workers work outside or at the nest entrance (foraging or guarding, Winston and Ferguson, 1985). The youngest bees perform house cleaning and capping. Brood and queen rearing occupy slightly older workers, nurse bees. Comb building and food processing are handled by middle-aged workers (who serve as a general reservoir of labor that be channeled into performing different tasks inside the nest, as needed). Finally, nest temperature regulation and ventilation, defense, and foraging occupy the oldest bees (Winston, 1992). The caste structure in honeybees is closely linked with the development of brood food glands (hypopharyngeal glands), mandibular glands, and wax glands (King, 1933; Simpson, 1960, 1966; Simpson et al., 1968; Wang and Moeller, 1969).

Colony Cycle: Migration, Swarming and Absconding

The annual colony cycle of honeybees in Thailand includes migration, swarming, and absconding. However, for beekeepers, the imported European honeybee, *A. mellifera* has been a popular option because this species has a prolific queen and both swarms and absconds less than native *Apis* species (Partarp and Verma, 1998). Migration from one habitat to another is normal part of seasonal cycle for *A. andreniformis, A. dorsata* and *A. florea*, (Wongsiri et al., 1996b). Migration may increase colony fitness by allowing the colony to move to an area with more food, enhanced outbreeding, and reduce parasite loads (Oldroyd et al., 1996). The migration period for *A. dorsata* in Thailand peaks in September to October (Wongsiri et al., 1996b). Absconding to escape predators (such as human honey hunters) can also occur.

Swarming is a natural mode of colony reproduction. If a colony is relatively safe from damage or destruction by its natural enemies, has an ample supply of forage, and queen and the workers have been performing their duties in an optimum manner, it will eventually outgrow its hive space. When this occurs, the colony is ready to reproduce itself by swarming. In temperate regions, natural food is most available to colonies in spring and summer, when warm ambient temperatures permit flights and active foraging (Partap and Verma, 1998; Wongsiri et al., 1996b). The colony is busily engaged in brood rearing during this period, until hive overcrowding and congestion signal the colony to swarm. During the cold autumn and winter months, however, colonies raise only a small amount of brood and depend on their stored food. Such a clearly defined annual cycle does not exist to the same extent in tropical regions, where colonies of indigenous *A. cerana* and introduced temperate races of *A. mellifera* rear brood whenever their food supply is plentiful (Partap and Verma, 1998).

Overcrowding of the hive can thus occur at almost any time, and swarming under tropical conditions occurs whenever the forage is seasonally plentiful. Under normal conditions, a temperate-zone colony of *A. mellifera* will swarm at least once per year. However, tropical species such as *A. cerana* may send out several successive swarms. *Apis cerana* is prone to swarm excessively during times of food abundance. Such swarming appears to depend primarily upon environmental conditions rather than swarm genotype (Punchichewa et al., 1998).

In preparation for swarming, a colony builds queen cells, normally just about the time the virgin queens begin to emerge and rears young queens. At the same time, the old queen receives less food and loses weight, which facilitates her departure flight. Before the new queens emerge, the colony's worker population leaves the parent hive in search of a new

Figure 12. The old open single comb nest of *A. dorsata* after absconding has occurred.

home site. At the next point, the bees will 'decide' who is to leave and who is to stay (Tew, 2006; Passino and Seeley, 2008). The swarming behavior may partially be the result of the evolution of absconding behavior or migratory behavior. Just as in absconding, the old queen goes with the swarm rather than the colony waiting around for the new queen to emerge and then leave with the new queen. It is commonly said that honeybees have a tropical ancestry where migratory and absconding behavior are much more prevalent. Whatever reason, the behaviors of swarming, supercedure, and absconding have obvious characteristics in common (Tew, 2006).

Foraging Behavior

Honeybees have sophisticated foraging coordination and communication (von Frisch, 1971; Suwannapong, 2000). This activity is only performed by workers, known as foragers or foraging bees. Some foragers specialize on pollen foraging and some on nectar foraging. Between these extremes, there are a large number of generalists who collect both (Fewell and Page, 1993). The range for the onset of foraging ranges from 18.3 days (Sakagami, 1953) to 37.9 days of age (Winston and Ferguson, 1985). This food consists of carbohydrates and proteins (nectar and pollen, Seeley, 1985). Under normal conditions, worker bees begin to forage when they are about 2 to 3 weeks old. Foraging is the last chore in the life of a worker. Part of the colony's stored honey is consumed by foraging bees who need fuel and therefore consume a certain amount of honey to ensure that she will have a sufficient energy supply for her round-trip journey (Akratanakul, 1976; Seeley, 1985). To obtain a full load of nectar and pollen (or both) in a single trip, she may have to visit several hundred flowers (Akratanakul, 1976). The amount of energy she expends, related to the amount of food she collects, is determined largely by such factors as the amount of nectar obtained per flower, floral density per unit area, the distance from the hive, and weather conditions (Akratanakul, 1976; Partap, 1992; Partap and Partap, 1997).

There are differences among flowering plant species with respect to nectar and pollen production. Not all plant species possess nectaries (glands secreting nectar) or have nectar that bees can reach with their proboscis (tongue) (Partap, 1992). Nectaries can be located in various areas of the flower and some species have extrafloral nectaries that may be visited by bees. In addition, some bees may perform nectar robbing, making a small hole at the base of a flower in order to obtain the nectar. In this case, the bee does not perform any pollination service for the “robbed” plant. A forager may prefer the nectar of one flower species. It is to her advantage to visit flowers producing greater quantities of nectar with a higher sugar concentration. The sugar concentration in the nectar of a given plant species may vary depending upon its location, time of day, and genotype. If nectar with a high sugar concentration is available, a forager of *A. mellifera* can carry as much as 70- 80 mg of nectar per load (Akratanakul, 1976; Partap, 1992; Partap and Partap, 1997).

Workers of all honeybee species carry nectar internally. Part of their alimentary canal is modified to form a “honey sac” or “honey stomach”. After returning to the hive, the forager regurgitates the nectar to one or more house bees, which then dehydrate the nectar and convert it into honey. They use the enzyme invertase, which splits sucrose in the nectar into fructose and glucose, the sugars predominant in honey. To dehydrate the nectar, house bees regurgitate a part of the nectar and hold the droplet in their mouthparts (Partap, 1992; Partap and Partap, 1997).

The entire body of a worker bee, particularly her thorax, is covered with fine, branched hairs, on which pollen grains are caught. She sometimes uses her mandibles to chew off the anthers, or deliberately rolls over the anthers to acquire the pollen. The tibiae of the bee's metathoracic legs are equipped with rows of short setae, which she uses to scrape the pollen from her body and to form it into pellets, sometimes regurgitating a slight quantity of nectar to provide moisture and adhesiveness. The pellets, attached to "pollen baskets" on the bee's rear tibiae, are carried back to the hive, where the load is deposited in a pollen-storage cell. Whereas cells containing ripe honey are capped, pollen-storage cells are not. The bees tightly pack pollen to about two thirds of the capacity of the cell and coat the top surface of the pollen in each cell with honey (Partap, 1992). This helps protect the pollen against spoiling (Partap, 1992; Partap and Partap, 1997; Verma and Partap, 1998).

In addition to collecting nectar and pollen, foragers can collect plant gum (propolis) and water (Fanesi et al., 2009; Marcucci, 1995; Bankova et al., 1983, 2000). Propolis, which is a resinous hive product exuded by certain plants, often to protect wounds on their surface and against the bees' enemies, is rich in tannin, flavonoids, aromatic acids, esters, aldehydes, ketones, fatty acids, terpenes, steroids, amino acids, polysaccharides, hydrocarbons, alcohols, hydroxybenzene, and several other trace compounds. Propolis exhibits antibiotic activity (Fanesi et al., 2009; Marcucci, 1995; Bankova et al., 1983, 2000). It is also an adhesive material, which the bees use in comb construction, to coat the interior of the hive, and to seal cracks. To collect propolis, a bee uses her mandibles to bite the substance from the plant surface and carries it back on her rear legs (much like pollen). In the hive, workers use their mandibles to remove the resin from the forager (Cheng and Wong, 1996).

The honeybee colony needs water for two purposes only: to cool the hive and to dilute the honey fed to the larvae (von Frosch, 1967; Seeley, 1996). Like nectar, water is collected by the forager through her proboscis and is carried back to the hive in her honey stomach. Water is regurgitated to the house bees on arrival. During the heat of the day, some foragers may switch from nectar to water collection, or they may prefer to collect nectar with a low sugar concentration and higher water. Interestingly, this process is mediated by the willingness of foragers to accept nectar from the forager. If there is need for more water, workers that unload nectar (food unloading bees) will preferentially unload bees with water or more dilute nectar (von Frosch, 1967). Bees bringing back sweeter nectar will wait for increasingly long periods and will therefore reduce their rate of recruitment for the sweeter nectar source (Seeley, 1996).

It has been also reported that *A. mellifera* foragers use 2-heptanone to mark previously visited flowers, thereby signaling nectar depletion to other bees (Engels et al., 1997; Giurfa, 1991). However, the four native Thai species do not appear to use aversive pheromone marking during foraging (Suwannapong, 2000; Suwannapong et al, 2010c). For example, they may revisit the same flower briefly after the first visit and continue to forage on the same flower simultaneously with several bees of their own species or other species. Suwannapong (2000) observed *A. florea,* two to three bees of *A. cerana*, one to two bees of *A. dorsata* and one to two bees of *A. andreniformis* visiting the same flower (Suwannapong, 2000). It is also possible that honeybees, like bumblebees can learn to associate floral depletion or floral reward using olfactory cues, cuticular hydrocarbon “footprints” deposited while walking on the food source (Leadbeater and Chittka, 2007). However, this remains to be investigated.

The mandibular gland of *A. mellifera*, the source of this putative food-marking pheromone is primarily 2-heptanone. However, the primary component of mandibular gland secretions in Thai honeybees is (Z)-11–eicosanol. In general, the ten most abundant components in the mandibular glands of all these species are 80% similar (Suwannapong, 2000).

Figure 13. *A. florea* foragers are foraging on palm flowers.

Table 1. Comparison of the main composition of mandibular gland pheromone of Thai honeybees (Suwannapong, 2000)

Pheromone	**A. an**	**A. c**	**A. d**	**A. f**	**A. m**
Z-11 eisosanol	+	+	+	+	+
1-butanol-3methyl-acetate	+	+	+	+	-
Dibutyle phthalate	+	+	+	+	+
Nonadecane	+	+	+	+	+
2-hexyl 1-decanol	+	+	+	+	+
Heneicosanol	+	+	+	+	+
Eicosane	+	+	+	+	+
1-octanol	+		+		
2-propyl 1-heptanol	+		+	+	+
2-butyl-1-octanol	+	+	+	+	+
Heneicosane		+			
Heptadecane				+	
Limonene	+	+	+	+	+
2-heptanone	Undetectable	undetectable	undetectable	undetectable	+

*A. an, *A. andreniformis*; *A.* c, *A.* cerana; *A.* d, *A. dorsata*; *A.* f, *A. florea* and *A.* m, *A. mellifera*

Thermoregulation

Honeybees are poikilothermic and have partial control over their internal body temperature. Unlike warm-blooded animals, honeybees can maintain lower body temperatures, but they can also elevate their body temperature through basking or muscular contractions (Woods et al., 2005). A populous honeybee colony can regulate the interior temperature of the hive, particularly within the area surrounding the developing brood (Nakamura and Seeley, 2006). In normal colonies, the brood-nest temperature is maintained at a remarkably constant 30-36 °C (Underwood, 1991). Thermoregulation by honeybee colony has a high energetic cost (Kronenberg and Heller, 1982; Heinrich, 1979; Heinrich and Esch, 1994). To generate more body heat, the worker bees will consume more food, especially honey: more heat is released because of the increased rate of food metabolism (Jones and Oldroyd, 2007).

By fanning their wings, evaporating the water regurgitated on worker mouthparts, and dispersing drops of water in empty cells, a honeybee colony can reduce its temperature. When water is available, a colony of *A. mellifera* can withstand external heat of up to 70° C. When the external temperature is low, bees reduce heat losses by clustering together, and the lower the temperature, the more compact the cluster (Jones and Oldroyd, 2007).

Figure 14. Thermoregulation in an *A. florea* colony. Although the fanning wings are not visible in this still photograph, the way that the wings are spread out is characteristic of fanning.

The survival ability of honeybee colonies during severe winter months depends on whether the colony has enough workers adequately provisioned with food. Insulating the hive wall and decreasing the volume of the hive can also improve the effectiveness of the colony's thermal regulation (Kronenberg and Heller, 1982; Heinrich, 1979; Heinrich and Esch, 1994). In the temperate regions, colonies of *A. mellifera* survive by forming clusters around the brood nest, the bees on the surface of the cluster and those within it exchange positions over time (Jones and Oldroyd, 2007). In this manner, *A. mellifera* colonies can survive temperatures as low as -40°C. The regulation of brood-nest temperature is not confined to the European races of *A. mellifera*. Tropical honeybee species and races can also regulate their brood-nest temperature to a certain extent, but they are able to survive only mildly cold temperatures, generally not below 0°C (Kronenberg and Heller, 1982; Heinrich, 1979; Heinrich and Esch, 1994).

3. HONEYBEE ANATOMY

Honeybees have many characteristics common to all insects. Insects have a hard outer covering called an exoskeleton that is made of a material called chitin that helps to protect the internal organs of the insect and prevent desiccation (Gary, 1992; Snodgrass, 1925). In order to grow, the insect must shed its exoskeleton. Its body can be divided into three sections: head, thorax, and abdomen. The head contains mouthparts, two compound eyes with three ocelli, and two antennae. The thorax contains the appendages for locomotion, the three pairs of legs and two pairs of wings. The abdomen contains the organs for digestion, reproduction, and defense (Gary, 1992; Snodgrass, 1925).

The Head of Honeybees

The head of the honeybee is triangular when viewed from the front. The two antennae arise close together near the upper center of the head. There are two compound eyes and three simple eyes located on the top of the head. The honeybee uses its proboscis, or long tongue, to feed on liquids and its mandibles to manipulate pollen and work wax in comb building (Gary, 1992; Snodgrass, 1925).

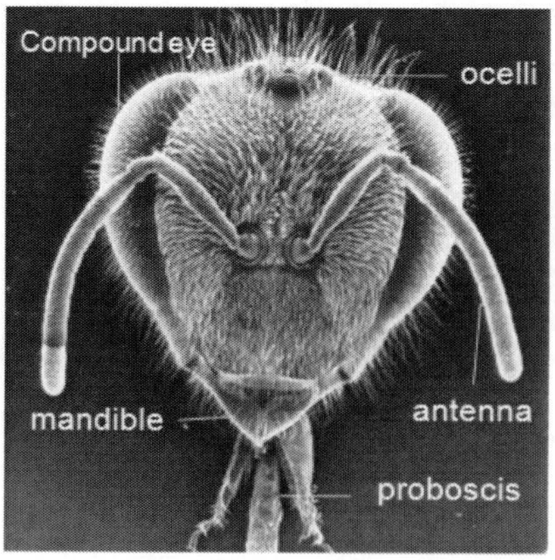

Figure 15. Scanning electron micrograph of the head of *A. florea* worker.

The Antenna

Honeybee antennae are sensitive to temperature, humidity, air pressure, odor, gustatory stimuli, near-field sound vibrations, substrate vibrations, and tactile contact. Odor perception is particularly important. Each antenna contains a few thousand of antennal sensilla distributed over the third segment of the antenna, known as a flagellum (Figs 15 and 16). The insect antenna functions primarily as an odor receptor and secondarily as a taste receptor. The antennal form and arrangement of sensilla appear to be well adapted to the pheromone perception needs of a particular species (Gupta, 1992; Payne et al., 1970).

Each antenna consists of a single long joint connected to a prominent knob, which is inserted into an antennal socket. The honeybee can turn the antenna in 360° direction at the base (Agren, 1977; Snodgrass, 1925; Chapman, 1982; 1988). Each antenna has a segmented scape, a pivoted pedicel and a flagellum, which is composed of 11 segments in females, queens and workers, and 12 segments in drones. Sensory organs or sensilla on the antennae of honeybee can be divided into seven different types. They are sensilla basiconica, sensilla campaniforme, sensilla placodea, sensilla trichodea type A, B, C, and D (Snodgrass, 1925). Sensilla placodea measure air pressure and have olfactory functions. They are located on the last eight segments of the antennae (Guirfa, 1991; Gupta, 1992; Naik et al., 1995). The sensilla placodea which such consist of oval cuticular plates with numerous pores and each innervated by 15 to 30 olfactory receptor neurons containing large numbers of branched dendrites (Esslen and Kaissling, 1976).

Honeybees from different castes have different roles in their colony and exhibit different external and internal morphology (Snodgrass and Erickson, 1992; von Frisch, 1971). This is especially true for the olfactory sensilla. Honeybee queens primarily use their antennal

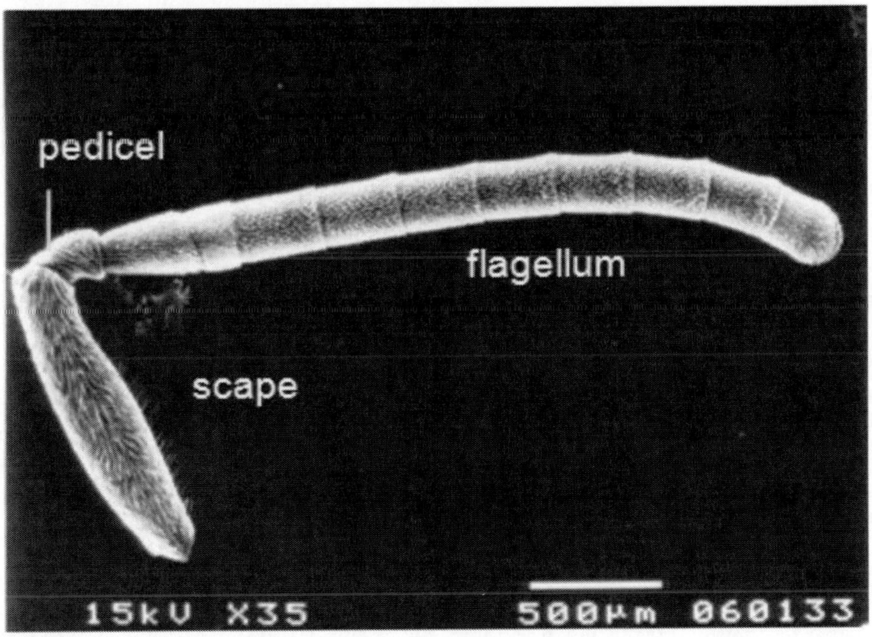

Figure 16. Scanning electron micrograph of the antenna of *A. dorsata* consists of scape, pedicel and the flagellum.

sensilla to detect or perceive odor within the colony. Workers use olfactory sensillae to detect colony odors such as brood pheromone and queen pheromone in addition odors obtained outside the nest such as floral odors, which they can learn to associate with rewarding resources (von Frisch, 1971). Drones use these sensilla to detect queen pheromone. Antennal sensilla also play an important role in detecting various types of odors in *A. dorsata* foragers. Their foraging occurs during daytime and continues after sunset, a nocturnal feature that is not reported in other honeybee species (Suwannapong and Wongsiri, 1999; Suwannapong et al., 2010c).

Antennal sensilla each have multiple microscopic pores that are 10 to 50 nm in diameter (Steinbrecht, 1996; 1997; 1998) and allow the passage of volatile molecules (Slifer et al., 1959). The number of pores per sensillum (3-18,000 pores) and the diameter of pores (between 10-100 nm) vary depending on type of sensilla and insect species (Kaissling, 1971; 1972; 1974). Associated with each cuticular pore is a pore cavity or kettle from which extend four to eight filaments or tubules into the lumen of the sensillum and terminate at the receptor membrane (Slifer et al., 1959; Schneider, 1962; Schneider and Steinbrecht, 1968; Kaissling, 1971). Trichogen cells form the pore tubules. They may be single walled with whole tubules, or double walled with spoke canals. Their surface can be pitted or grooved longitudinally.

Most of the multiporous sensilla (MP) can easily be identified as multiporous pitted (MPP) or multiporous grooved (MPG). The MPP cuticles have many round holes or slits at the surface (Steinbrecht, 1997). The lumen is filled with sensillum lymph secreted by the trichogen (Tr) and tormogen (To) support cells. The thecogen support cell acts as a glial cell for the olfactory neuron (Ernst, 1969; Ernst, and Boeckh, 1983; Farbman, 1992). This fluid contains a high concentration of odorant-binding proteins (OBPs), bathing the dendrites of the olfactory neurons. One to four olfactory neurons project dendrites into a single sensillum (Farbman, 1992). There are four types of olfactory sensillae: sensilla trichodea, sensilla basiconica, sensilla placodea, and sensilla coeloconica (Schneider and Steinbrecht, 1968). Each olfactory sensillum is connected to a single or multiple bipolar receptor neurons (ORNs) and auxiliary cells, trechogen (ensheaths the neuron and the peg), trichogen (synthesizes the peg) and tormogen cells (secrete the surrounding socket). Each sensilla cell envelopes the ORNs in each sensillum (Shanbhag et al., 1999, 2000; Steinbrecht, 1997).

The Compound Eye

The honeybee eye is made up a large number of ommatidia or facets. Each ommatidium is composed a crystal line lens (the front surface of which makes up a single facet) that usually including light focusing elements (lens and a transparent crystalline cones). Within each ommatidium, light is focused onto eight light sensing cells (retinal cells) arranged in a radial pattern like sections of an orange (Giurfa, 1991; Giurfa et al., 1995). The pigment cells ensure that only light entering the ommatidium roughly parallel to its long axis reaches the visual cells and triggers nerve impulses. Thus, each ommatidium contributes information about only one small area in the field of view, just like a single pixel in a CCD of a digital camera. Each facet in a compound eye corresponds to one ommatidium and takes in one small part of the insect's vision. The brain then takes the image from each tiny lens and creates one large mosaic-like picture. Workers of *Apis mellifera* have about 4,000-6,000 ommatidia but drones have from 7,000-8,600, presumably because drones need better visual ability during mating (Giurfa, 1991; Giurfa et al., 1995; Giurfa et al., 1996a; Giurfa et al., 1996b;

Suwannapong and Wongsiri, 1999). There are approximately 6300 ommatidia in *A. dorsata* and *A. florea*, respectively. As in most insects, bee eyes are not designed to see high-resolution images like human eyes, but rather they see a mosaic lower resolution image. However, their compound eyes are better at motion detection than most camera lens eyes. Honeybees can adjust their light sensitivity and in dim light can adapt their eyes by concentrating the visual pigments of their ommatidia into the lower ends of the pigment cells. This shift enables light entering a single ommatidium at an angle to pass into and stimulate adjacent ommatidia. When multiple ommatidia respond to a single area in the visual field, the image becomes coarser and has reduced resolution (Giurfa, 1991; Giurfa et al., 1995; Giurfa et al., 1996a; Giurfa et al., 1996b; Suwannapong and Wongsiri, 1999).

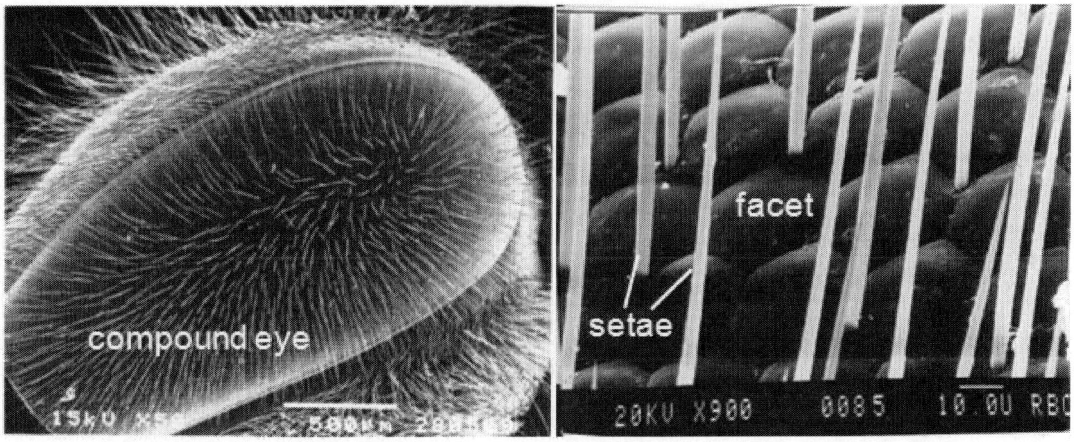

Figure 17. Scanning electron micrograph of the compound eyes of *A. dorsata* worker (left) and their facets (right).

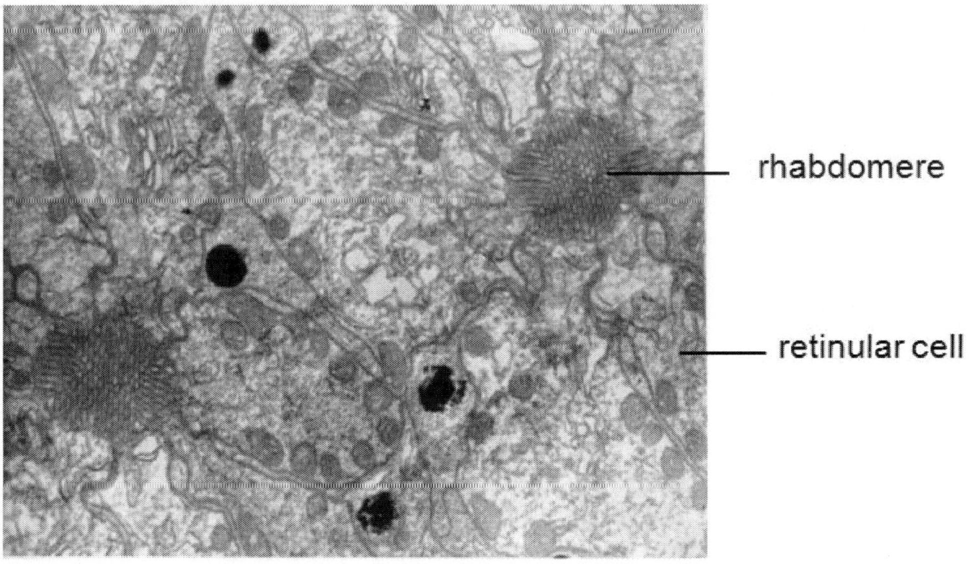

Figure 18. Transmission electron micrograph of the eight retinular cells surrounded the rhabdomere of *A. florea* workers.

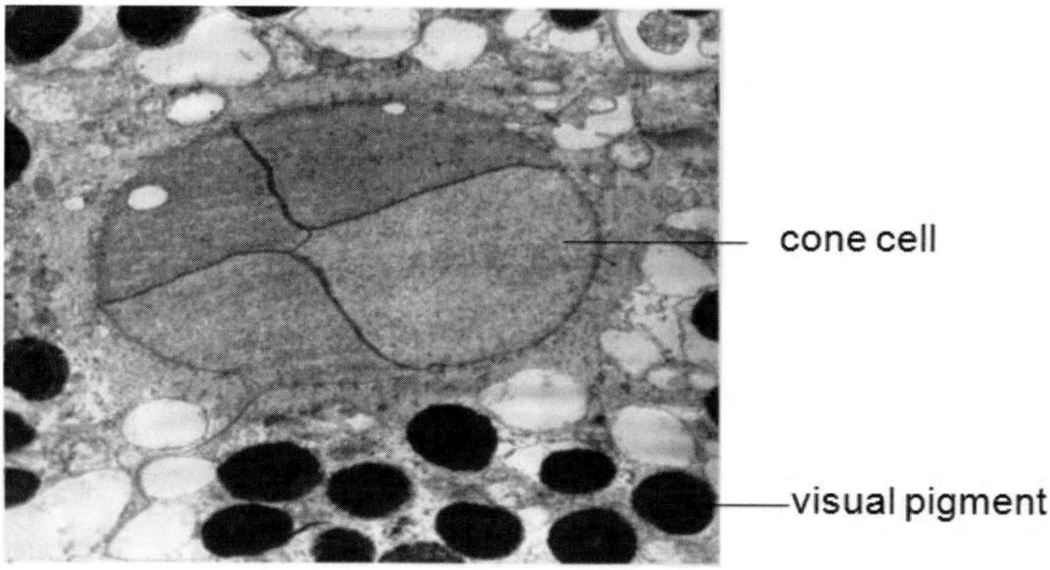

Figure 19. Transmission electron micrograph of the crystalline lens of *A. dorsata* workers.

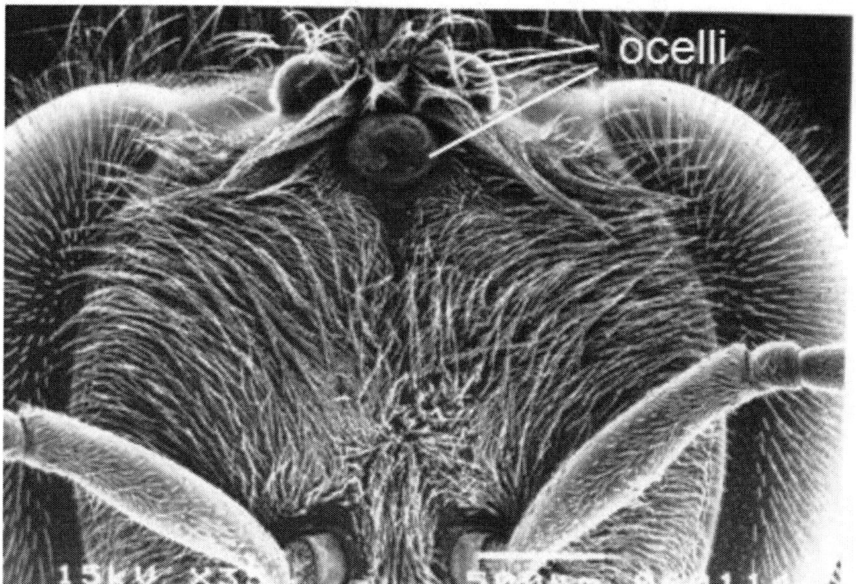

Figure 20. Scanning electron micrograph of the three ocelli of *A. dorsata* workers.

Honeybees have trichromatic color vision. Each ommatidium consists of four cells that respond best to yellow-green light (544 nm), two that respond maximally to blue light (436 nm) and two that respond best to ultraviolet light (344 nm). This system enables the honeybee to distinguish colors, and this has been amply demonstrated in behavioral discrimination experiments (Chapman, 1988; Giurfa, 1991; Giurfa et al., 1996a, 1997). Although honeybees perceive a fairly broad color range, they strongly differentiate six major categories of color:

yellow, blue-green, blue, violet, ultraviolet, and also a color known as "bee purple", a mixture of yellow and ultraviolet (Chapman, 1988; Giurfa, 1991; Giurfa et al., 1996a). Bees see red poorly. Differentiation is not equally good at all wavelengths and is best in the blue-green, violet, and bee purple colors. In addition, honeybees can discriminate various shapes and patterns, inability useful in recognizing different flowers and in local landmark orientation (Giurfa et al., 1995, 1996b). Honeybees can easily differentiate between solid and broken patterns, but show a preference for broken figures (Guirfa et al., 1995). Honeybees also have three smaller eyes in addition to their two compound eyes. These simple eyes that are called ocelli (singular: ocellus) and are located near the top of a bee's head. The ocelli only provide information about light intensity. They cannot resolve images (Chapman, 1988; Giurfa, 1991; Giurfa et al., 1996a, 1996b).

Table 2. Number of ommatidia in honeybee eyes

Honeybee species	Number of facet	Size of facet
A. florea	7933 7760 queen	Un publish data (Suwannapong)
A. cerana	8160 7036 queen	Un publish data (Suwannapong)
A. dorsata	6200 (worker) 5700 (queen)	Rasmidatta et al., 1999 Suwannapong, 1999
A. mellifera	4500	Gould and Gould, 1988
A. andreniformis	7780	Un publish data (Suwannapong)

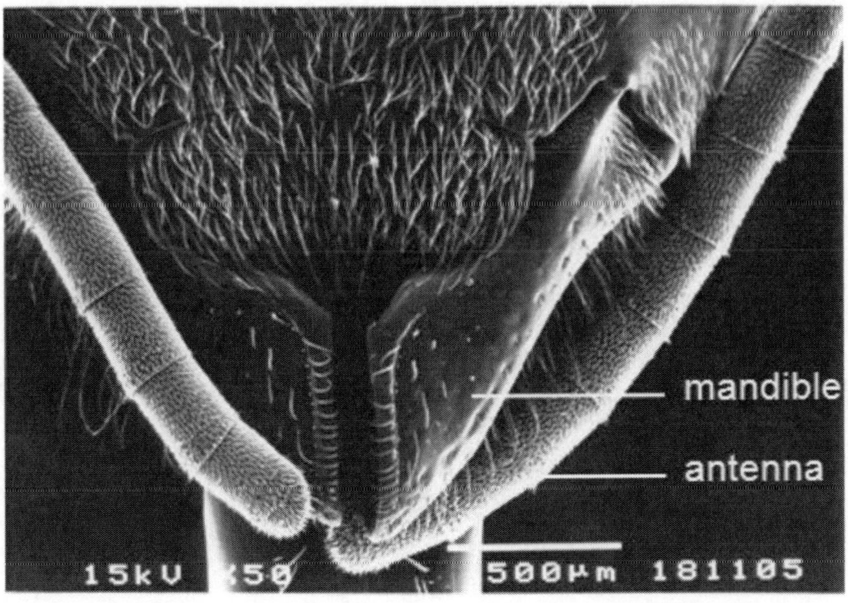

Figure 21. Scanning electron micrograph of the mandibles of *A. dorsata* workers.

The Internal Organ of the Head

The main internal organs in the head are the brain, subesophageal ganglion, and the esophagus (food canal). The brain has a large area for receiving inputs from the two compound eyes, called optic lobes. The next largest input is from the antennal lobes. One important region in the middle of the brain is called the "mushroom body" because its cross section resembles a mushroom. This area is important for learning and memory formation (both consolidation memory and long term memory) (Snodgrass and Erickson, 1992).

Endocrine organs are also attached to the nerve cord, very close to the esophagus. One is called the corpora allata (in Latin it means the body beside the food canal). The corpora allata is the only source of a key important hormone, juvenile hormone, which is involved in the queen-worker differentiation and worker division of labor. The second key endocrine organ is called the corpora cardiaca (the body near the heart). This neurohemal organ stores and releases another hormone (PTTH, prothoracicotropic hormone). PTTH can stimulate the production of ecdysone in the prothoracic glands, located in the thorax (Snodgrass, 1956; Snodgrass and Erickson, 1992).

Lastly, there are several exocrine glands inside the head: mandibular, hypopharyngeal and salivary glands. Mandibular glands are a simple sac-like structure attached to each of the mandibles. In the queen, this is the source of the powerful queen pheromone (Boch and Shearer, 1971; Plettner et al., 1993; 2000). In young workers, the gland produces a lipid-rich white substance that is mixed with the secretion of hypopharyngeal glands to make royal jelly or worker jelly and fed to the queen or other workers. In old *A. mellifera* workers, (foragers) the gland also produces 2-heptanone, a component of the alarm pheromone (Boch and Shearer, 1971; Plettner et al., 1993; Snodgrass, 1956; Snodgrass and Erickson, 1992).

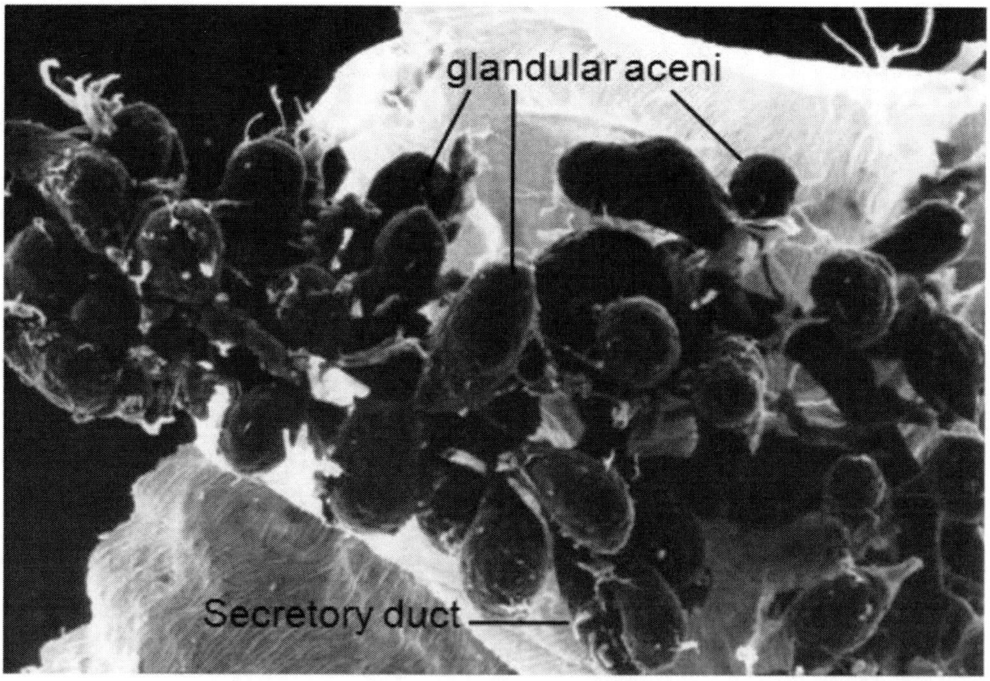

Figure 22. Scanning electron micrograph of the hypopharyngeal gland of an *A. andreniformis* worker.

Other species of honeybees produce different compounds in their mandibular glands (see above). The hypopharyngeal glands produce protein-rich secretions in nurse bees, but produce invertase (an enzyme that breaks down sucrose into fructose and glucose) in foragers. The glands consist of a central duct with many of small grape-like spheres (acini, singular: acinus). The secretion flows to the mouth through the long duct. The glands are large (hypertrophied) in nurse bees (Deseyn, and Billen, 2005; Kubo et al., 1996; Ohashi et al., 2000; Suwannapong et al., 2010a). There is also a pair of salivary glands inside the head. The glands produce saliva, which is mixed with wax scales to change the physical properties of wax, making it easier to work (Snodgrass, 1956; Snodgrass and Erickson, 1992).

Hypopharyngeal Glands

Hypopharyngeal glands (HPGs) of honeybees are age-dependent structures that change with the size of acini and are correlated with social behavior (Feng et al., 2008; Kubo et al., 1996; Ohashi et al., 1999; Ohashi et al., 2000; Suwannapong et al., 2010a). Gland protein concentration increases progressively in nurse bees, and this has been correlated with the appearance of enriched protein granules in the cytoplasm. The glands are composed of several secretory units, each opened into a secretory duct that passed through the mouthparts. In pupae, the secretory cells are irregular in shape with low concentrations of proteins and carbohydrates, while the glands of nurse bees and foragers are fully developed with numerous secretory vesicles (Suwannapong et al., 2007). It has been also reported that the hypopharyngeal glands of *A. cerana* and *A. mellifera* workers had significantly larger than those of *A. florea* and *A. andreniformis* either guards or forgers, but without substantial differences between these two species after counting for caste (Suwannapong et al., 2007; Suwannapong et al., 2010a).

The structure of honeybee HPGs depends on the development and age of individuals, which corresponds with age-specific tasks and is known as age polyethism (Deseyn and Billen, 2005; Robinson, 1987; 1994). Studies have shown that the histochemical structure of carbohydrate and protein in *A. cerana* and *A. mellifera* was correlated with honeybee age-specific tasks of the colony. Young worker nurse bees care for and feed their brood with royal jelly that is synthesized and secreted from the hypopharyngeal glands (Deseyn and Billen, 2005; Feng et al., 2008; Kubo et al., 1996; Ohashi et al., 1999). These hypopharyngeal glands were strongly positive to PAS and Ninhydrin Schiff's reagent reactions in this study (Suwannapong et al., 2007; Suwannapong et al., 2010). However older workers, when they became guards, had less positive staining to PAS as compared to nurse bees. This may be related to the development of the hypopharyngeal glands, which are fully developed when young workers take care of the brood by synthesizing and secreting royal jelly. Older workers no longer feed the brood, and thus gland atrophy is expected (Suwannapong et al., 2010a).

The development of HPGs in dwarf honeybee workers, *A. andreniformis* and *A. florea* primarily depends on age. These glands begin to differentiate at pupal stage and are largely undeveloped at emergence (Suwannapong et al., 2007). When workers become nurse bees, they perform brood rearing that is associated with HPGs development. The size of HPGs is correlated with glandular production and generally increases with age from 6 to 18 days in nurse bees (Deseyn and Billen, 2005; Hrassnigg and Crailsheim, 1998). HPGs synthesize and secrete proteinaceous substances and royal jelly that are fed to the queen and brood (Deseyn and Billen, 2005). The highest rate of protein synthesis occurs during nursing ages from 8 to

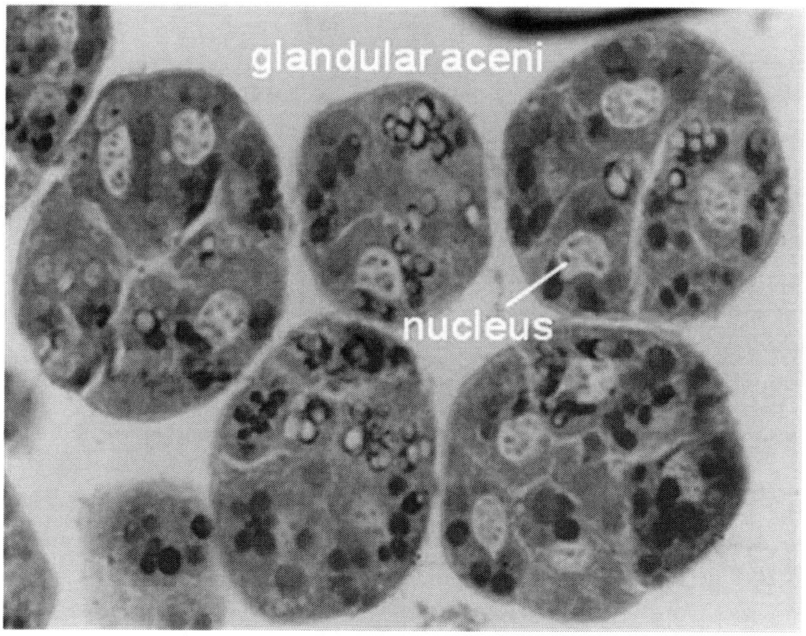

Figure 23. Light micrograph of the hypopharyngeal gland of an *A. andreniformis* worker stained with PAS and counterstained with light green (200X).

16 days (Ohashi et al., 2000; Knecht and Kaatz, 1990). In bees older than 18 days (guards and foragers), the HPGs decrease considerably in size and secrete enzymes such as α-glucosidases, leucine arylamidase and invertase (Kubo et al., 1996; Ohashi et al., 1999; Feng et al., 2008). Forager gland size is reduced and correlated with gland activity (Deseyn and Billen, 2005; Furi et al., 1982; Ohashi et al., 2000).

Moreover, other studies have reported that the secretory units of the HPGs were filled with numerous vesicles that gave a strong positive staining with PAS and Ninhydrin Schiff's reagent (Suwannapong et al., 2010a). This indicates that the glands of nurses, guards, and foragers in four species of honeybees (i.e., *A. andreniformis, A. florea, A. cerana,* and *A. mellifera,*) play an important role not only in secretion of carbohydrate rich substance but also in the secretion of enzymes for converting nectar to honey (Suwannapong et al., 2007; Suwannapong et al., 2010a; Simpson et al., 1968; Brouwers, 1982). However, there were also differences found in structure of the glands between the hive cavity nest honeybees, *A. cerana* and *A. mellifera,* and the single open nest honeybees, *A. andreniformis* and *A. florea.* The structure of the extracellular space between adjacent cells of *A. andreniformis* and *A. florea* was wider than that of *A. cerana* and *A. mellifera.* In addition, the secretory units of hypopharyngeal glands of *A. mellifera* from this study were different from results in the study of Deseyn and Billen (2005) who showed that the volume of acini decreased in foragers or displayed degenerative structure. This was not found by Suwannapong et al. (2010a). Hypopharyngeal glandular size is known to be positively correlated with gland activity and is influenced by larval feeding (Hrassnigg and Crailsheim, 1998; Ohashi et al., 2000; Crailsheim and Stolberg, 1989; Free, 1981; 1987). These glands gradually decrease in size when honeybees become guards, cease feeding, and begin defending the colony (Deseyn and Billen, 2005). However, the hypopharyngeal gland size of foragers was significantly larger than that of guards indicating that glandular development corresponds well with total protein synthesis in the hypopharyngeal glands at different adult life stages. Lastly, a number of reports indicate that the HPGs produce enzymes that are used to hydrolyze nectar into honey, including amylase, α-glucosidases, glucosidase oxidase, galactosidase, esterase, leucine arylamidase, and invertase (Hrassnigg and Crailsheim, 1998; Kubo et al., 1996; Li et al., 2008).

The Thorax

The thorax is the middle part of the bee. The thoracic contains the thoracic glands, which are derived from the cocoon-spinning gland of the larva and are well developed in workers, queens and drones (Snodgrass and Erickson, 1992; Snodgrass, 1956). Primarily, the thoracic is the anchor point for a bee's locomotory appendages for walking and flight. Three pairs of legs arise from the thorax: prothoracic (closest to the head), mesothoracic (center), and metathoracic (hind) legs. Honeybee metathoracic legs are modified to be pollen baskets. There are two also two pairs wings also located on the thorax of adult bees (Snodgrass, 1956).

Legs: The honeybee has three pairs of segmented legs. The legs of the bee are primarily used for walking. However, the legs have specialized areas such as the antennae cleaners on the prothoracic legs, and the pollen baskets on the metathoracic legs. These pollen baskets are also used to transfer propolis (Snodgrass and Erickson, 1992). Three pairs of legs divided into six segments by joints. The “foot” or tibia, of the insect has claws and a smooth pad to enable them to cling to surfaces. The mesothoracic leg has brushes for cleaning the thorax, and long spines at the end that bees use to loosen pollen from the pollen baskets and to clean the wings and the small breathing pores or spiracles (Snodgrass and Erickson, 1992; Snodgrass, 1956). In addition, the mesothoracic legs remove wax scales from the abdomen.

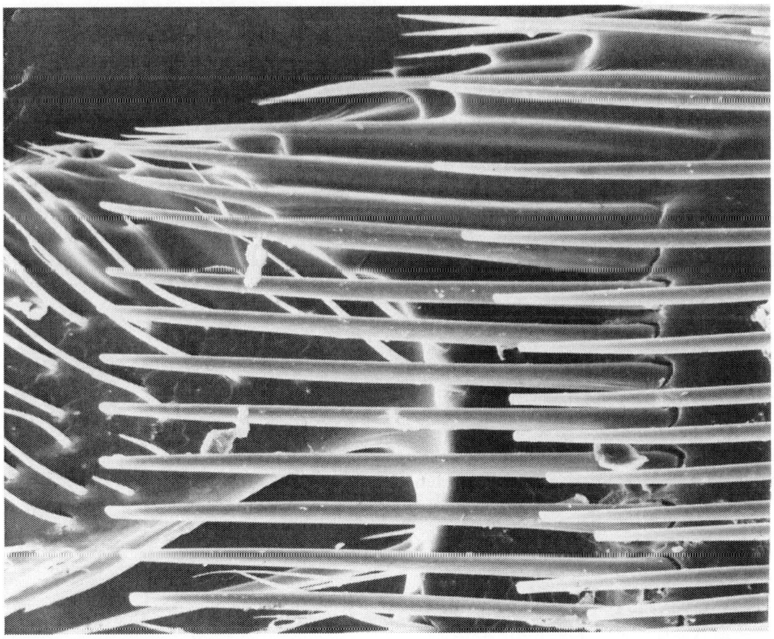

Figure 24. Scanning electron micrograph of the pollen basket of *A. cerana*.

The metathoracic legs differ from the other legs in their larger size and broad flattened form. These legs differ in size in the queen, worker, and drone. Only the worker collects pollen. The legs have long curved hairs. The space enclosed by these hairs is called the pollen basket. This consists of a smooth, somewhat concave surface on the outer metathoracic leg that is fringed with long, curved hairs to hold the pollen in place. This structure is known as a corbiculum and has 10 transverse rows of stiff hairs projecting backwards (Snodgrass and Erickson, 1992; Snodgrass, 1956). The deep notch between the upper and middle portion of the metathoracic leg transfers the pollen from brushes on the ventral section of each leg to the medial (inner) section of each leg. In this notch, there are short stiff spines called a rake. Bees use the rake by rubbing the leg on one side against the other leg.

The basitarsal hairs of *A. florea* are white while in *A. andreniformis* they are black. Drones of *A. florea* and *A. andreniformis* have a distinctive long inner lobe on the hind tibia (Maa, 1953; Wongsiri et al., 1996a; Wu and Kuang, 1987).

Wings: The honeybee has two sets of flat, thin, membranous wings, strengthened by veins (Crushman, 2010; Snodgrass and Erickson, 1992). The fore wings are much larger and stronger than the hind wings, and the two wings of each side work together in flight since small hooks called hammulae connect them. The bottom wing has hooks on the top edge. During flight, the front wings are drawn over the hind wings and held together by the hooks. Flight results from a propeller-like twist given to each wing during the upstroke and the down stroke. The wing vein measurement is very important for calculating the cubital index, the ratio of two wing vein segments. For instance, in Figure 25 points A, B & C should be judged to the centroid of the vein junctions concerned. The distance "AB" divides the distance "BC". There is a tool for doing this known as the Herold Fan named after the individual who devised this method (Crushman, 2010).

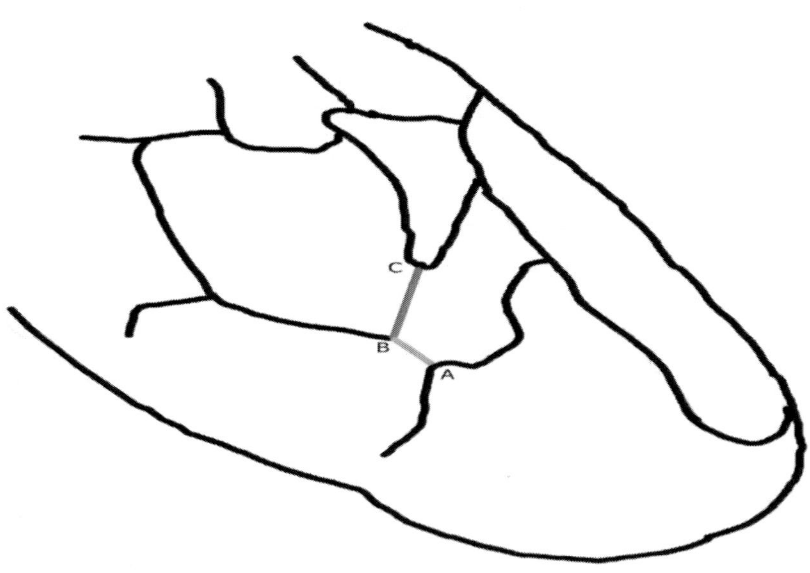

Figure 25. Wing venation of *A. florea* worker.

The pattern of the veins of the fore wings and the cubital index is consistent for a given race of bee. The cubital index of *A. florea* is 2.78 (Wongsiri et al., 1990) and 2.86 (Rinderer et al., 1995). The wing length is 6.26 ±0.10 mm (Seeley, 1982). In *A. andreniformis*, the cubital index ranges from 6.28-6.37 (Rinderer et al., 1995; de Guzman et al., 1993). *Apis dorsata* workers have a rusty brown pubescence and dark tinge to the wings that are approximately 12.34 ± 0.34 mm in length (Seeley, 1982). The wing length of *A. cerana* is 7.54± 0.14 mm (Tingek et al., 1996) and the cubital index is 4.40 (Seeley, 1982). In *A. mellifera*, the forewing length is 7.64-9.70 mm and the cubital index is 2.30 (Seeley, 1982).

Honeybee species are clustered into three groups base on body and wing size. The largest size is *A. dorsata* group with a forewing length of 12-15 mm. Honeybees with a forewing length of 7-10 mm are a medium size of *A. mellifera* group while the smallest species, *A. florea* have fore wing length are from 5-7 mm (Maa, 1953; Ruttner, 1988).

The Abdomen

The honeybee abdomen is composed of nine segments and contains digestive organs, wax and some pheromone glands, reproductive organs, and the stinger which is a modified ovipositer that can both sting and deposit eggs in adult females. The body color is a distinctive characteristic among honeybee species and subspecies (Woyke, 1977). *Apis florea* has bright red-brown color on the first and the second abdominal segments, but *A. andreniformis* has a chestnut brown color, and other segments are yellow and banded in queen and workers, but are black in drones.

Figure 26. *Apis andreniformis, A. cerana, A. dorsata, A. florea,* and *A. mellifera* showing body color.

The gene responsible for body color expression of *A. florea* is designated Fl (Woyke, 1998). The pattern of yellow and black between queen and workers are different. The body color of *A. andreniformis* workers after the second segment is yellow although they are the darkest of all five honeybee species of Thailand. Queen and drones are black. In *A. dorsata* the workers are yellow and black on abdomen segments (Sakagami et al., 1980): however, the queen and drones are brown color. The gene governing body color in this species is designated Do (Woyke, 1998). *Apis cerana* workers are yellow while queen and drone are brownish- black. The gene control color expression of this species is designated Ce (Woyke, 1998). The body color of *A. mellifera* is similar in workers, queen and drones, has yellow and black color abdomen, however the black area is different among subspecies which is control by Gene Y (Woyke, 1977; 1998).

Stinger: The stinger is located in a chamber at the end of the abdomen, from which only the sharp pointed shaft protrudes. The stinger is modified from an egg-laying organ, known as the ovipositor. However, the stinger can also be used to inject venom. Only females have a stinger/ovipositor. When the stinger is not in use, it is retracted within the sting chamber in the abdomen. The shaft of the sting is a hollow tube, like a hypodermic needle. The tip is barbed so that it sticks into the flesh or exoskeleton of the victim. The hollow needle has three sections. The top section is called the stylet and has ridges while the bottom has two pieces called lancets (Snodgrass and Erickson, 1992; Snodgrass, 1956).

When the stinger penetrates the skin, the two lancets move back and forth on the ridges of the stylet so that the whole apparatus is driven deeper into the skin. The enclosed tube of the two adjacent lancets creates the poison canal. In front of the shaft is the bulb. The ends of the lancets within the bulb are enlarged, and, as they move, they force the venom into the poison canal, like miniature plungers. The venom comes from two acid glands that secrete

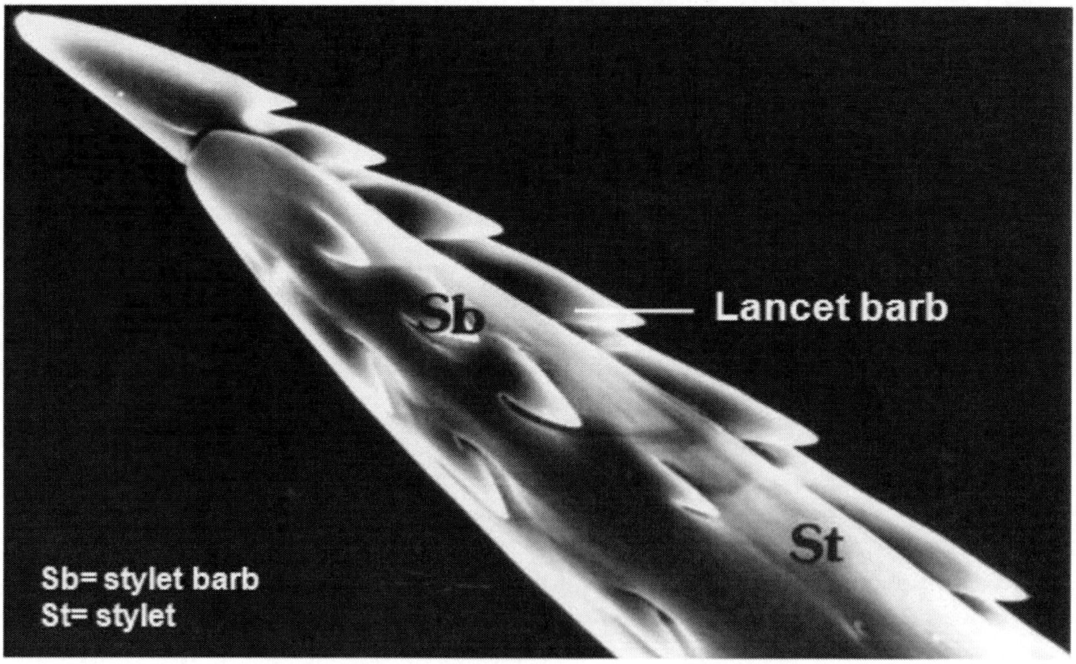

Figure 27. Scanning electron micrograph of the sting apparatus of *A. cerana*.

into the poison sac. During stinging, the contents of the alkaline gland are dumped directly into the poison canal where they mix with the acidic portion. When a honeybee stings another animal, the sting becomes embedded. In its struggle to free itself, a portion of the stinger is left behind in the victim. This often damages the honeybee enough to kill her (Snodgrass and Erickson, 1992; Snodgrass, 1956).

Interestingly, although separated from the bee, the stinger continues to contract reflexively, continuously pumping venom into the wound for several seconds. The stinger of the queen is longer than that of the worker, and more solidly attached within the sting chamber. The lancets are fewer and the barbs are smaller than those found in a worker's stinger (and therefore less likely to remain inside the victim such as another queen), but the poison glands are well developed and the poison sac is very large (Snodgrass and Erickson, 1992). Queens typically only use their stinger upon other queens in the competition over who will take over the colony (see above section on castes and colony re-queening) (Snodgrass and Erickson, 1992; Snodgrass, 1956).

4. Honeybee Pheromones

Pheromone Glands and Pheromone Production

Communication among insects is extremely important for their survival, especially for social insects that live in complex colonies. Honeybees are well known their chemical communication and use of olfactory cues (Free, 1987; Free et al., 1983; Leal 2010). For example, honeybees can navigate to food sources by detecting floral scents and by orienting towards food-marking pheromones or, possibly, cuticular hydrocarbons (cues) deposited as footprint (Balerrama et al., 1996). Honeybees smell or detect pheromone with their antennae using odorant-binding proteins in sensillum lymph. They produce volatile and non-volatile chemicals as signal molecules from their exocrine glands to communicate with others of the same species or with other species (Kaissling, 1987). These signaling chemicals are often called semiochemicals (Kaissling, 1972).

The secretory product of worker mandibular glands consists of 10- and 8-carbon acids that have an oily appearance. In young workers, this gland produces a lipid-rich white substance that is mixed with the secretion of the hypopharyngeal glands to make royal and worker jelly that is fed to the queen or workers. In old workers (foragers), the gland also produces 2-heptanone, a volatile substance that accumulates in the central reservoir, the amount of which progressively increases with increasing age (Engels et al., 1997). This compound can repel guard bees, and is a potential component of alarm pheromone. On guards, 2-heptanone has been reported to have either an attractive or a repulsive effect, according to the season. A foraging bee may mark a nectar-depleted flower with 2-heptanone (Balerrama et al., 1996; Boch and Shearer, 1962; 1971; Blum, 1969; 1982; 1992; Blum et al., 1978; Crewe and Hastings, 1976; Engels et al., 1997; Guirfa, 1991; Suwannapong, 2000). In foraging bees, 2-heptanone can have a temporary, repulsive effect on the visitation of flowers. Suggesting that, it is "forage marking" pheromone (Vallet et al., 1991). Mandibular glands of Thai honeybees mainly produce (z) -11-eicosanol instead of 2-heptanone (Suwannapong, 2000, Suwannapong et al., 2010a).

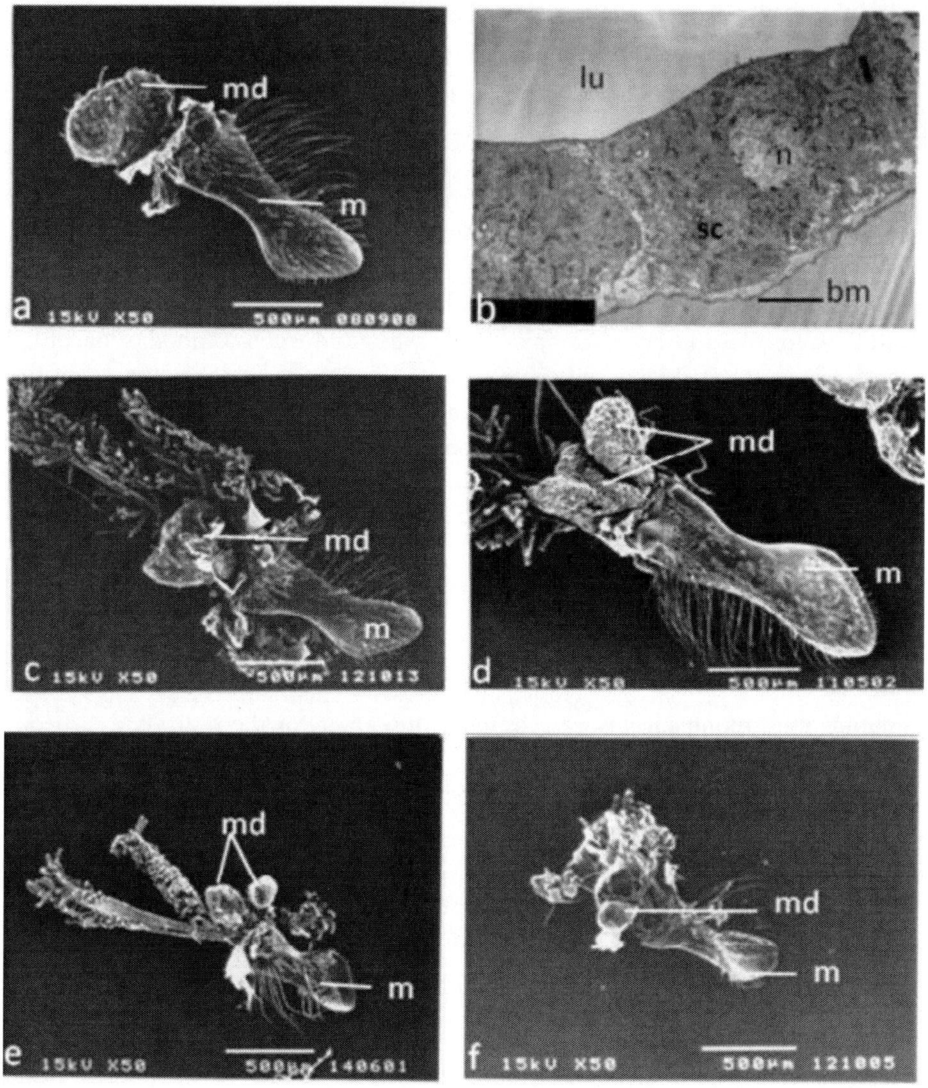

Figure 28. (a) Scanning electron micrographs of one mandibular gland of *A. mellifera*, (b) the duct cells of that mandibular gland, (c) one mandibular gland of *A. cerana*, (d) a mandibular glands of *A. dorsata*, (e) mandibular glands of *A. andreniformis*, and (f) mandibular glands of *A. florea*; bm, basement membrane; lu, lumen; m, mandible; md, mandibular gland; n, nucleus; sc, secretory cell (Suwannapong and Benbow, 2011).

Maschwitz (1964) suggested that mandibular glands produce alerting pheromones (in Blum, 1969). Shearer and Boch (1965) identified 2-heptanone from mandibular gland secretions. When filter paper with 2-heptanone was placed at the hive entrance, guard bees were alerted and attacked the paper. This is consistent with suggestions by Boch and Shearer (1962) that 2-heptanone has two functions: as an alarm pheromone and as a foraging repellant pheromone. However, it is possible that cuticular hydrocarbons deposited by bees walking on a depleted food source are sufficient to indicate that food sources are depleted. It is possible that 2-heptanone is a repellent at high concentrations and an attractant at low concentrations

(Boch and Shearer, 1971; Kerr et al., 1974; Vallet et al., 1991). The conflicting hypotheses of 2-heptanone's function on a depleted resource (repellent or attractant) require additional study.

In *A. florea* and *A. cerana*, the response of guards towards 2-heptanone was different from the response of foraging bees (Suwannapong et al., 2010; Suwannapong et al., 2011). The response of antennal sensillum of *A. florea* to low concentrations of 0.1 % 2-heptanone was higher than the response to higher concentrations of 5% and 10% 2-heptanone (Suwannapong et al., 2011). In contrast, *A. cerana* foragers exhibited a stronger response to a 10% 2-heptanone compared to that of 0.1 and 1.0%. The membrane potential of foragers following exposure to 2-heptanone was higher than that of guards in both *A. florea* and *A. cerana* (Suwannapong et al., 2010c; Suwannapong et al., 2011). One of the components of sting pheromone, isoamyl acetate (IPA), releases strong alarm behavior in bees. Aggressive behavior can be observed when mandibular glands or crushed heads of worker bees are presented at the hive entrance (Shearer and Bosch, 1962; 1965). IPA is 20-70 times more effective as an alarm pheromone than 2-heptanone (Boch and Shearer, 1971; Boch et al., 1975).

All species of *Apis* have alarm pheromones and the compounds are generally similar among honeybee species with the exception of *A. laboriosa*, the giant Himalayan honeybee (Vander Meer et al., 1998; Harborne, 1993). Africanized bees secrete alarm pheromones with the same concentration of isopentyl acetate as other bee species, but with more 2-nonanol and decyl acetate. These differences may leads to a more aggressive response to alarm pheromone. Honeybee species that have open nests, such as *A. florea*, *A. andreniformis*, and *A. dorsata* tend to have alarm pheromones that persist longer, such as 2-decenyl acetate (Vander Meer et al., 1998). Workers of Thai *Apis* species also have these compounds, along with 9-HDA and ODA, which are normally not present in *A. mellifera* worker glands. Queens and workers of each different *Apis* species have different combinations of mandibular compounds (Plettner et al., 1982; 1996). The sting glands of *A. dorsata* and *A. florea* have an additional pheromone besides isopentyl acetate (IPA), 2-decen-1-yl-acetate (2-DA). Upon presenting this pure compound, these Thai species exhibited a prolonged alarm reaction as compared to the reaction for pure IPA. A mixture of IPA and 2DA had a similar effect on the behaviour and reaction time, as did sting extracts (Koeniger et al., 1979).

5. POLLINATION BY THAI HONEYBEES

Thailand is tropical country that has four main regions: (1) the northern mountainous region, (2) the north-east (including the semi-arid Korat plateau, the most desolate and least-visited part of the country), (3) central Thailand (including the fertile plains surrounding the Chao Phraya River which is the country's most populous region and its rice basket), and (4) the southern region which stretches for hundreds of miles along the Malay peninsula. Thailand has a tropical climate divided into three seasons: cool in November to February, hot in March to May, and rainy in June to October. The seasons are more extreme in the northern regions (Maksong, 2008).

In Thailand, more than a thousand plant species, including agricultural crops and native plants are pollinated by insects. Of these, several hundred plant species are visited by honeybees. For example, honeybee pollination increases the productivity of crops such as corn, sunflower, lychee, mango, rambutan, longan, sesame, and durian. This section provides an overview of the plants upon which Thai honeybees forage, how to use pollen analysis to identify these plants, how to manage bee plants to the benefit of bees, and when different species flower (Maksong, 2008).

Honeybee Pollination in Thailand

Honeybees play are major agricultural pollinators around the world (McGreger, 1976; Crane, 1991; Free 1993; Partap and Verma, 1994) play an important role in tropical ecosystems, such as in Thailand. All species of honeybees in Thailand tend to be good generalist pollinators for native plants. Honeybees typically do not nectar rob, their body size and proboscis lengths are suited for pollinating many types of crops and they can forage in a wide variety of weather conditions.

Thai honeybee species forage by visiting plants for nectar and pollen. As the bees forage for nectar, pollen sticks to the fuzzy hairs that cover their bodies. Some of this pollen rubs off on the next flower they visit, fertilizing the flower. Some plants will not produce fruit at all without the help of honeybees. This floral fidelity of bees is due to their preference for nectars having sugar contents and pollens with higher nutritive values. All species of honeybees in Thailand are very good pollinators for native plants due to their related morphological structure of the organs that fit and provide other functions important for pollination, such as a body covered with hairs that help carry nectar and pollen. Further the bees do not injure the plants, as the body size and proboscis length are very much suitable for many Thai crops (Pyramarn and Wongsiri, 1986).

The availability of natural insect pollinators in Thailand is decreasing rapidly as a result of increased and continued use of pesticides. There is timely need for better management of hive honeybees such as *A.* cerana and *A. mellifera* in rare pollinator areas to increase fruit production. Information on the role of honeybees in pollination leads to increase in the quality and yield of crops that has been reported worldwide (McGreger, 1976; Crane, 1991; Free 1993; Partap and Verma, 1994).

Categorization of Bee Flora

Nectar content, odor, color, and shape of flowers affect honeybee foraging behavior, which is somewhat different among species. Honeybees forage on a variety of plant species to collect nectar and pollen (McGregor, 1976). However, not all plant species are available in one locality. A plant that produces nectar and pollen prolifically in one area may not yield the same amount of nectar and pollen in another area (Erdtman, 1966, 1969; Latif et al., 1960; Singh, 1981).

Bee flora or bee plants are the plants at which bees collect pollen and nectar. There are three types of bee flora: plants that only supply nectar, plants that only supply pollen, and plants that provide both (Allen et al., 1998; Baker, 1971; 1983; Bhattacharya, 2004; Crane et al., 1989; Partap, 1997). Some plants provide only resin, but these are less common. Floral nectar provides energy for flight activity, foraging activity and other activity in the colony. Honeybees also convert the nectar into honey and store in honey storage area of the comb.

Pollen provides protein, lipids, minerals, and vitamins (Gary, 1975; 1992). Pollen from different plant species differs in nutritive value and attractiveness to honeybees (Baker, 1971; 1983; Erdtman, 1966, 1969; Shuel, 1992).

The five species of honeybees in Thailand differ somewhat morphology, differences that affect their foraging preferences. Nectar content, odor, color and shape of flowers affect to honeybee foraging behavior and they are different among honeybee species (Akratanakul, 1976; Maksong, 2008; Pyramarn and Wongsiri, 1986). In Thailand, several plant species are most visited by these five species. These plants consist of agricultural and horticultural crops such as maize, rice, cucurbitaceous plants, bean, carrots, cabbage, litchi, mango, papaya and wild plants and forest trees such as rubber tree, Eucalyptus, teak, Burmese Ebony, and jambolan plum (Akratanakul, 1976; Maksong, 2008; Pyramarn and Wongsiri, 1986). Studies of bee flora can be carried out by several techniques such as pollen load analysis, melissopalynology (identification bee flora from honey), identification from the midgut, and from observations of foraging activity on flowers of different local plants (Akratanakul, 1976; Maksong, 2008; Pyramarn and Wongsiri, 1986).

The success of beekeeping essential depends on the abundance and management of bee flora in an area. However, a plant that produces nectar and pollen prolifically in one area may not yield the same amount of nectar and pollen in another area (Latif et al., 1958; Singh, 1981). The species of plants visited by Thai honeybees are shown in Table 3. There are more than 30 species of plants visited by *A. andreniformis* in Thailand such as Anacardium occidentale L., Antigonon leptopus Hook., Balakara baccata Roxb., Brassica chinensis Jusl var., Castanopsis acuminatissima Rehd., Chrysal, Cocos nucifera L., Coriandrum sativum L., Conyza sumatrensis Retz., Cucurbita citrillus L. Cucumis sativus Linn, Cuphea hyssopifola H.B.K., Dimocarpus longan Lour., Eugenia javanica and Mimosa pigra (Maksong, 2008).

The plants visited by *A. florea* include more than 40 species such as Mimosa pigra, Callistemon viminalis, Vetchia merrillii (Becc.) H.E. Mosre, Cocos nucifera L., Melampodium divaricatum, Zea mays L., Cuphea hyssopifola H.B.K., Dimocarpus longan Lour., Durio zibethinus L. Eugenia javanica, Eupatorium odoratum L., Euphoria longana Lamk., Fragaria ananassa Guedes, Hopea odorata Roxb (Maksong et al., 2011). *Apis dorsata* reportedly uses fewer food plants than *A. florea*. Only 38 species are reportedly used by *A. dorsata*: Ageratum conyzoides L., Amomum xanthioides Wall., Anacardium occidentale L., Blumea balsamifera L. DC., Bidens biternata Merr. and Sherff., Celosia argentea, Cinnamomum kerrii Kosten, Citrus aurantifolia Swing., Citrus maxima (J. Burman) Merr., Cocos nucifera L. (Maksong 2008).

There are more than 68 plant species visited by *A.* Cerana. These include Aeschynomene Americana L., Ageratum conyzoides L., Amomum xanthioides Wall., Anacardium occidentale L., Antigonon leptopus Hook. Balakara baccata Roxb., Bidens biternata Merr. & Sher, Brachiaria ruziziensis Germain & Evrard, Castanopsis acuminatissima Rehd., Cinnamomum kerrii Kosten, Coccinia grandis CL.Voigt, Cocos nucifera L., Coffea Arabica L., Conyza sumatrensis Retz. The number of bee *florea* of the introduced honeybee species in Thailand are more than 54 species such as Ageratum conyzoides L. , Durio zibethinus L., Euphoria longana Lamk., Fragaria × ananassa Guedes, Leersia hexandra Sw., Macadamia integrifolia maiden & Betche, Mikania cordata Roxb., Mimosa pigra, Musa acuminata Colla., Nephelium lappaccum L., Ocimum basillicum L., Oryza sativa L., Oxalis acetosella L. Prunus mume Sieb., Psidium guajava L., Sesamum indicum L., Schoenoplectus juncoides (Roxb.) Palla, Raphanus sativus L. (Maksong, 2008).

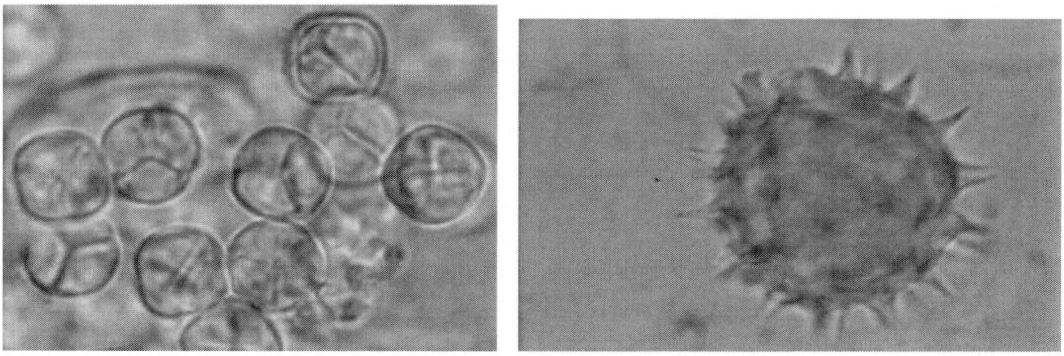

Figure 29. Pollen of bee flora, Mimosa pudica L. (left) and Tagete erecta L. (right) from the midgut of *A. dorsata*.

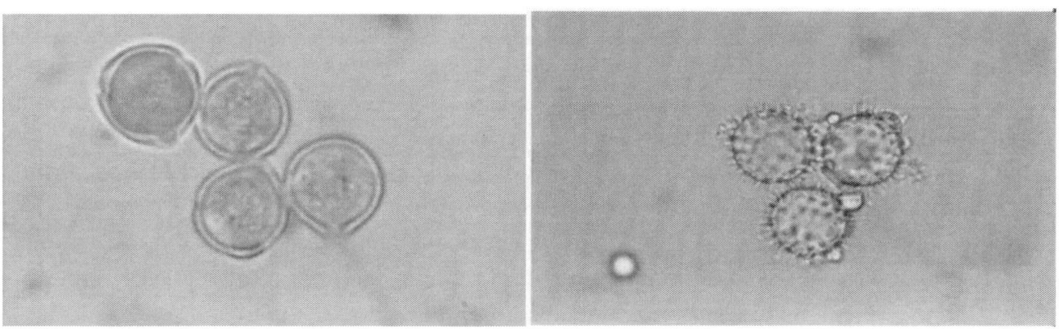

Figure 30. Pollen of bee flora, Muntingia calabure L. (left) and Melampodium divaricatum (right) from the midgut of *A. andreniformis*.

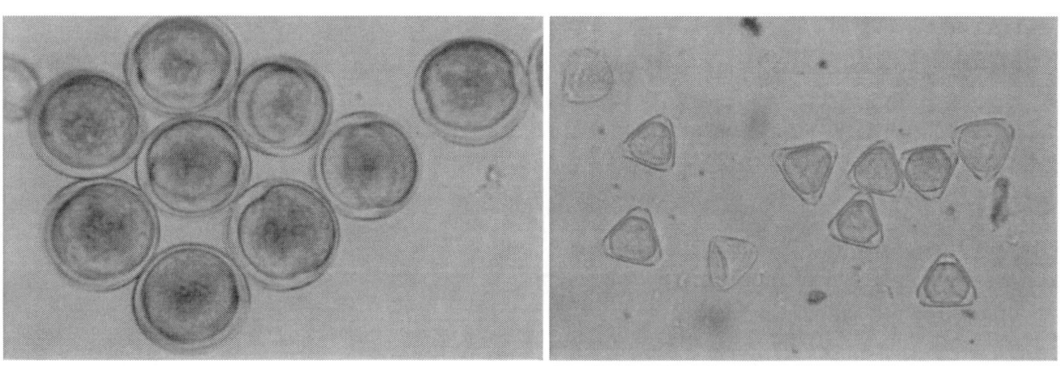

Figure 31. Pollen of bee flora, Nymphaea nouchali Borm. F (left) and Callistemon viminalis (right) from the mid gut of *A. florea*.

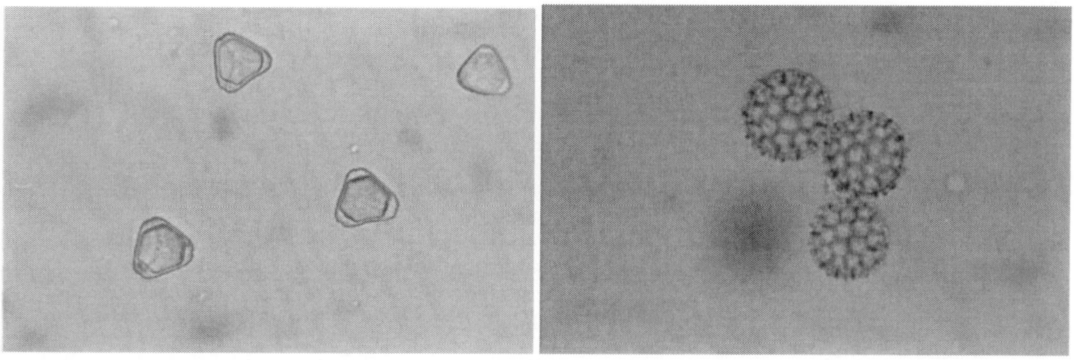

Figure 32. Pollen of bee flora, Syzygium malaccense L. (left) and Gomphrena globosa (right) from the mid gut of *A.* cerana.

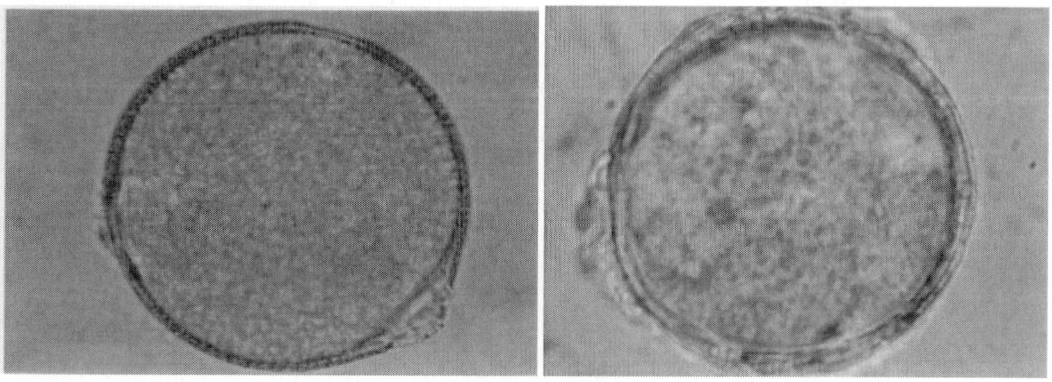

Figure 33. Pollen of bee flora, Luffa cylindrical Roem (left) and Coccina grandis CL.Voigt (right) from the mid gut of *A.* cerana.

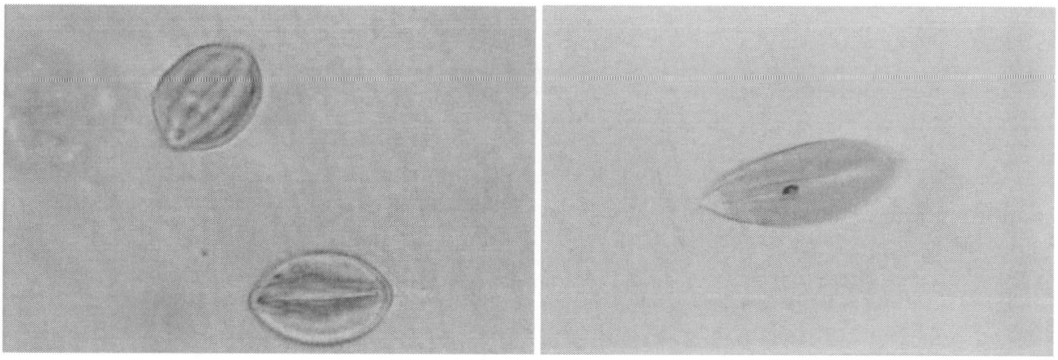

Figure 34. Pollen of bee flora, Euphorbia millii Desmoul. (left) and Vetchia merrillii (Becc.) H.E. Mosre (right) from the mid gut of *A. mellifera.*

Table 3. Bee flora of honeybees in Thailand study by pollen load and honey analysis

Plant species	Honey bee species			
	A. cerana	A. *dorsata*	A. *florea*	A. *mellifera*
Aeschynomene Americana L. **Common names:** American jointvetch, joint-vetch, shyleaf, deervetch	+	-	-	-
Ageratum conyzoides L. **Common names:** Catinga de Bode, Mexican ageratum, Erva de Sao Joao, Aru batu, Bandotan, Berokan, Rumput tahi ayam, Rompesaraguelo, Wedusan	+	+	-	+
Amomum xanthioides Wall. **Common names:** Bastard Cardamom , Tavoy Cardamom	+	+	-	-
Anacardium occidentale L. **Common names:** cashew	+	+	+	+
Antigonon leptopus Hook. **Common names:** Chain of Love; Coral Creeper; Coral Vine; Mexican Rose.	+	-	+	-
Balakara baccata Roxb. **Common names:** Pho bai (Thai)	+	-	+	-
Blumea balsamifera (L.) DC. **Common names:** Blumea camphor, Ngai Camphor, Sambong	+	+	-	-
Bidens biternata Merr. & Sherff. **Common names:** Spanish Needles, yellow flowered blackjack, black jack, five leaved blackjack, beggar ticks	+	+	+	+
Brachiaria ruziziensis Germain & Evrard **Common names:** Ruzi grass, Congo grass, Congo signal grass, prostrate signal grass	+	-	-	+

Brassica chinensis Jusl var. **Common names:** Chinese cabbage, Bai-Cai, Sawi-Putih, Bok Choy, Pak-Choi	+	-	+	+
Callistemon lanceolatus DC. **Common names:** Bottle brush plant	-	-	+	-
Castanopsis acuminatissima Rehd. **Common names:** Castanopsis chestnut, Gon, Ko duel	+	-	+	+
Ceiba pentandra (L.) **Common names:** Kapok tree, silk cotton tree, ceiba de lana, bois coton, kapokier, pacae, sumauma, kankantri.	-	-	-	+
Celosia argentea **Common names:** Will cockcomb, Cockcomb	+	+	-	-
Chrysalidocarpus lutescens H. Wendl **Common names:** Butterfly palm, Cane palm, Madagascar palm, Golden feather palm, yellow palm, Bamboo palm, Areca palm.	+	-	+	-
Cinnamomum kerrii Kosten **Common names:** -	+	+	-	-
Citrus aurantifolia Swing. **Common names:** Common lime, Lime		+	-	+
Citrus maxima (J. Burman) Merr. **Common names:** Pomelo	+	+	-	+
Citrus eticulate Blanco **Common names:** orange	-	-	-	+
Coccinia grandis CL.Voigt **Common names:** Ivy gourd	+	-	+	+
Cocos nucifera L. **Common names:** Coconut palm	+	+	+	+
Coffea Arabica L. **Common names:** Kofi, coffee, koffie, Brazilian coffee	+	-	-	+

Table 3. (Continued)

Coriandrum sativum L. **Common names:** Chinese Parsley	-	-	+	+
Conyza sumatrensis Retz. **Common names:** fleabane, tall fleabane, broad-leaved fleabane, white horseweed, Sumatran fleabane, Guernsey fleabane	+	-	+	+
Cosmos sulphureus Cav. **Common names:** Yellow cosmos, Orange cosomos	+	-	-	-
Crataeva magna Lour. **Common names:**-	-	+	-	-
Croton oblongifolius Roxb. **Common names:** Oblong-leaved croton	+	-	-	+
Cucurbita citrillus L. **Common names:** Pumpkin	+	+	+	-
Cucumis sativus Linn **Common names:** Cucumber	-	-	+	-
Cuphea hyssopifola H.B.K. **Common names:** Mexican Heather, False Heather, Hawaiian Heather	-	-	+	-
Dalbergia oliveri Gamble ex Prain **Common names:** Burma Rosewood, Tamalan (Thai), Mai Kham Phii)Lao	-	+	-	-
Datura metel L. **Common names:** angel's trumpet, devil's trumpet, metel.	+	-	-	+
Dillenia ovata Wall. **Common names:** Ovate dillenia, Kadah Simpoh, Simpoh Beludu	-	-	-	+
Dimocarpus longan Lour. **Common names:** longan	+	+	+	+
Diospyros glandulosa Lacc. **Common names:** Kluai-ruesi (Thai)	-	+	-	+

Diospyros areolata King& Gamble **Common names:** phlap, kayu arang, maphlap(Thai)	-	+	-	-
Duabanga grandiflora Walp. **Common names:** lampatti, lamphu-pa (Thai)	-	+	-	-
Duranta erecta L. **Common names:** Golden Dewdrop, Pigeon Berry, Skyflower	-	-	-	-
Durio zibethinus L. **Common names:** Durian	+	+	+	+
Elaeagnus latifolia L. **Common names:** Oleaster, silverberry	+	-	-	+
Erythrina suvumbrans Merr. **Common names:** thong-lang (Thai), December-tree	-	+	-	-
Eugenia javanica **Common names:** wax apple, love apple, java apple, chomphu (Thai)	+	+	+	
Eupatorium odoratum L. **Common names:** Siam Weed, Christmas Bush, Common Floss Flower	+	+	+	+
Euphorbia millii Desmoul **Common names:** Crown-of-thorns, Christ Plant	+	-	-	-
Euphoria longana Lamk. **Common names:** longan, longyan	+	+	+	+
Fragaria × ananassa Guedes **Common names:** Strawberry	+	-	+	+
Gmelina arborea Roxb. **Common names:** Beechwood, Gmelina, Goomar teak	+	+	-	+
Gomphrena globosa L. **Common names:** Globe amaranth, Bachelor's bottons, Gomphrena	+	-	-	-
Helianthus annuus **Common names:** Sunflower	-	-	+	+

Table 3. (Continued)

Hopea odorata Roxb. **Common names:** Takian, white thingan	-	+	-	-
Jacaranda filicifolia D.Don **Common names:** Blue Jacaranda	-	+	-	-
Leersia hexandra Sw. **Common names:** Swamp rice grass, swamp cut grass, Southern cutgrass	+	-	-	+
Leucaena leucocephala de Wit. **Common names:** White Leadtree, Jumbay, White Popinac	+	-	+	+
Litchi chinensis Sonn **Common names:** lychee	+	-	+	+
Luffa cylindrica Roem **Common names:** sponge gourd, Smooth loofah, Loofah	+	-	-	-
Lxora stricta Roxb. **Common names:** Chinese ixora, Needle flower, Jungle flame	+	-	-	-
Macadamia integrifolia maiden & Betche **Common names:** macadamia nut, Australian brush nut, Bopple nut	+	-	-	+
Mangifera indica L. **Common names:** Mango, Mangot, Manga	+	-	+	-
Melampodium divaricatum **Common names:** butter daisy, melampodium	-	-	+	-
Mikania cordata Roxb. **Common names:** heartleaf hempvine	+	+	+	+
Mimosa diplotricha C. Wright. **Common names:** Giant sensitive plant, creeping, nila grass	+	+	+	-

Mimosa pigra **Common names:** Giant Sensitive Tree, bashful plant, catclaw mimosa	+	+	+	+
Mimosa pudica L. **Common names:** Sensitive plant, Sleeping grass, shame plant	+	+	+	+
Muntingia calabura L. **Common names:** strawberrytree, Jamaican cherry, takhop farang (Thai)	+	+	+	+
Musa acuminata Colla. **Common names:** Dwarf Cavendish Banana	+	+	-	+
Musa sapientum L. **Common names:** Banana	+	-	-	+
Nephelium lappaccum L. **Common names:** Rambutan	+	-	-	+
Nymphaea nouchali Borm. F **Common names:** Lotus	+	-	+	-
Ocimum sanctum L. **Common names:** Holy Basil	-	-	+	-
Ocimum basillicum L. **Common names:** Sweet Basil	+	-	-	+
Oryza sativa L. **Common names:** Rice	-	-	-	+
Oxalis acetosella L. **Common names:** Wood Sorrel, Shamrock	-	-	-	+
Passiflora laurifolia L. **Common names:** passion fruit, bell apple, yellow granadilla	+	+	-	-
Pithecellobium dulce (Roxb.) Benth. **Common names:** monkeypod	+	-	+	-
Portulaca oleracea L **Common names:** Verdolaga, Pigweed, Little Hogweed	-	-	+	-

Table 3. (Continued)

Prunus cerasoides D.Don **Common names:** sour cherry, wild Himalayan cherry	-	-	-	+
Prunus mume Sieb. **Common names:** Japanese Apricot, Black plum, Mume	-	-	+	+
Psidium guajava L. **Common names:** guava, goiaba, guayaba, djamboe	+	+	-	+
Pyrostegia venusta (Ker-Crawl.) **Common names:** flame flower, flame vine, orange creeper	-	-	-	+
Raphanus sativus L. **Common names:** cultivated radish, jointed charlock, wild radish	+	-	+	+
Schoenoplectus juncoides (Roxb.) Palla. **Common names:** Sedge, Rock bulrush	+	-	-	+
Sesamum indicum L. **Common names:** Sesame seed, benne seed, til	-	-	+	+
Shorea siamensis Miq. **Common names:** Dark Red Meranti, Red Lauan		+	-	+
Solanum torvum SW. **Common names:** Turkey berry	+	-	-	+
Spilanthes paniculata Wall. Ex DC. **Common names:** phak khraat (Thai),yari sennichimodoki (Japanese)	+	-	+	-
Synedrella nodiiflora (L.) Gaerth. **Common names:** Synedrella, cerbatana	+	-	+	-

Syzygium malaccense L. **Common names:** Mountain Apple, Malaysian Apple, 'Ohi'a 'Ai, Rose Apple, Malay Apple, Pomerac, Otaheite-apple	-	+	-	-
Tagete erecta L. **Common names:** Marigold	+		-	-
Tamarindus Indica L. **Common names:** Tamarind	+	+	+	+
Vetchia merrillii (Becc.) H.E. Mosre **Common names:** Manila palm, Christmas palm, Merrill palm	+	-	+	-
Wedelia trilobata (L.) Hiteh. **Common names:** Climbing wedelia, Creeping daisy, Singapore daisy	+	+	+	-
Wrightia religosa **Common names:** Water Jasmine, Wild Water Plum	-	-	-	-
Wrightia arborea (Dennst.) Mabb. **Common names:** Woolly Dyeing Rosebay	+	-	-	-
Zea mays L **Common names:** Corn	+	+	+	+
Zizyphus mauritiana Lamk. **Common names:** Indian jujube, common jujube	+	-	-	+

Table 4. Nectar, pollen and Nectar and pollen source plants of Thai honeybees

Number	Plant species	Nectar source	Pollen source
1	Ageratum conyzoides L.	+	+
2	Amomum xanthioides Wall.	+	+
3	Balakara baccata Roxb.	+	+
4	Blumea balsamifera (L.) DC.	+	+
5	Bidens biternata Merr. & Sherff.	+	+
6	Brachiaria ruziziensis Germain&Evrard	-	+
7	Brassica chinensis Jusl var.	+	+
8	Castanopsis acuminatissima Rehd.	-	+

Table 4. (Continued)

9	Ceiba pentandra (L.)	+	+
10	Cinnamomum kerrii Kosten	+	+
11	Citrus aurantifolia Swing.	+	+
12	Citrus maxima (J. Burman) Merr.	+	+
13	Coccinia grandis CL.Voigt	+	+
14	Cocos nucifera L.	+	+
15	Coffea Arabica L.	+	+
16	Coriandrum sativum L.	+	+
17	Conyza sumatrensis Retz.	+	+
18	Crataeva magna Lour.	+	+
19	Croton oblongifolius Roxb.	+	+
20	Cuphea hyssopifola H.B.K.	+	-
21	Dalbergia oliveri Gamble ex Prain	+	+
22	Datura metel L.	+	+
23	Dillenia ovata Wall.	+	+
24	Dimocarpus longan Lour.	+	+
25	Diospyros glandulosa Lacc.	+	+
26	Diospyros areolata King & Gamble	+	+
27	Duabanga grandiflora Walp.	+	+
28	Elaeagnus latifolia L.	+	+
29	Erythrina suvumbrans Merr.	+	+
30	Eucalyptus camaldulensis	+	+
31	Eugenia javanica	+	+
32	Eupatorium odoratum L.	+	+
33	Euphoria longana Lamk.	+	+
34	Fragaria × ananassa Guedes	+	+
35	Gmelina arborea Roxb.	+	+
36	Hopea odorata Roxb.	+	+
37	Jacaranda filicifolia D.Don	+	+
38	Leersia hexandra Sw.	-	+
39	Leucaena leucocephalade Wit.	-	+
40	Litchi chinensis Sonn	+	+
41	Macadamia integrifolia maiden & Betche	+	+
42	Mangifera indica L.	+	+
43	Mikania cordata Roxb.	+	+
44	Mimosa diplotricha C. Wright.	+	+
45	Mimosa pigra	+	+
46	Mimosa pudica L.	-	+

47	Muntingia calabura L.	+	+
48	Musa acuminata Colla.	+	+
49	Musa sapientum L.	+	+
50	Ocimum sanctum L.	+	+
51	Oryza sativa L.	-	+
52	Oxalis acetosella L.	+	+
53	Passiflora laurifolia L.	+	+
54	Prunus cerasoides D.Don	+	+
55	Prunus mume Sieb.	+	+
56	Psidium guajava L.	+	+
57	Raphanus sativus L.	+	+
58	Schoenoplectus juncoides (Roxb.) Palla.	-	+
59	Shorea siamensis Miq.	+	-
60	Solanum torvum SW.	+	+
61	Spilanthes paniculata Wall. Ex DC.	+	+
62	Synedrella nodiiflora (L.) Gaerth.	+	+
63	Wedelia trilobata (L.) Hiteh.	-	+
64	Wrightia arborea (Dennst.) Mabb.	+	+
65	Zea mays L	-	+
66	Zizyphus mauritiana Lamk.	+	-

6. BEEKEEPING IN THAILAND

Introduction

Beekeeping is an important component of agriculture and rural development programs in many Asian countries (Ahmad, 1992; Partap, 1992; Partap and Verma, 1998). It is mainly conducted using the European honeybee, *A. mellifera* and the Asiatic honeybee *A.* cerana. The benefits provided by beekeeping include enhanced nutritional, economic and ecological security to rural communities of Asia (Partarp and Verma, 1998). Thai people have learned how to hunt honey from wild bee colonies. However, their methods tend to destroy the colony. It is important that education is enhanced that can teach sustainable ways of harvesting honey. This training should teach beekeepers the following basic activities: checking general bee colony health, checking for the general presence of the queen, how to establish new colonies, control colony pests and diseases, moving bee hives to food sources, feeding colonies (if necessary), routine maintenance of bee hives, and how to harvest hive products.

History

The first report of Western-style beekeeping in standard Langstroth hives was recorded in 1940 by Professor Supachai Wattana (Wongsiri et al. 2000). He imported foreign honeybees to study at Chulalongkorn University in Bangkok. In 1953, Professor Saman Worakitta (who at the time served as the Dean of Agriculture at Kasetsart University) introduced European honeybees from Australia and raised them on campus. However, the operation was not successful (Wongsiri et al. 2000). Later, Thailand established a cooperative agreement with Taiwan to exchange and share knowledge related to beekeeping. Taiwan sent bee experts to advise farmers in northern Thailand: however, beekeeping was still quite limited. Until about 1976-1979, private companies hired specialists from Taiwan for management and operation of beekeeping in Thailand (Wongsiri et al. 2000).

Since 1980, the Thai government has recognized the importance of bees to the national economy. And as a result the government has a policy to encourage and promote agricultural apiarists. The Ministry of Agriculture and Cooperatives is an agency responsible for promoting and educating farmers on beekeeping. New apiarists can obtain information about beekeeping from government resources, take training courses, and receive advice provided by organizations distributed throughout the country that include the following:

1. Department of Agriculture
2. Department of Agricultural Extension
3. Agricultural Extension and Development Center Chiang Mai (Beekeeping)
4. Agricultural Extension and Development Center Phitsanulok (Beekeeping)
5. Agricultural Extension and Development Center Khon Kaen (Beekeeping)
6. Agricultural Extension and Development Center Chanthaburi (Beekeeping)
7. Agricultural Extension and Development Center Chumphon (Beekeeping)

Many universities in Thailand also have researchers who study honeybees, educate, and advise beekeepers. Some research units have their own apiary. Others cooperate with beekeepers and local communities. To receive information about bee biology and beekeeping you can go to any of the universities shown below:

1. Bee Biology Research Unit, Department of Biology, Faculty of Science, Chulalongkorn University, Bangkok, Thailand 10330.
2. Department of Entomology, Faculty of Agriculture, Kasetsart University, Bangkok, Thailand 10900.
3. Department of Agricultural Technology, Faculty of Science, Ramkhamhaeng University, Bangkok, Thailand 10900.
4. Department of Biology, Faculty of Science, Burapha University, Chon Buri, Thailand 20131.
5. Department of Entomology, Faculty of Agriculture, Khon Kaen University, Khon Kaen, Thailand 40000.
6. Department of Entomology, Faculty of Agriculture, Chiang Mai University, Chiang Mai, Thailand 50000.

7. Department of Agricultural Science, Faculty of Agriculture Natural Resources and Environment, Naresuan University, Phitsanulok, Thailand 65000.

There are also beekeeping associations with regional and local branches, which help beekeepers: the Beekeeper Association of Northeastern Thailand, the Beekeeper Association of Thailand (http://thailandbee.net), and the Thai Organic Beekeeper Association (http://www.thaiorganicbee.net). Some private bee farms, such as Supha Bee Farm in Chiang Mai (http://www.suphabeefarm.com) are good beekeeping resources.

Status of Beekeeping in Thailand

Beekeeping is chiefly conducted to produce honey, but some beekeepers focus on producing royal jelly, beeswax, bee colonies, or queens. Some beekeepers provide bees to pollinate crops. Each type of operation requires specific experience and management techniques, and thus beekeepers often specialize on one type of production. Only two-thirds of honeybee colonies in Thailand are used for crop pollination. Primarily, these colonies are used to pollinate longan, lychee, sesame, sunflower and bitter weed (Pyramarn and Wongsiri, 1986). The number of beekeepers rearing *A. mellifera* across the country according to information provided by the Agricultural Extension and Development Center is 887 as shown in Table 5.

The cavity nesting honeybee *A.* cerana and the introduced *A. mellifera* are the primarily species used for beekeeping. In Thailand, there are more than a thousand beekeepers keeping more than two hundred thousand colonies of *A. mellifera. Apis* cerana has been recognized as native to eastern Asia including Laos, Thailand, and Malaysia. Most races of *A.* cerana are slightly smaller than *A. mellifera* and have smaller colonies. Unlike *A. mellifera, A.* cerana provides fewer products for beekeepers. *Apis* cerana does not gather propolis; however, it is more resistant to mites than *A. mellifera*. In nature, *A.* cerana build their nest in logs or other cavities. Many local beekeepers keep this species in traditional beehives. They drill out the core of tree trunks and cover both ends with wooden lids. Three to five inches from the

Table 5. Number of beekeepers and honeybee colonies according to center

Province	No. of beekeepers	Total Colonies	Average
Agricultural Extension and Development Center Chiang Mai			
Chiang Mai	127	31,170	245
Chiang Rai	62	21,270	343
Lamphun	75	18,870	252
Lampang	27	1,510	56
Phayoa	12	6,870	573
Phrae	106	33,450	316
Nan	54	7,000	130
Total	**463**	**120, 140**	**259**

Table 5. (Continued)

Agricultural Extension and Development Center Phitsanulok			
Chai Nat	1	150	150
Kamphaeng Phet	1	20	20
Nakhon Sawan	1	150	150
Phetchabun	2	1,300	650
Sukhothai	1	100	100
Phichit	12	1,590	133
Phitsanulok	24	6,750	281
Uthai Thani	4	180	45
Uttaradit	55	12,051	219
Total	**101**	**22,291**	**221**
Agricultural Extension and Development Center Khon Kaen			
Khon Kaen	20	2,920	146
Udon Thani	32	5,586	175
Roi Et	4	1,470	368
Nong Khai	3	126	42
Loei	17	3,349	197
Chaiyaphum	6	580	97
Maha Sarakham	4	400	100
Sri Sa Ket	5	420	84
Amnat Charoen	3	510	170
Nong Bua Lam Phu	9	1,468	163
Nakhon Ratchasima	12	1,530	128
Buri Ram	4	160	40
Surin	1	850	850
Ubon Ratchathani	1	88	88
Total	**121**	**19,457**	**161**

Agricultural Extension and Development Center Chanthaburi			
Chanthaburi	18	1,480	82
Sa Kaeo	3	210	70
Saraburi	35	970	28
Lop Buri	27	12,540	464
Kanchanaburi	12	1,180	98
Total	**95**	**16,380**	**172**
Agricultural Extension and Development Center Chumphon			
Chumphon	24	1,970	82
Surat Thani	21	354	17
Nakhon Si Thammarat	18	323	18
Trang	3	57	19
Phattalung	24	72	3
Songkhla	12	720	60
Pattani	3	255	85
Satun	2	210	105
Total	**107**	**3,961**	**37**

bottom of the hive, they make a hole for bee entrance. However, most honeybee farms use movable-frame Langstroth-style hives for *A.* cerana. Management of *A.* cerana is relatively simple. Beekeepers need to provide sufficient food and protect the colony from its enemies (ants, wasps) to minimize absconding. Major beekeeping areas for *A.* cerana are in Southern Thailand: Chumphon, Surat Thani, Nakhon Si Thammarat, Trang, Phattalung, Songkhla, Pattani, and Satun.

Conserving *A.* Cerana and others Native Species in Thailand

At present, conservation of native Thai honeybee species is of primary importance. In terms of practicality for beekeepers, *A.* cerana is the best choice when compared to *A. dorsata*, *A. florea* and *A. andreniformis* because swarming and absconding is less frequent in *A.* cerana. However, *A.* cerana populations are declining because of competition with *A. mellifera*. The conservation of *A.* cerana is being promoted by the Agricultural Extension and Development Center of Thailand, particularly in southern areas such as Chumphon and Samui Island. A key point is to convince Thai farmers to think about crop pollination in addition to honey production (which is traditional and for which *A. mellifera* is superior). *Apis* cerana may be better suited for pollinating certain kinds of crops. This is an area of current Thai research. In addition, Thai honeybee researchers and agriculturalists are educating people to preserve natural bee habitat and increase honeybee habitat be planting bee flora.

Figure 35. A beekeeper wears a tight suit with veil to protect her from being stung by bees.

Figure 36. Beekeeper is removing *A.* cerana bood frame from coconut truck hole to keep in the box hive.

Beehive

A beehive is an enclosed structure in which beekeepers keep bees. Traditionally, beehives are often made from wood. A modern beehive is generally a box of wood with a bottom board, a brood box or brood chamber, a honey box (top box where most of the honey is stored), and frames (where bees build combs for egg laying and honey storage). In addition, a queen excluder made of slotted zinc, plastic, or wire keeps the queen in the brood chamber, an inner cover or crown board (which prevents the hive lid from sticking due to wax or propolis collected by bees), and the lid. However, in Thailand, beehives are generally built in one piece: the base, brood box, and honey box are united together. This box contains 8 to 9 removable frames. This hive is then set on a stand, which is usually 0.5-0.6 m in height to separate the hive from damp ground and to exclude ants (with oil or similar material applied to the legs). Hive stands can be made from wood or metal Wongsiri et al., 2000).

7. Honeybee Pathogens, Parasites and Predators

Introduction

Four main factors contribute to honeybee decline in Thailand: honeybee pests and diseases, deforestation, pesticides, and human management and honey hunting practices. Over the past two decades, these four factors have affected *A. mellifera* and the native species, *A. dorsata*, *A. florea* and *A. andreniformis* (Wongsiri et al. 2000).

Honeybee Mites

Parasites such as bee mites have been spread, in part, by human management practices because beekeepers move hives for commerce and pollination (Anderson, 1999; Oldroyd and Wongsiri, 2006). Such parasites are now a global problem, causing significant reduction in bee populations (particularly *A. mellifera*) and problems for fruit and vegetable producers who rely on bee pollination. The outbreak of parasite and disease has the potential to destroy beekeeping in Thailand (Wongsiri et al., 2000; Oldroyd and Wongsiri, 2006). The commercial beekeepers are usually preferring to switch to *A. mellifera* colonies to explore different bee flora leading to an increased exchange of disease and parasite within bee species and colonies (Boecking et al., 2000). In Asian countries including Thailand, honeybees are kept almost exclusively for honey production, in fact, pollination of crop plants by honeybees is much more important than honey production. It is estimated that honeybees account for 80 per cent of all crop pollination (Robinson et al., 1989). Mites are the largest and most diverse group of honeybee parasites. *Apis dorsata*, *A.* cerana, *A. florea*, *A. andreniformis*, and *A. mellifera* are parasitized by a wide variety of ectoparasitic mites, particularly Acar*Apis* woodi (Acarine tracheal mites), Varroa jacobsoni, V. destructor, Tropilaelaps clareae, T. koenigerum, T. mercedesae, and T. thaii. Consequently, mites constitute a major threat to all beekeeping in Thailand (Wongsiri et al., 2000; Oldroyd and Wongsiri, 2006).

Table 6. Mesostigmatic mites parasitizing bees, arranged according to host species (modified from Koeniger, 1996)

Honeybee species	Mite species	Reference
A. *andreniformis*	Euvarroa sinhai Euvarroa wongsirii	Delfinado-Baker, Baker and Phoon, 1989 Lekprayoon and Tangkanasing, 1991
A. *florea*	Euvarroa sinhai Tropilaelaps clareae	- Delfinado-Baker, Baker and Phoon, 1989
A. cerana	Tropilaelaps clareae Varroa jacobsoni Varroa underwoodi Acarapi woodi	Delfinado-Baker, Baker and Phoon, 1989 - Delfinado-Baker , Baker and Phoon, 1989
A. koschevnikovi	Varroa rindereri Varroa jacobsoni	de Guzman and Delfinado-Baker,1996 Delfinado-Baker, Baker and Phoon, 1989
A. *dorsata*	Tropilaelaps clareae Tropilaelaps koenigerum	Delfinado-Baker, Underwood and Baker, 1985
A. *laboriosa*	Tropilaelaps clareae Tropilaelaps koenigerum	- Delfinado-Baker, Baker and Phoon, 1989
A. *mellifera*	Tropilaelaps clareae Varroa jacobsoni Euvarroa sinhai Acarapi woodi	Delfinado-Baker, Baker and Phoon, 1989 Delfinado-Baker, Baker and Phoon, 1989 Koeniger, Koeniger, de Guzman and Lekprayoon, 1993

The relationship between different bee species and their acarine parasites is still being explored (Otis, 1991; Smith et al., 1991). Recently, several new mites have discovered parasitizing older and newly recognized honeybee species (see Table 7). *Apis andreniformis* and *A. florea* are attacked by Euvarroa spp. (Lekprayoon and Tangkanasing, 1991; 1993). *Apis* cerana, *A.* koschevnikovi, and *A. mellifera* are parasitized by different Varroa species (de Guzman and Delfinado-Baker, 1996). Finally, *A. dorsata* and *A. laboriosa* are parasitized by Tropilaelaps species (Delfinado-Baker et al., 1985; 1987). This pattern also seems to apply to Varroa rindereri rindereri (Anderson, 1999; Lekprayoon and Tangkanasing, 1991; 1993).

More and perhaps unexpected associations between honeybees and parasitic mites will likely be discovered. More species of Varroa and Tropilaelaps are likely to be found in Southeast Asia. In addition, significant differences have been found between E. sinhai from India and from Thailand (Morin and Otis, 1993). In Borneo, de Guzman and Delfinado-Baker (1996) reported finding multiple Varroa species similar to V. underwoodi on *Apis* nuluensis.

Tropilaelaps Clareae

This species of mite was originally discovered in the rat (Delfinado and Baker, 1961), but is also found on five species of honeybees (see below), is primarily found on *A. dorsata*, and occur on *A. mellifera* (Laigo and Morse, 1969). Currently, T. clareae is restricted to Asia and ranges from Iran to Papua New Guinea, including India, Pakistan, Philippines, Nepal, and Burma (Mattheson, 1993; 1996). Tropilaelaps koenigerum has been reported in Sri Lanka and Nepal (Delfinado-Baker, 1985), Borneo (Koeniger et al., 2002), and Thailand (Tangjingjai et al., 2003). Tropilaelaps clareae is now found in five honeybee species: *A. mellifera*, *A. dorsata*, *A.* cerana, *A. florea*, and *A. laboriosa* (Aggarwal, 1988). After the European honeybee (A. *mellifera*) was introduced to Asia and subsequently to Thailand, this species was parasitized by T. clareae, resulting in a significant annual decrease in commercial honey production (DeJong, 1990; 1997). Interestingly, T. clareae is more harmful to the exotic species, *A. mellifera* than to its native host, *A. dorsata* (Eickwort, 1988). Because of the importance of this parasite, we have included detailed information about its biology.

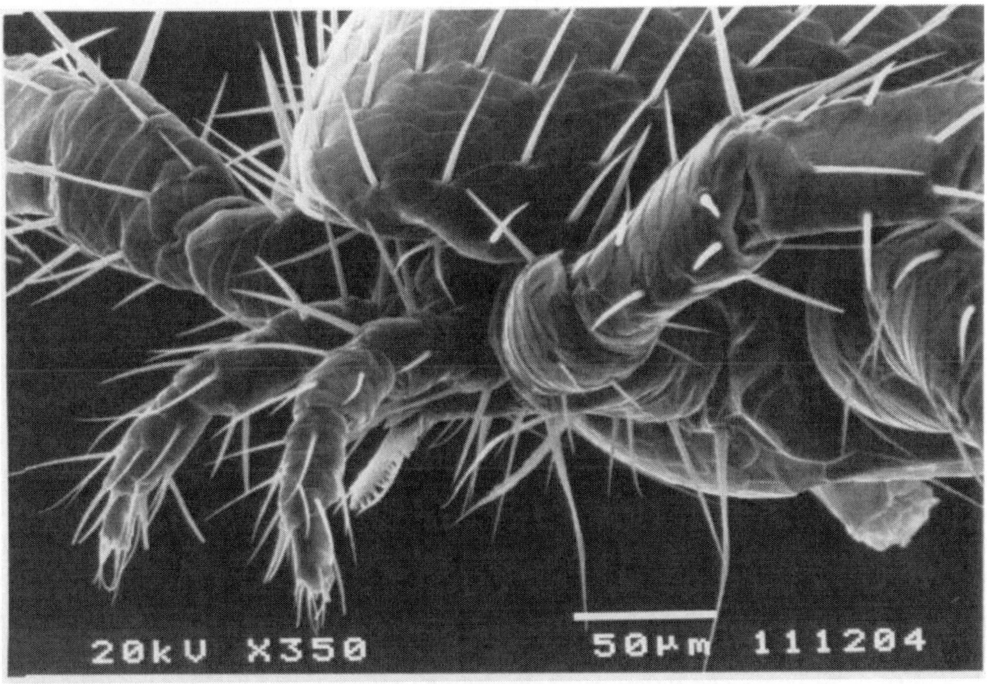

Figure 37. Scanning electron micrograph shows the mouth part of T. clareae.

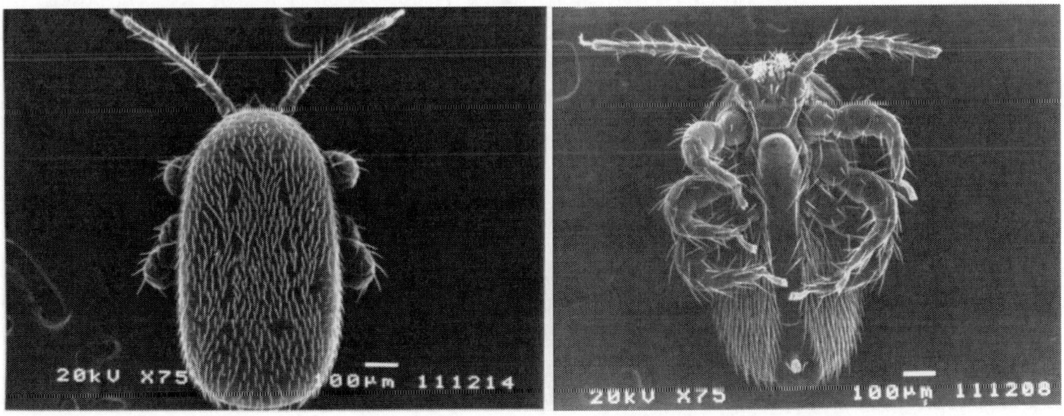

Figure 38. Scanning electron micrograph of the dorsal surface (left) and ventral surface (right) of T. clareae.

Taxonomy of Tropilaelaps clareae

Kingdom Animalia
Phylum Arthropoda
Class Arachnida
Subclass Acari
Order Parasitiformes
Suborder Mesostigmata
Family Laelapidae
Genus Tropilaelaps
Species T. clareae
(Delfinado and Baker, 1962)

General Biology and Life Cycle of Tropilaelaps Clareae

Females are medium sized (< 1 mm), elongated, and light reddish brown. Males are similar but less sclerotized (Lekprayoon and Tangkanasing, 1991; 1993; Sammataro, 1997). They are visible to the naked eye. The foundress mite places three to four eggs on mature bee larvae shortly before they are capped and the progeny feed only on developing bee larvae. The mite requires about one week to develop, and the adults, including the foundress mite, emerge with the adult bee and search for new hosts. This short life cycle is one reason why T. clareae populations increase more rapidly than those of Varroa species. Tropilaelaps clareae out competes the latter when both infest the same colony of *A. mellifera* (Sihag, 1988). Nevertheless, populations of both mites can survive in the same apiary for 12 months, probably because their niches are not completely congruent (Rath et al., 1995). Like Varroa, female T. clareae are dispersed by bees, but can only survive in this dispersal phase for a short time. Gravid female mites die within two days unless they deposit their mature eggs (Woyke, 1987; 1994a, 1994b). Even adult T. clareae chelicerae (jaws) cannot pierce the integument of adult bees. The mouthparts are stubby, with an apically bidentate fixed upper digit, and a longer, unidentate and pointed moveable digit. This piercing-grasping structure is more suitable to piercing soft brood tissue. This implies that Tropilaelap can feed only on soft tissues, such as honeybee brood (Griffiths, 1988). However, the tearing-sawing jaws of Varroa can penetrate the harder adult *Apis* exoskeleton (Griffiths, 1988).

Pest Status: Symptoms and Distribution

Tropilaelap clareae causes severe damage to *A. mellifera*, *A.* cerana, *A. dorsata*, and *A. florea*. Honeybees infested by this parasite may become unable to fly. Heavily infested bees may crawl on the hive floor or form a parasitized cluster in the hive. Bee lifespans are shortened by heavy mite infestations, creating a condition called acarine disease or acariosis (Anderson and Morgan, 2007). This also leads to reduced brood production. However, this does not generally cause colony death, only reduced colony health, and productivity because bees in the warm Thai climate can produce brood thoughout the year, unlike bees in colder climates (Anderson and Morgan, 2007).

Varroa Jacobsoni

Varroa jacobsoni is another important and dangerous ectoparasite of Thai honeybees and can feed on the bodily fluids of larvae, pupae, and adult bees. This parasite has a strong economic impact and places stress on the Thailand's commercial and wild honeybee populations. The pathology it causes is commonly called varroasis (also seen as varroatosis or varrosis). The varroa mite was first discovered in Southeast Asia around 1904. Since then, they have spread worldwide and from their original host, *A.* cerana (Sasagawa et al., 1999) to European honeybees (*Apis mellifera*). The name Varroa destructor has been proposed (D. Anderson, personal communication) and may be recognized if several species and strains are lumped under the name V. jacobsoni (Anderson, 1999; de Guzman et al., 1997; 1998; De Jong and Goncalves, 1999). Varroa became an economic concern in Japan and China in the 1950s and 1960s, in Europe in the late 1960s and 1970s, and in Israel and North America in the 1980s. Because it is a significant pest of honeybees, we provide more details about its biology below.

Taxonomy of Varroa mite

Kingdom Animalia
Phylum Arthropoda
Class Arachnida
Subclass Acari
Order Parasitiformes
Suborder Mesostigmata
Family Varroidae
Genus Varroa
Species V. jacobsoni
(Anderson and Trueman, 2000)

General Biology of Varroa

Adult females are a reddish-brown color. Mites are dorsoventrally compressed, allowing them to fit beneath a bee's abdominal sclerites (Yoder et al., 1999). The average female mite is approximately 1.1 mm in length and 21.6 mm in width and approximately 0.14 mg in mass (Sammataro, 1997; Sammataro et al., 1994). Adult males are smaller and lighter in color than adult females. The mite feeds on the honeybee haemolymph by making a soft hole through the honeybee's exoskeleton between the soft intersegmental tissues of the bee's exoskeleton. Mites have modified chelicerae that contain a moveable digit that is like a saw blade. It can pierce and tear open the host's integument (Delfinado and Baker, 1987).

Varroa mites reproduce on a 10-day cycle. Females ride on adult bees during dispersal (phoresy). These mites prefer young ''house'' bees to older workers, probably because of the lower titer of the Nasonov gland pheromone geraniol in older workers, which strongly repels the mite (Hoppe and Ritter, 1989). The female mite enters a honeybee brood cell, one to two days before capping and produces its first egg 60 hours after the cell is sealed (Ifantidis, 1983). These eggs then hatch to typically produce one haploid male and several females. Mites go through the following instars: pharate larvae, mobile protonymph, pharate deutonymph, mobile deutonymph, pharate adult, and adult (De Jong, 1997, Donze and Guerin, 1997). The young mites hatch at about the same time and leave the cell with the host.

Young females mature in 6.5 -6.9 days (De Jong, 1997). When the young bee emerges from the cell after pupation, the Varroa mites also leave and spread to other bees and larvae. They can survive off of the host for 18 -70 hours, depending on the substrate (de Guzeman et al., 1993). The mites preferentially infest drone larvae over workers because they contain higher quantities of fatty acid esters (Le Conte et al., 1989), more aliphatic alcohols and aldehydes, and a larger amount of hemolymph (Donze et al., 1998).

There are several signs of infestation. The pale or dark red-brown mites can be easily seen on white pupae. Infested drone or worker brood develops into disfigured, stunted adults with deformed legs and wings (De Jong, 1997; Gerson et al., 1988; Le Conte et al., 1989). Additionally, workers bees are seen discarding infested larvae. Because mite populations increase in proportion to the available bee larvae, Varrao can destroy bee colonies within months (Gerson et al., 1988). Varroa mites attack adults and brood, weakening and shortening the life span of infested bees. Losses due to these parasitic mites are sometimes attributed to standard winter mortality or queenlessness. Beekeepers in Thailand tend to treat Varroa using natural products such as extracts of the snake root plant. However, commercial produces such as Coumaphos®, Bayer Bee Strips® or CheckMite® are also used (De Jong, 1997; Gerson et al., 1988; Le Conte et al., 1989).

Acar*Apis* Woodi

This mite lives inside a bee's tracheal tubes and was discovered and first named Tarsonemus woodi (Rennie, 1921, Rinderer et al., 1999). However, it was later renamed Acar*Apis*, from Acarus, mite, and *Apis*, bee (Hirst, 1921). The disease was then called Isle of Wight disease. This mite is found worldwide and is a serious pest of *A.* cerana and *A. mellifera*. Beekeepers have reported heavy colony mortality due this mite in *A.* cerana colonies in southern Thailand and *A. mellifera* in northern Thailand. Beekeepers suspect that *A.* cerana colonies are more susceptible to this mite than *A. mellifera* colonies (Partap and Verma, 1998).

Taxonomy of Tracheal mite

Kingdom Animalia
Phylum Arthropoda
Class Arachnida
Subclass Acari
Order Trombidiformae
Family Tarsonemidae
Genus Acar*Apis*
Species *A.* woodi
(Anderson and Trueman, 2000)

General Biology and Life Cycle

The mite's entire life cycle is spent within the thoracic tracheal system of adult honeybees, except for brief migratory periods (Sammataro and Needham, 1996; Smith et al., 1991). Mites are occasionally found in air sacs in the thorax and abdomen. Female tracheal mite length ranges from 120-190 μm and the width ranges from 75-84 μm (Delfinado-Baker et al., 1989). Females each weigh 5.5-10.4 mg. Male range in length from 125-136 μm, in width from 60-77 μm, and weigh 2.61-10.4 mg. Males complete their development in 11-12 days. Females complete development in 14-15 days. Like other members of the prostigmatic Heterostigmata, *A.* woodi has a foreshortened life cycle. It has only three apparent stages: egg, larva, and adult. However, the mite has an apdous nymphal instar that remains inside the larval skin (Lindquist, 1986).

These mites feed on bee hemolymph, which they obtain by piercing the tracheae with their closed-ended, sharply pointed stylets, operated by internal chitinous levers (Hirschfelder and Sachs, 1952). Once a trachea is pierced, the mites' mouth, located just below the stylets, is pressed against the wound, and the mite sucks host hemolymph into its pharynx.

Mites are spread within the colony as a result of bee-to-bee contact. Mated female mites leave the breathing tubes where they develop and climb to the tip of a body hair. As bees come in contact with one another, the mites attach themselves to the hairs of a passing bee and enter the tracheae through the thoracic spiracles. Dispersing female mites are attracted to air expelled from the prothoracic (first thoracic) spiracle of young bees (Hirschfelder and Sachs, 1952), as well as to specific hydrocarbons from the cuticle of callow bees, less than four days old (Phelan et al., 1991). This mite is less attracted to older bees and prefers drones to workers (Royce and Rossignol, 1991). If the mite does not locate a new host within 24 hours, it will die. Once the female mite enters honeybee's spiracle, she lays 4-8 eggs within a couple days. The mite's small size is critical to its survival. The tiny mites can hide under the flat lobe that covers the bee's first thoracic spiracle. Mites begin to disperse when the host bee is more than 13 days old. Dispersal peaks when they are 15- 25 days old. Mites leave the tracheae after the death of the bee. Drifting bees between hives and swarms from infested colonies can spread the mite the apiary (Publication 1753, Extension Service of Mississippi State University, cooperating with U.S. Department of Agriculture).

Pest Status: Symptoms and Distribution

It is difficult to determine if tracheal mites are present in honeybee colonies: the parasites are very small, and they infest the host bees internally. Adult bees infested by *A.* woodi show no noticeable signs, but their lifespan is shortened. Infested bees are whitish in color, and have a shiny cuticle with a few long fine hairs on the body and legs (Fyg, 1964; Morse, 1978). The only reliable diagnostic method is the microscopic examination of dissected tracheae. If present, the mites are usually found within the trachea closest to the bees' thoracic spiracles. The infested tracheae display a color darker than normal. Infested queens can live for many years (Fyg, 1964). Morse (1978) estimated that mite infestation reduces colony size by approximately five percent. Honey production and pollen collecting are correspondingly reduced. In addition, honeybees infested by this parasite may become unable to fly (Fyg, 1964; Morse, 1978).

The Parasitic Mite, Euvarroa

Euvarroa sinhai is a parasite of *A. florea*, and ranges from Iran through India and Sri Lanka (Sammataro et al., 2000). The mite infests capped drone brood (Mossadegh and Komeili, 1986), but can be reared in the laboratory on *A. mellifera* worker brood (Mossadegh, 1990). Development requires less than one week, and each female produces four to five offspring. Mites disperse by riding on drones and workers. The female mite overwinters in the colony, probably feeding on the clustering bees. Colony infestation by E. sinhai is somehow

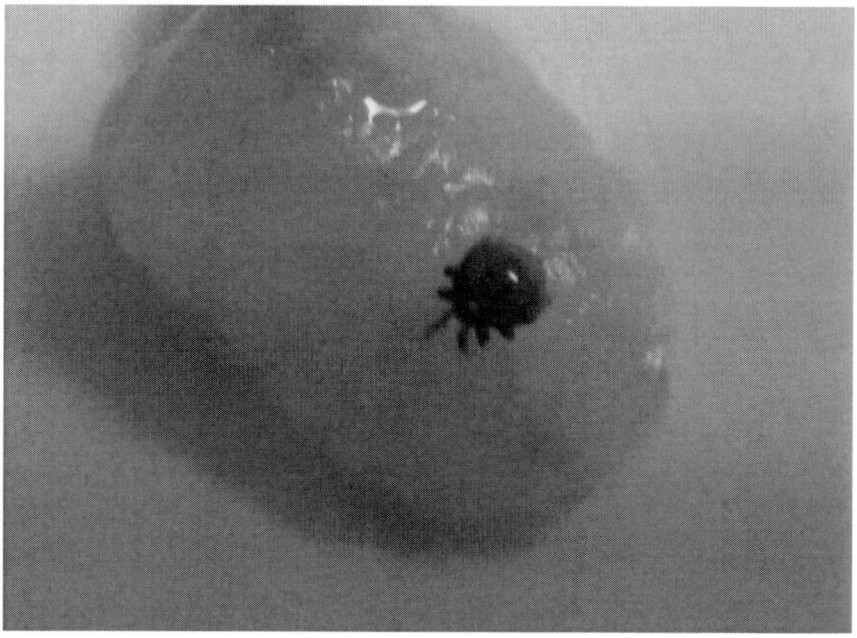

Figure 39. Euvarrao singhi feeding on drone larva of *A.* florae.

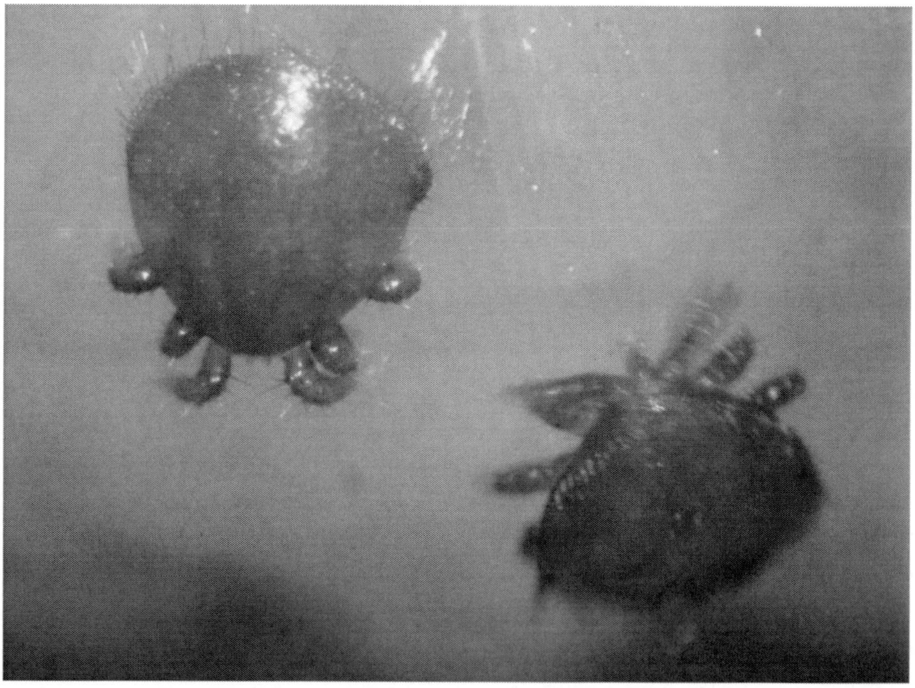

Figure 40. Euvarrao singhi collected from *A. florea.*

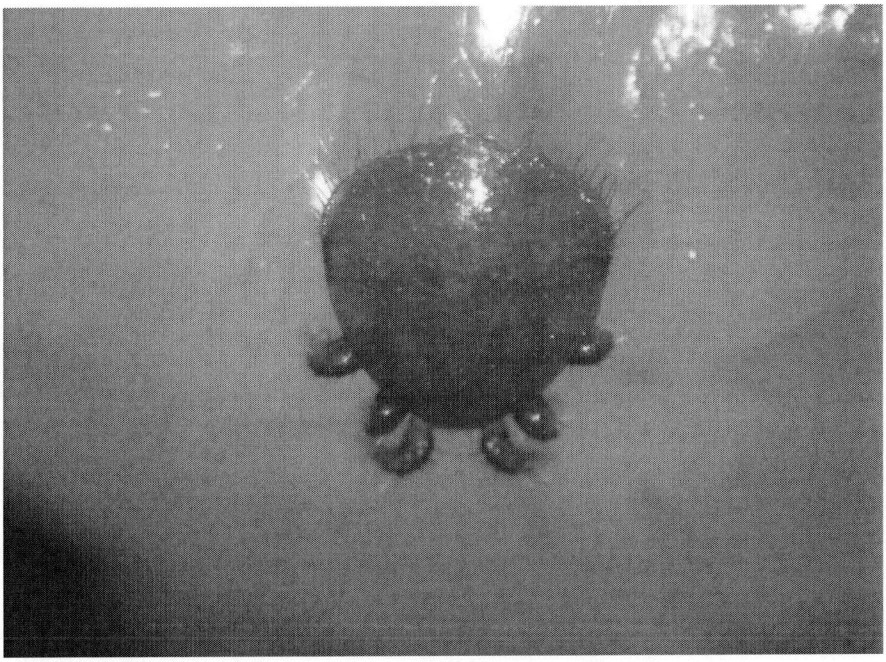

Figure 41. Euvarrao singhi collected from an *A. florea* worker.

hindered by the construction of queen cells (Aggarwal and Kapil, 1988), and its population growth is inhibited in the presence of T. clareae and V. jacobsoni (Sihag, 1988). Transfer experiments (Koeniger et al., 1993) confirmed that E. sinhai can survive and thus potentially cross-infest *A. mellifera* and *A.* cerana. The closely related Euvarroa wongsirii parasitizes drone brood of *A. andreniformis* in Thailand and Malaysia. Its biology appears similar to E. sinhai and it can live for at least 50 days on worker bees outside the nest (Morin and Otis, 1993).

Viral Diseases

Honeybees are subject to many viruses (Allen and Ball, 1996; Ball and Bailey, 1997), five of which are associated with Varroa and one with tracheal mites. Several viral diseases affect *A.* cerana. Thai honeybees suffer from the Kashmir bee virus (KBV) (Anderson, 1991; Bailey, 1962; Ball and Bailey, 1997). This disease is similar in size to several picornavirus-like agents. KBV may be activated in the presence of Varroa, multiplying to lethal levels. Another important virus is *Apis* iridescent virus (AIV). Recent reports state that AIV has been causing serious damage to commercial colonies of *A.* cerana in northern India and Pakistan, the virus being associated with "clustering disease" (Aemprapa and Wongsiri, 2000). The bees are unusually inactive and frequently form small, detached clusters of bees that do not fly. Many individual bees are observed crawling on the ground and are lost. At first, these symptoms were associated with the presence of the tracheal mite Acar*Apis* woodi on some diseased bees, but it was later shown that AIV is the major causative agent (Anderson, 1991; Ball and Bailey, 1997).

Table 7. Viral infections in Thai bees (modified from Allen and Ball, 1996; Anderson, 1991)

Honeybee species	Viruses	Reference
Apis andreniformis	-	-
cerana	Thai sac brood virus Deformed wing virus *Apis* iridescent virus Kashmir bee virus	Abrol and Bath, 1990; Bailey et al., 1982; Oldroy and Wongsiri, 2007
dorsata	Thai sac brood virus	Abrol and Bath, 1990; Oldroy and Wongsiri, 2007; Vermat et al., 1990
florea	Black queen cell virus	Abrol and Bath, 1990; Oldroy and Wongsiri, 2007; Verma et al., 1990
mellifera	Acute bee paralysis virus Kashmir bee virus Sac-brood virus (SBV) Deformed wing virus	Sanpa and Chantawannakul, 2009

Thai Sac-brood virus (TSBV) is a spherical virus that infects the cytoplasm of fat cells of honeybee larvae (Aemprapa and Wongsiri, 2000; Ball and Bailey, 1997). It was found among colonies in mountainous areas of Northern Thailand, for example in *A.* cerana indica collected from Doi pui, Chiang Mai province (Aemprapa and Wongsiri, 2000). In 1990, there was an outbreak in the southern Thailand in Chumporn province (Jarungjit et al., 1990). The pupae turn into sac-like structures filled with lemon-colored liquid at the posterior end. Later, the larvae change their appearance from yellowish to brownish to black color. No discernible foul odor can be detected. Many Indian bee colonies were destroyed by TSBV in South India during early 1990's. TSBV was first reported in *A.* cerana colonies in Thailand and has since been found other Asian countries. Its natural range may cover the entire Asian continent. In Thailand, the disease is found in colonies experiencing stress: lack of food, excessive humidity, low worker population, poor-laying queens, etc. TSBV is also called Chinese sacbrood virus (CSBV), which occurs in *A.* cerana colonies in China (Allen and Ball, 1996; Anderson, 1991; Yan et al., 2009). Molecular data from TSBV from India (Genbank Accession # EU156753) and CSBV from China (Genbank Accession# AF469603) suggest that these viruses are very closely related and can be regarded as different varieties of the same virus (Hepburn and Radloff, 2010).

Bacterial Diseases

There are two main bacterial diseases found in Asian honeybees: American Foul Brood disease (AFB) caused by Paenibacillus larvae (Nakamura, 1996; Oldroyd and Wongsiri, 2006) and European Foul Brood (EFB) caused by Melissococcus pluton (Bailey and Collins, 1982). Because EFB is a stress-related disorder, colonies that are heavily infested with Varroa are susceptible to EFB. Colonies of *A.* cerana are occasionally infested with bacterial diseases such as AFB and EFB (Bailey and Collins, 1982). Other microbial diseases have also been reported.

American Foul Brood (AFB)

American foul brood is caused by the spore forming gram-positive bacterium, Paenibacillus larvae ssp, (formerly classified as Bacillus larvae). It is the most widespread and destructive of the bee brood diseases (Genersch, 2005; Genersch et al., 2006). Paenibacillus larvae are rod-shaped bacteria, which are visible only under a compound microscope. Ingesting spores that are in their food infects larvae up to 3 days old, with larvae less than 24 hours most susceptible. Spores germinate in the gut of the larva and the vegetative form of the bacteria begins to grow, taking its nourishment from the larva. Spores will not germinate in larvae over 3 days old. Infected larvae normally die after their cell is sealed. The vegetative form of the bacterium will die but not before it produces many millions of spores. Each dead larva may contain as many as 100 million spores (Alippi et al., 1995). AFB is spread worldwide in *A. mellifera* and *A.* cerana (Hansen et al., 1999).

European Foul Brood (EFB)

Melissococcus plutonius is a bacterium that infests the mid-gut of an infected bee larva. European foulbrood is less deadly to a colony than American foulbrood. Melissococcus plutonius does not form spores, though it can overwinter on comb. Symptoms include dead and dying larvae which can appear curled, brown or yellow, melted or deflated with tracheal tubes dried out and rubbery (Bailey and Ball, 1991; Shimanuki, 1990). European foulbrood is often considered a "stress" disease, a disease that is dangerous only if the colony is already under stress for other reasons. An otherwise healthy colony can usually survive European foulbrood. An outbreak of the disease may be controlled chemically with oxytetracycline hydrochloride, but honey from treated colonies could have chemical residues from the treatment (Waite et al., 2003). The 'Shook Swarm' technique involves replacing all brood frames with new frames in a single operation, thus removing all potentially diseased equipment and minimizing disease transfer. The advantage is that chemicals are not used. Prophylactic treatments are not recommended as they may lead to resistant bacteria (Waite et al., 2003).

Fungal Diseases

Nosema Disease (Nosemosis)

Nosema is a fungus in the class Microsporidia (Nosematidae), a large group of obligate intracellular parasites that are highly widespread in nature. They frequently infect insects, including honeybees. This parasite enters via the ventriculus of honeybees (Bailey, 1952a; 1952b; Chen et al., 2008, 2009; Fries et al., 1996). Nosema is also associated with black queen-cell virus. There are two species of Nosema: Nosema *Apis* (Hassanein, 1952, 1953; Higes et al., 2007; Rinderer and Elliott, 1977) and N. ceranae (Fries et al., 1996). The differences between these two species can be detected in their SSUrRNA sequences. Moreover, their spores also differ. Spores of N. ceranae have shorter and fewer polar filament coils than those of N. *Apis* (Fries, 1989a; 1989a).

Nosema spore infections are found only in adult bees. Infected workers, drones, and queen bees become weak and suffer early mortality (Fries et al., 1996; Webster, 1993; Webster et al., 2004). Queens suffer damage to reproductive organs and, consequently, the colony bee population can drop dramatically. Nurse bees infected with Nosema do not fully develop their hypopharyngeal glands, reducing the production of royal jelly and brood food. Infected foraging bees have reduced foraging activity (Anderson and Giacon, 1992; Clark, 1978, 1980; Goodman, 2007; Hassanein, 1951; Malone and Gatehouse, 1998; Malone et al., 1995; 2001). Many compounds have been tested against Nosema, but currently the only effective product is the antibiotic fumagillin (Moffet et al., 1969). This product inhibits the development of N. *Apis* in honeybees (Katznelson and Jamieson, 1952; Liu, 1973).

Recently, N. *Apis* has become more widely distributed by cross infection from *A. mellifera* to *A.* cerana. The closely related N. ceranae was first found in *A.* cerana Fabricius, 1793 by Fries et al. (1996). Recently, N. ceranae was found in managed *A. mellifera* Linneaus, 1758 colonies (Huang et al., 2005; Higes et al., 2006; Huang et al., 2007, Klee et al., 2007). Nosema ceranae also infects three native species of Thai honeybees: *A. dorsata*, *A.* cerana and *A. florea* (Suwannapong et al., 2010b; Suwannapong et al., 2011).

Nosema invades and destroys cells in the bee gut, resulting in drooping wings, lack of hair, dysentery marked by brown fecal marks in the comb and early mortality. Nosema infections are acquired by the uptake of spores during feeding or grooming (Bailey, 1969; Ellis and Munn, 2005; Fries, 1983, Fries et al., 1992, 2006; Huang et al., 2007; Matheson, 1993). The parasite invades the posterior region of the ventriculus, giving rise to large numbers of spores within a short period of time. Nosema levels generally increase when bees are confined (Goodman, 2007; Higes et al., 2007). The spores are also transmitted among bees via the ingestion of contaminated comb material and water, and by trophallaxis via honey stores. Crushed infected bees may also play a role in disease transmission. The pathological consequences of N. ceranae in *A. mellifera* are not well known, however they proved highly pathogenic in infected *A. florea* (Suwannapong et al., 2010b; Suwannapong et al., 2011). Newly reported N ceranae has recently jumped hosts to *A. mellifera* Linneaus, 1758 (Higes et al., 2006). However, in *A. mellifera* colonies, Nosema is normally only a problem when the bees cannot leave the hive to eliminate waste (Higes et al., 2006).

Nosema in Thailand

Nosema ceranae was first observed in Thailand by Suwannapong et al in 2007 in *A.* cerana and can now be found throughout the country infecting *A. florea*, *A. dorsata* and *A. andreniformis* (Suwannapong et al., 2010b, 2011). Nosema spores can be found in pollen removed from pollen storage area in honeybee comb of *A.* cerana, *A. florea* and *A. dorsata*. The presence of Nosema spores in corbicular pollen may be due to self-contamination during the process of pollen collection. As the forager bee moistens the surface of its body with its protruding tongue and brushes the collected pollen, the spores inside its body become mixed with the pollen, via an unknown mechanism. Spores can either come directly from the intestines after regurgitation or be present in the saliva. Nosema spores also are found in the honey of four species, *A.* cerana, *A. florea*, *A. dorsata* and *A. mellifera*. Experimental cross infection of *A.* cerana and *A. florea* with N. ceranae isolated from both *A.* cerana and *A. florea* workers has been reported by Suwannapong et al. in 2010, where they reported that the parasite rapidly divided in the honeybee midgut.

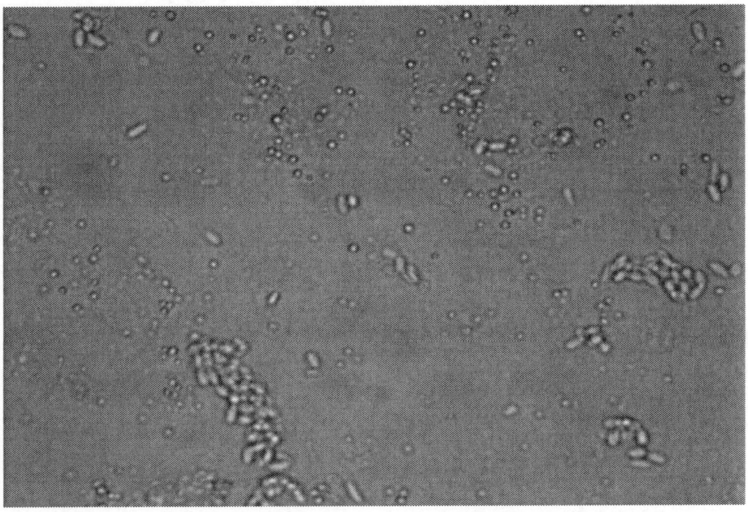

Figure 42. Light microscope image of Nosema ceranae spores.

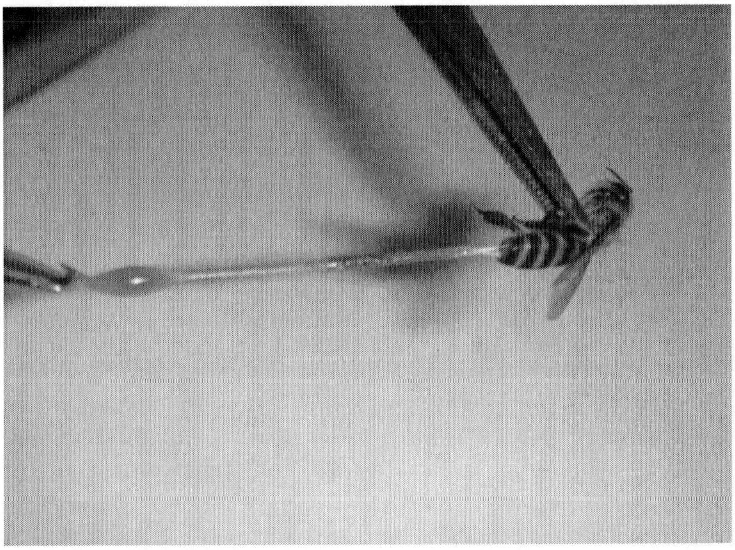

Figure 43. Removing the midgut of Nosema infected *A. florea* worker.

Control of Nosema Disease

Nosema spores may be killed by heating hive equipments or tools to a temperature of at least 60°C for 15 minutes. Another treatment involves heating the equipment to 49°C for 24 hours. This is best conducted in a room where the temperature is uniform and thermostatically controlled (Morse and Shimanuki, 1990). Fumigation with acetic acid is effective, especially when the bees are transferred as early as possible in the season from contaminated equipment to fumigated equipment. An efficient method is to intersperse absorbent material with acetic acid between groups of hive bodies containing the combs (Bailey and Ball, 1991; Shimanuki et al., 1992). Fumigation with ethylene oxide (ETO) has also been demonstrated to kill spores on combs (100 mg ETO/l for 24 hours at 37.8 °C). However, there are a number of safety issues associated with the use of ETO (Shimanuki et al., 1992). Fumagillin has been found to be effective against N. *Apis* (Katznelson and Jamieson, 1952). In addition, this chemical inhibits DNA replication of the microsporidian without affecting the DNA of the host cell (Hartwig and Przelecka, 1971; Liu, 1973). Fumagillin's activity remains high in honey kept at 4°C for several years and for at least 30 days at 30°C (Furgala and Sugden, 1985).

Several natural compounds have been examined for potential efficacy against Nosema: thymol, vetiver essential oil, lysozyme, and resveratrol. Lysozyme was not effective. However, resveratrol and thymol have high potential for the control of Nosema (Maistrello et al., 2008). Resveratrol (trans- 3,5,4'-trihydroxystilbene) is a phytoalexin produced by certain plants in response to infections caused by phytopathogens. It is known for its anti-cancer and anti-inflammatory effects (Fremont, 2000). Recent studies have shown that resveratrol can inhibit the development of the microsporidian Encephalitazoon cunicoli in vitro experiments (Leiro et al., 2004).

Thymol (3-hydroxy-p-cymene) is a constituent of the essential oil found in thyme and other plant species. It has been shown to suppress N. vespula disease in Helicoverpa armigera caterpillars. Some evidence suggests that thymol may suppress Nosema disease in honeybee colonies (Rice, 2001; Yucel and Dogaroglu, 2005). It also inhibits the growth of pathogenic bacteria and fungi such as Salmonella typhimurium, Staphylococcus aureus (Juven et al., 1994), Aspergillus flavus (Mahmoud, 1999), and Cryptococcus neoformans (Viollon and Chaumont, 1994). In apiculture, it is known to suppress the parasitic mite Varroa destructor (Chiesa, 1991). Recent research has shown that thymol fed orally to adult bees is not toxic (Ebert et al., 2007).

Other treatments also show promise. Protofil is a natural product obtained from plants through hydro-alcoholic extraction. Protofil prevents the development cycle of N. *Apis*, inhibits intestinal pathogens, and stimulates the digest enzymes. In general, it impairs the development of bee colonies (Chioveanu et al., 2004). The addition of acetic acid into winter food may have positive effects in preventing different diseases. An experiment in Norway found that acetic acid in food reduced the occurrence of chalk brood, but these results should be replicated (Pederson, 1976). Laboratory experiments in Belgium suggested that acidified food decreases the development of N. *Apis* in the midgut, but field studies performed in France demonstrated no impact of acidified food on Nosema development (Chioveanu et al., 2004). The chemical composition of food may have an impact on of N. *Apis* the spore germination. Changing the chemical environment (i.e. lowering the pH) may reduce spore germination (Crane, 1975).

Wax Moths

Two varieties of wax moth infest honeybee colonies: the lesser wax moth Aphomia sociella which has a length of 10-13mm, a wingspan of 11-14 mm and is silvery-grey to buff in color with a yellow head. When seen, it flies, runs very quickly or holds onto the comb vibrating its wings. Each female can lay 250-300 eggs hatching into larvae that are similar in appearance to Greater Wax Moth larvae but not as large being up to 20mm in length. Though larvae consume honey, pollen and wax they are not found in comb occupied by bees and do not damage hive components. Lesser Wax Moth larvae are unable to compete with Greater

Wax Moth larvae because the latter will eat them (Gambino, 1995). In the live bee situation the best preventative against wax moths is strong healthy colonies. If not controlled wax moth infestations can rapidly multiply, (which is mottled grey in color and 0.5-1.9 cm long) is found worldwide. In southern India, it causes severe damage in the plains and lower altitudes but it is rare at high altitudes. The greater wax moth is a natural scavenger of honeybee comb and their contents. Beeswax combs are vulnerable to wax moth damage anytime they are unprotected by bees: whether in a weak and declining colony or in shed storage. It is one of the most important enemies of the bee colony, causing serious damage particularly to weak colonies where the number of bees is not sufficient enough to cover all the combs. They will not attack the bees directly, but feed on the wax used by the bees to build their honeycomb. Their full development requires access to used brood comb or brood cell cleanings for protein (Ali et al., 2009; Gambino, 1995).

A strong hive generally needs no treatment to control wax moths. The bees themselves will kill and clean out the moth larvae and webs. Damaged comb may be scraped out and replaced by the bees. The moths prefer to mate and lay their eggs at night. In Thailand, wax moths occur throughout the year and are found in all species of honeybees. In nature, wax moths are valuable members of the ecosystem because they clean abandoned cavities of old comb (potentially contaminated with disease) and render it clean for the next occupying swarm (Ali et al., 2009; Gambino, 1995).

Life Cycle of Wax Moths

Wax moth development goes through three consecutive stages: egg, larva and pupa. This sequence is only interrupted if the temperature is too low or when there is no food. Therefore, the cycle can last between 6 weeks and 6 months depending on temperature and food. According to the literature, over-wintering can take place as egg, larva or pupa. The females start laying eggs between days 4 and 10 after emergence (Shimanuki, 1981). At dusk, the females attempt to enter the beehive to lay their eggs. Normally, females lay their eggs into crevasses and gaps. If the colony is strong enough to repel the wax moth, the moths lay their eggs outside in cracks in the wood. This puts them out of reach of the bees and prevents their destruction. After hatching, the young larva immediately searches for a comb to feed and to build the silk-lined feeding tunnels. Speed of growth is directly dependent on temperature and food supply. Under ideal conditions, larval weight can double daily during the first 10 days. Newly hatched larvae are white, but successive instars are medium to dark gray on the top with creamy white undersides. The larval head capsule is brown. Heat, which is created by this rapid growth, can increase the temperature in the spun silk nests far above the ambient environmental temperature (Ali et al., 2009; Gambino, 1995; Shimanuki, 1981).

The larvae feed on wax, impurities in wax, the cocoons of bee larvae, and remnant pollen. Larvae that have been reared exclusively on pure wax (foundation and fresh comb), do not complete their development. Dark, old combs that contain many old bee cocoons provide the most food to wax moth larvae. The larvae grow rapidly and will migrate toward the edges of the frames or corners of the supers to spin a cocoon and pupate. At the end of the larval stage, the larva spins a very strong silk cocoon on a firm support. Frequently the larva spins its cocoon in a hollow it had bored into the wood (Ali et al., 2009; Gambino, 1995; Shimanuki, 1981).

Figure 44. Larvae of wax moth in an *A. florea* nest.

Figure 45. Larvae and pupae of wax moths collected from an *A. andreniformis* nest.

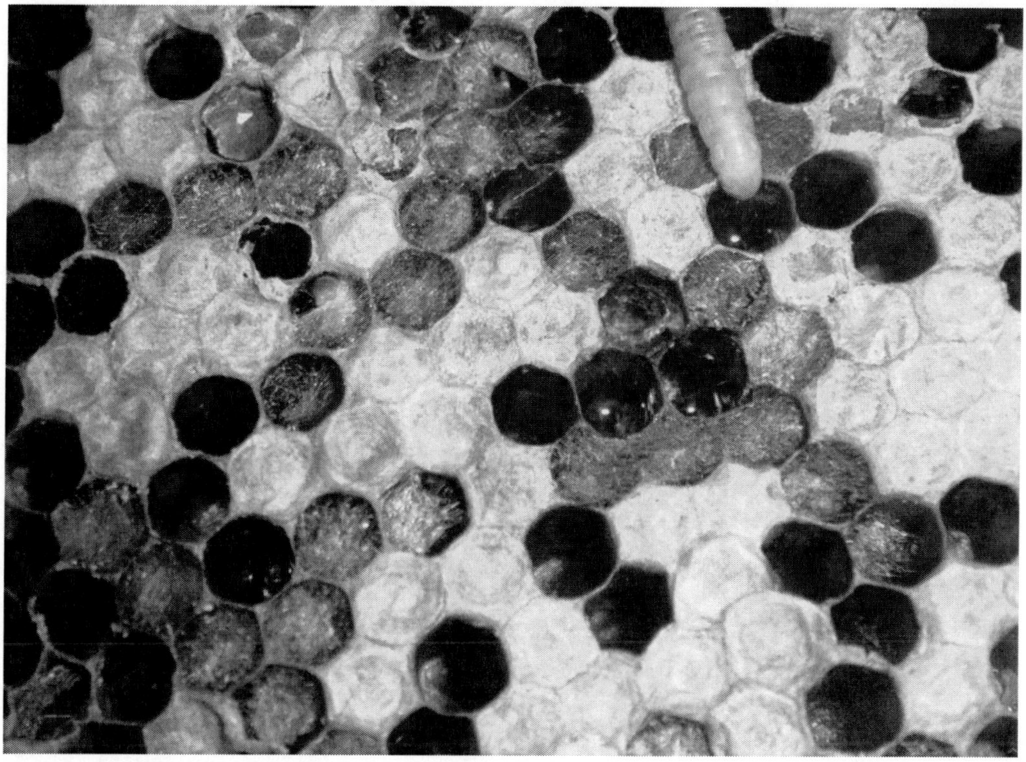

Figure 46. *Apis* cerana brood comb infested by wax moths.

Damage occurs as the larvae burrow into the comb feeding on the wax, larval skins, pollen and honey. As the larvae chew through the comb they spin a silk lined tunnel through the cell walls and over the face of the comb. These silk threads can tether emerging bees by their abdomens to their cells and they die of starvation because they are unable to escape from their cell (Ali et al., 2009; Shimanuki, 1981). This phenomenon is termed galleriasis. In severe infestations, the wax comb, wooden frames, and sides of the hive bodies can be heavily damaged. After hatching, the larvae, if not removed, by house bees can destroy a hive that is in weak. Adult wax moths cause no direct damage because their mouthparts are atrophied. They do not feed as adults. Only larvae feed and destroy combs. However, adult wax moths and larvae can transfer disease pathogens. In colonies infested with foulbrood, the feces of wax moths contain large amounts of Paenibacillus larvae spores (Ali et al., 2009)

The most effective method for preventing wax moth damage is to maintain strong colonies. The bees will remove the moth larvae and repair damage as it occurs. Stored equipment can be protected against wax moths by fumigating it with para-dichlorobenzene crystals or by stacking honey supers in a criss-cross fashion in open sheds. The penetrating air and daylight discourage colonization by moths. Some beekeepers store supers in enclosed barns with a lighted bug-zapper running constantly to kill emerging adult moths. This practice can eventually eradicate moths from the room. Control of wax moths by other means includes the freezing of the comb for at least 24 hours (Ali et al., 2009; Gambino, 1995; Shimanuki, 1981).

Figure 47. The Asian giant hornet, Vespa mandarinia and *A. florea* nest.

Wasps, Hornets and Ants

Ants are the most common predators of honeybees in Thailand. They will collect honey, brood, and adults (dead or alive). Black ants (Camponotus compressus) and red ants (Dorylus labiams) are dangerous enemies of bees (Akratanakul, 1976). They attack weak colonies and carry away the honey, pollen and the brood. Strong colonies are able to withstand the ants. In weak colonies, ant attack will result kill the colony. Beekeepers sometimes kill ant nests in the proximity of the apiaries. A common practice is to rub the hive stand posts with engine oil or any lubricants. A more reliable method of defense is to place the hive stand posts in plastic pots or cans filled with water or oil. Liquids require replenishment regularly and removal of all vegetation which can be the foundations of bridges that can be crossed by ants (Akratanakul, 1976; Hirai et al., 1981).

Wasps and hornets are also enemies of bees. The yellow-banded hornet, Vespa cincta F., is a large wasp with a broad transverse band on its abdomen. They attack a weak colony en masse, using their strong mandibles to bruise the guardian bees at the hive entrance. Hornets drop the dead and dying bees to the ground. The colony under attack will eventually lose its defenders. Hornets will then invade the hive and carry away honey and brood to store in their nest (Akratanakul, 1976; Hirai et al., 1981).

The Asian giant hornet, Vespa mandarinia is a relentless hunter that preys on Thai honeybees. They often attack honeybee nests with the goal of obtaining the honeybee larvae. A few scouts will cautiously approach the nest, giving off pheromones that lead other hornets to the hive's location. The hornets can devastate a colony of honeybees. A single hornet can kill as many as 40 honeybees per minute. It takes only a few of these hornets a few hours to kill thousands of bees. *Apis mellifera* has relatively small stings that do little damage to hornets, since they are five times the size of honeybees and twenty times their weight. The honeybees make futile solo attacks without mounting a collective defense, and are easily killed individually by the hornets. Once a hive is emptied of all defending bees, the hornets feed on the honey and carry the larvae back to feed to their own larvae (Hirai et al., 1981).

Adult hornets cannot digest solid protein, so the hornets do not eat their prey, but chew them into a paste and feed them to their larvae (Hunt et al., 1982). The larvae produce a clear liquid, Vespa amino acid mixture, which the adults consume. Larvae of social Vespidae produce these secretions. The exact amino acid composition varies considerably among species (Hunt et al., 1982). In many parts of Asia, including Thailand hornets, hornets are reported to be serious pests of honeybees. Making efforts to catch hornets that come near hive entrances can prevent serious destruction. However, the best way of dealing with the problem is reduce hornet nest habitat by removing nearby trees until the hornet population is much reduced (Akarakul, 1976). This method, however, is not practical for forest beekeepers.

REFERENCES

Aemprapa, S. and Wongsiri, S. (2000). Thai sac brood virus situation in Thailand. Asian beekeeping: progress of research and development, *Proceeding of Fourth Asian Apicultural Association International Conference*, Kathmundu, March 23-28, 1998, pp. 57-59.

Aggarwal, K. and Kapil, R. P. (1988). Observations on the effect of queen cell construction on Euvarroa sinhai infestation in drone brood of *Apis florea*. In *Africanized Honey Bees and Bee Mites*, ed. G.R., Needham, R.E., Page Jr, M., Delfinado-Baker, C. E., Bowman, pp. 404-8. Chichester, UK: Ellis Horwood.

Agren, L. (1977). Flagellar sensilla of two species of Andrena (Hymenoptera: Andrenidae), *Journal Insect Morphology Embryology* 7: 73-79.

Ahmad, R. (1992). *Present status of beekeeping in Pakistan. Honeybees in Mountain Agriculture*. New Delhi: Oxford and IBH. pp. 211-220.

Akratanakul, P. (1976). Honeybees in Thailand. *American Bee Journal* 116:120-121.

Ali, *A*. D., Bakry, N. M., Abdellatil, M. *A*. and Sawaf, K.E.I. (2009). The control of the greater wax moth, Galleria mellonella L. by chemicals. *Journal of Applied Entomology*, 74(1-4): 170-177.

Alippi, *A*. M., Albo, G. N., Marcangeli, J., Leniz, D. and Noriega, *A*. (1995). The mite Varroa jacobsoni does not transmit American foulbrood from infected to healthy colonies. *Experimental Applied Acarology* 19:607-13.

Allen, M. D. (1955). Observation of honeybees attending their queen. *Journal of Animal Behavior* 3: 66-69.

Allen, M. R. and Ball, B. V. (1996). The incidence and world distribution of honey bee viruses. *Bee World* 77:141-62.

Allen, W. G., Peter, B., Bitner, R., Burquezs, *A*., Buchmann, S. L., Cane, J., Cox, P. *A*., Dalton, V., Feinsinger, P., Ingram, M., Inouge, D., Jones, E. E., Kennedy, K., Kevan P., Koopowitz, H., Medellin, R., Medellin, M. S. and Nabnam, G. P. (1998). The potential consequences of pollinator declines on the conservation of biodiversity and stability of food crop yields, conserv. *Biology* 12:8-17.

Anderson, D. L. (1991). Kashmir bee virus- A relatively harmless virus of honey bee colonies. *American Bee Journal* 131:767-770.

Anderson, D. L. (1999). Genetic and reproductive variation in Varroa jacobsoni. *Proc. XIII Int. Congr. IUSSI, Adelaide*, p. 33.

Anderson, D. L. and Giacon, H. (1992). Reduced pollen collection by honey bee (Hymenoptera: Apidae) colonies infected with Nosema *Apis* and sacbrood virus. *Journal of Economic Entomology* 85(1): 47-51.

Anderson D. I. and Morgan M. J. (2007). Genetic and morphological variation of beeparasitic Tropilaelaps mites (Acari: Laelapidae): new and re-defined species. *Experimental and Applied Acarology* 43(1): 1-24.

Anderson, D. L. and Trueman, J. W. H. (2000). Varroa jacobsoni (Acari; Varroidae) is more than one species. *Experimental Applied Acarology* 24: 165-189.

Anderson, R. H. (1963). The laying worker in the Cape bee, *Apis mellifera* capensis. *Journal of Apicultural Research* 2: 85-92.

Bailey, L. (1955a). The infection of the ventriculus of the adult honeybee by Nosema *Apis* (Zander). *Parasitology* 45: 86-94.

Bailey, L. (1955b). The epidemiology and control of Nosema disease of the honeybee. *Annuals of Applied Biology* 43: 379-389.

Bailey, L. (1962). *Bee Diseases*. United Kingdom: Harpenden.

Bailey, L. (1969). The signs of adult bee diseases. *Bee World* 50: 66-68.

Bailey, L. and Ball, B. (1991). *Honey Bee Pathology*. United Kingdom: Academic Press.

Bailey, L. and Collins, M. D. (1982). Reclassification of Streptococus pluton (White) in a new genus Melissococus pluton. *Journal of Applied Bacteriology* 53: 215-217.

Baker, R. J. (1971). The influence of food inside the hive on pollen collection by a honeybee colony. *Journal of Apicultural Research* 10: 23-26.

Baker, H. G. and Baker, I. (1983). *A Brief historical review of chemistry of floral nectar. The Biology of Nectaries*. New York: Columbia University. pp. 29-52.

Balderrama, N., Nunez, J., Giurfa, M., Torrealba, J., De Albornoz, E. G. and Almeida, L. C. (1996). A deterrent response in honeybee (*Apis mellifera* L.) foragers: dependence on disturbance and season. *Journal of Insect Physiology* 42: 463–470.

Ball, B. V. and Bailey, L. (1997). Viruses. In *Honey Bee Pests, Predators, and Diseases* (3rd ed). In: R. M. Morse and Flottum, P. K. (eds.). 2:13–31. Medina, OH: Root.

Bankova , V., De Castro, S. L. and Marcucci, M. C. (2000). Propolis: recent advances in chemistry and plant origin. *Apidologie* 31: 3–15.

Bankova, V. S., Popov, S. S. and Marekov, N. L. (1983). A study on flavonoids of propolis. *Journal of Natural Products* 46: 471-474.

Beekman, M., Gloag, R. S., Even, N., Wattanachaiyingchareon, W. and Oldroyd, B. P. (2008). Dance Precision of *Apis* florae – Clues to the Evolution of the Honeybee Dance Language. *Behavioral Ecology and Sociobiology* 62 (8): 1259-1265.

Benholf, L. M. (1925). The moults of the honeybee. *Journal of Economic Entomology* 18: 380-384.

Bhattacharya, *A.* (2004). Flower visitor and fruitset of Anacardium occidentole. *Annales Botanici Fennici* 41: 385-392.

Billen, J., Evershed, P. J. and Morgan, E. D. (1984). Morphological comparison of Dufour glands in workers of Acromyrmex octospinosus and Myrmica rubra. *Entomological Experimental Applied* 35: 205-213.

Blum, M. S. (1969). Alarm pheromones. *Annual Review of Entomology* 14: 57-81.

Blum, M. S. (1982). Pheromonal bases of insect sociality: communications, conundrums and caveats. In Les MediateursChimiques. Les Colloques de l'INRA: Versailles. pp. 149-162.

Blum, M. S. (1992). *Honeybee pheromones,* The Hive and the Honey Bee. Michigan: Dadant & Sons. pp. 373-400.

Blum, M. S., Fales, H., Tucker, K. W. and Collins, *A.* M. F. (1978).Chemistry of sting apparatus of worker honeybee. *Journal of Apicultural Research* 17: 218-221.

Boch, R. and Shearer, D. *A.* (1962). Identification of geraniol as the active compound in the Nasonoff pheromone of the honeybee. *Nature* 194: 704-706.

Boch, R. and Shearer, D. *A.* (1971). Chemical releaser of alarm behaviour in the honeybee *Apis mellifera. Journal of Insect Physiology* 17: 2277-2285.

Boch, R., Shearer, D. *A.* and Young, J. C. (1975). Honey bee pheromones: field tests of natural and artificial queen substance. *Journal of Chemical Ecology* 1: 133-148.

Boecking, O., Bienefeld, K. and Drescher, W. (2000). Heritability of the Varroa-specifichygienic behaviour in honey bees (Hymenoptera: Apidae). *Journal of Animal Breeding and Genetics* 117(6): 417-424.

Brouwers, E. V. M. (1982). Measurement of hypopharyngeal gland activity in the honeybees. *Journal of Apiculture Research* 21: 193-198.

Brouwers, E. V. M. (1983). Activation of the hypopharyngeal glands of honeybees in winter. *Journal of Apicultural Research* 22: 137-141.

Butler, C. G. (1975). *The honey bee colony: life history*, The Hive and Honey Bee. Hamilton, Illinois: Dadant & Sons. pp. 39-74.

Butler, C. G., Fletcher, D. J. C. and Watler, D. (1970). Hive entrance finding by honeybee (*Apis mellifera* L.) foragers. *Animal Behaviour* 18: 78-91.

Butler, W., Anderson, E. and Holzer, G. (1964). Pheromone of the honeybee: biological studies of the mandibular gland secretion of the queen. *Journal of Apicultural Research* 30: 650-657.

Chapman, R. F. (1998). *The Insects: Structure and Function*. New York: Cambridge University Press.

Chen, Y., Evans, J. D., Smith, I. B. and Pettis, J. S. (2008). Nosema ceranae is a long-present and wide-spread microsporidrian infection of the European honey bee (*Apis mellifera*) in the United States. *Journal of Invertebrate Pathology* 97: 186-188.

Chen, Y., Evans, J. D., Zhou, L., Boncristiani, H., Kimura, K., Xiao, T., Litkowski, *A.* M. and Pettis, J. S. (2009). Asymmetrical coexistence of Nosema ceranae and Nosema *Apis* in honey bees. *Journal of Invertebrate Pathology* 101: 204-209.

Cheng, P. C. and Wong, G. (1996). Honey bee propolis: prospects in medicine. *Bee World* 77(1): 8-15.

Chiesa, F. (1991). Effective control of varroatosis using powdered thymol. *Apidologie* 22: 135–145.

Chioveanu, G., Ionescu, D. and Mardare, *A.* (2004). Control of nosemosis- the treatment with "Protofil". *Apiacta* 39: 31-81.

Clark, T. B. (1978). Honey bee spiroplasmosis, a new problem for beekeepers. *American Bee Journal* 118(23): 18-19.

Clark, T. B. (1980). A second misrosporidian in the honeybee. *Journal of Invertebrate Pathology* 35: 290-294.

Collins, *A.* M. and Kubasek, K. J. (1982). Field test of honey bee (Hymenoptera, Apidae) colony defensive behavior. *Annals of the Entomological Society of America* 75: 385-387.

Collins, *A.* M., Rinder, T. E., Daly, H. V., Harbo, J. R., and Pesante, D. (1989). Alarm pheromone production by two honeybee (*Apis mellifera*) types. *Journal of Chemical Ecology* 15: 1747-1756.

Crailsheim, K. (1992). The flow of jelly with in a honeybee colony. *Journal of Comparative Physiology* 162: 681-689.

Crailsheim, K. (1998). Trophallactic interactions in the adult bee (*Apis mellifera* L.). *Apidologie* 29: 189-204.

Crailsheim, K. and Stolberg, E. (1989). Influence of diet, age and colony condition upon intestinal proteolytic activity and size of the hypopharyngeal glands in the honeybee (*Apis mellifera* L). *Journal of Insect Physiology* 35: 595-602

Crane, E. (1975). *The word's beekeeping– past and present*. The Hive and the Honey Bee. Hamilton, Illinois: Dadant & Sons. pp. 1-18.

Crane, E. (1990). *Managing other bees for honey production*, Bees and Beekeeping Science, Practice and World Resource. New York: Cornell University. pp. 274-284.

Crane, E. (1991). *Apis* species of tropical Asia as pollinators and some rearing methods for them. *Acta Horticulture* 288: 29-48.

Crane, P. R., Friis, E. M. and Pedersen, K. R. (1989). Reproductive structure and function in Cretaceous Chloranthaceae. *Plant Systematics and Evolution* 165: 211-226.

Crewe, R. M. and Hastings, H. (1976). Production of pheromone by workers of *Apis mellifera* adansonii. *Journal of Apicultural Research* 6: 17-28.

Danforth, B. N., Sipes, S., Fang, J. and Brady, S. G. (2006). The history of early bee diversification based on five genes plus morphology. *Proceedings of the National Academy of Sciences of U.S.A.* 103 (41): 15118-23.

De Guzman, L. I. and Delfinado-Baker, M. (1996). A new species of Varroa (Acari: Varroidae) associated with *Apis* koschevnikovi (Hymenoptera: Apidae) in Borneo. International Journal of Acarology 22: 23-27.

De Guzman, L. I. and Rinderer, T. E. (1998). Distribution of the Japanese and Russian genotypes of Varroa jacobsoni. *Honeybee Sci.* 19: 115-119.

De Guzman, L. I. and Rinderer, T. E. (1999). Identification and comparison of Varroa species infesting honey bees. *Apidologie* 30: 85-95.

De Guzeman, L. I., Rinderer, T. E. and Beaman, L. D. (1993). Survival of Varroa jacobsoni Oud. (Acari: Varroidae) away from its living host *Apis mellifera* L. *Experimental Applied Acarolology* 17:283-290.

De Guzman, L. I., Rinderer, T. E. and Stelzer J. *A.* (1997). DNA evidence of the origin of Varroa jacobsoni Oudemans in the Americas. *Biochemical Genetic* 34:327-35.

De Guzman, L. I., Rinderer, T. E., Stelzer, J. *A.* and Anderson, D. (1998). Congruence of RAPD and mitochondrial DNA markers in assessing Varroa jacobsoni genotypes. *Journal of Apicultural Research* 37:49-51.

De Jong, D. (1990). Mites: Varroa and other parasites of brood. In: *Honey Bee Pests, Predators and Diseases* (Morse, R.A. and Nowogrodski, R., eds.). Cornell University Press, Ithaca, pp. 200-218.

De Jong, D. (1997). Mites: varroa and other parasites of brood. pp. 281–327

De Jong, D. and Goncalves, L. S. (1999). The Africanized bees of Brazil have become tolerant of Varroa. p. 131.

Delfinado-Baker, M. and Aggarwal, K. (1987). A new Varroa (Acari: Varroidae) from the nest of *Apis* cerana (Apidae). *International Journal of Acarology* 13: 233-237.

Delfinado, M. D. and Baker, E. W. (1961). Tropilaelaps, a new genus of mite from the Philippines (Laelapidae: Acarina). *Fieldiana-Zoology* 44(7): 53-56.

Delfinado, M. D., Baker, E. W. and Phoon, *A.* C. C. (1989). Mites (Acari.) associated with bees (Apidae) in Asia, with description of new species. *American Bee Journal*, 129(9): 612-613.

Delfinado-Baker, M. D., Underwood, B. *A.* and Baker, E. W. (1985). The occurrence of Tropilaelaps mites in brood nests of *dorsata Apis* laborious and *A.* in Nepal with description of the nymphal stages. *American Bee Journal* 125 (10): 703-706.

Deseyn, J. and Billen, J. (2005). Age-dependent morphology and ultrastructure of the hypopharyngeal gland of *Apis mellifera* workers (Hymenoptera, Apidae). *Apidologie* 36: 49-57.

Dietz, *A.* (1992). *Honey bees of the world*, The Hive and The Honey Bee. Michigan: Dadant & Sons. pp. 23-71.

Donze, G. and Guerin, P. M. (1997). Time-activity budgets and space structuring by the different life stages of Varroa jacobsoni in capped brood of the honey bee, *Apis mellifera. Journal of Insect Behaviour* 10:371–93.

Donze, G., Schnyder, C. S., Bogdanov, S., Diehl, P. *A.* and Guerin, P. M. (1998). Aliphatic alcohols and aldehydes of the honey bee cocoon induce arrestment behavior in Varroa jacobsoni (Acari: Mesostigmata), an ectoparasite of *Apis mellifera. Arch. Insect Biochemistry Physiology* 37:129-45.

Durham, O. C. (1953). Pollen identification. A Manual of Clinical Allergy. Philadelphia and London: N. B. Saunders. pp. 112-158.

Dyer, F. C. (2000). Individual cognition and group movement: insights from social insects. In: *Group Movement in Social Primates and Other Animals: Patterns, Processes, and Cognitive Implications*. (Ed. by P. Garber and S. Boinski). Chicago: University of Chicago Press.

Dyer, F. C. (2002). Biology of the dance language. *Annual Review of Entomology* 47: 917-949.

Dyer, F. C. and Seeley, T. D. (1991). Dance dialects and foraging range in three Asian honey bee species. *Behavioral Ecology and Sociobiology* 28: 227-233.

Ebert, T. *A*., Kevan, P. G., Bishop, B. L., Kevan, S. D. and Downer R. *A*. (2007). Oral toxicity of essential oils and organic acids fed to honey bees (*Apis mellifera*). *Journal of Apiculture Research* 46: 220-224.

Eickwort, G. C. (1988). *The origins of mites associated with honey bees*, pp. 327-338. In G. R. Needham, R. E. Page Jr, M. Delfinado-Baker and C. E. Bowman (eds.) Africanized honey bees and bee mites. Chichester, Harwood, 572 p.

Elias, T. S. and Gelband, H. (1975). Nectar: its production and function in trumpet creeper. *Science* 189: 289-291.

Ellis, J. D. and Munn, P. *A*. (2005). The worldwide health status of honey bees. *Bee World* 86: 88-101.

Engels, W., Rosenkranz, P., Adler, *A*., Taghizadeh, T., Lubke, G. and Francke, W. (1997). Mandibular gland volatile and their ontogenetic pattern in queen honeybees, *Apis mellifera* carnica. *Journal of Insect Physiology* 43: 307-313.

Erdtman, G. (1966). *Angiosperm (An introduction to palynology I)*. Pollen Morphology and Plant Taxonomy. New York: Hafner. pp. 89-95.

Erdtman, G. (1969). *An introduction to the study of pollen grains and spores*. Handbook of Palynology. New York: Hafner. pp. 65-78.

Ernst, K. D. (1969). Die Feinstruktur von Riechsensillen auf der Antenne des Aaskäfers Necrophorus. *Z. Zellforsch.* 94: 72-102.

Ernst, K. D. and Boeckh, J. (1983). A neuroanatomical study on the organizationof the central antennal pathways in insects. *Cell Tissue Research* 229: 1-22.

Esslen, J. and Kaissling, K. E. (1976). Zahl und Verteilung antennaler Sensillen bei der Honigbiene (*Apis mellifera* L.). *Zoomorphology* 83: 227-251.

Farbman, *A*. I. (1992). *Structure of olfactory mucous membrane*. In: P. W. Barlow, D. Bray, P. B. Green, and J. M. W. Slack (Eds), Cell Biology of Olfaction, Developmental and Cell Biology Series 27. Cambridge: Cambridge University Press.

Farnesi, *A*. P., Aquino-Ferreira, R., De Jong, D., Bastos, J. K. and Soares, *A*. E. E. (2009). Effects of stingless bee and honey bee propolis on four species of bacteria. *Genetics and Molecular Research* 8(2): 635-640.

Feng, J.L.M., Zhang, Z. and Pan, Y. (2008). Identification of the proteome complement of hypopharyngeal glands from two strains of honeybees (*Apis mellifera*). *Apidologies* 39(2): 199-214.

Ferguson, *A*. W. and Free, J. B. (1979). Production of forage-marking pheromone by the honeybee. *Journal of Apicultural Research* 18: 128-135.

Ferguson, *A.* W. , and Winston, M. L. (1988). The influence of wax deprivation on temporal polyethism in honeybee (*Apis mellifera* L.) colonies. *Canadian Journal of Zoology* 66: 1997-2001.

Fewell, J. H. and Jr. Page, R. E. (1993). Genotypic variation in foraging responses to environmental stimuli by honey bees, *Apis mellifera. Experientia* 49:1106-1112.

Forsgren, E. and Fries, I. (2005). Acidic- Benzoic feed and Nosema disease. *Journal of Apiculture Science* 49: 2.

Free, J. B. (1967). Factors determining the collection of pollen by Honeybee foragers. *Animal Behaviour* 15: 134-144.

Free, J. B. (1981). Biology and behaviour of the honey bee *Apis florea*, and its possibilities for beekeeping. *Bee World* 62: 46-59.

Free, J. B. (1987). *Pheromone of Social Bees*. London: Chapman and Hall. pp. 218.

Free, J. B. (1993). *Insect pollination of crops*. London: Academic Press. pp. 684.

Free, J. B., Furguson, *A.* W. Simpkins, R. J. and AL-Sa'ad, B. N. (1983). Effect of honeybees Nasanoff and alarm pheromone components on behaviour at the nest entrances. *Journal of Apicultural Research* 22: 214-223.

Free, J. B., Williams, I. H. Pickett, J. *A.*, Ferguson, *A.* W. and Martin, *A.* P. (1982). Attractiveness of (z)-1-eicosan-1-ol to foraging honeybees, *Apis mellifera. Journal of Apicultural Research* 21: 151-156.

Free, J. B., Ferguson, *A.* W. and Simpson, J. R. (1988). Honeybee response to chemical components from the worker sting apparatus and mandibular glands in field tests. *Journal of Apicultural Research* 28: 7-21.

Fremont, L. (2000). Biological effect of resveratrol. *Life Science* 66: 663-673.

Fries, I., Ekbohm, G. and Villumstead, E. (1983). Nosema *Apis*, sampling techniques and honey yield. *Journal of Apicultural Research* 23(2): 102-105.

Fries, I. (1997). Protozoa. In: R. *A.* Morse (Ed.), Honey Bee Pests, Predators and Diseases (3rd) (pp. 57-76). Ohio: *A.*I. Root.

Fries, I. (1997). Infectivity and multiplication of Nosema *Apis* Z. in the ventriculus of the honey bee. *Apidologie* 19: 319-328.

Fries, I. (1989a). Comb replacement and Nosema disease (Nosema *Apis* Z.) in honey bee colonies. *Apidologie* 19: 343-354.

Fries, I. (1989b). Observation on the development and transmission of Nosema *Apis* Z. in the ventriculus of the honeybee. *Journal of Apicultural Research* 28: 107-117.

Fries, I., Feng, F., Dasilva, *A.*, Slemenda, S. B. and Pieniazek, N. J. (1996). Nosema ceranae (microspora, Nosematidae) morphological and molecular characterization of a Microsporidian parasite of the Asian honey bee *Apis* cerana (Hymenoptera, Apidae). *European Journal of Protistology* 32: 356-365.

Fries, I., Granados, R. R. and Morse, R. *A.* (1992). Intracellular germination of spores of Nosema *Apis* Z. *Apidologie* 23: 61-71.

Fries, I., Martín, R., Meana, *A.*, García-Palencia, P. and Higes, M. (2006). Natural infections of Nosema ceranae in European honey bees. *Journal Apicultural Research* 45: 230-233.

Furgula, B. and Mussen, E. C. (1990). *Protozoa*. In R. *A.* Morse, and R. Nowogrodzk (Eds.), Honey bee pests, predators, and diseases (pp. 48-58). Ithica and London: Cornell University Press.

Furgala, B. and Sugden, M. *A.* (1985). Residual activity of bicyclohexylammonium Fumagillin In sucrose and high fructose corn syrup stored at two temperatures. *American Bee Journal* 125: 47-48.

Furi, P., Wille, H., Gerig, L. and Luscher, M. (1982). Change in weight of pharyngeal glands and haemolymph titres of juvenile hormone, proteins and vitellogenin in worker honeybees. *Journal of Insect Physiology* 14: 39-59.

Fyd, W. (1959). Normal and abnormal development in the honeybee. *Bee World* 40: 57-66: 85-96.

Fyg, W. W. (1964). Anomalies and diseases of the queen honey bee. *Annual Review of Entomology* 9: 207-224.

Gambino, P. (1995). Dolichovespula (Hymenoptera: Vespidae), hosts of Aphomia sociella (L.) (Lepidoptera: Pyralidae). *Journal of the New York Entomological Society* 103 (2): 165-169.

Gary, N. E. (1963). Observation of mating behavior in the honeybee. *Journal of Apicultural Research* 2: 3-13.

Gary, N. E. (1975). *Activities and behavior of honeybee*. The Hive and the Honeybee. Hamilton, Illinois: Dadant & Sons. pp. 185-225.

Gary, N. E. (1992). *Activities and behavior of honeybee*. The Hive and the Honeybee. Hamilton, Illinois: Dadant & Sons. pp. 269-371.

Genersch, E. (2005) Development of a rapid and sensitive RT-PCR method for the detection of deformed wing virus, a pathogen of the honey bee (*Apis mellifera*), *Vet. J.* 169: 121-123.

Genersch, E., Yue, C., Fries, I. and de Miranda, J. R. (2006). Detection of deformed wing virus, a honeybee viral pathogen in bumble bee (Bombus terristris) and Bombus pascoorum) with wing deformities. *Journal of Invertebrate Pathology* 91: 61-63.

Gerson, U., Lensky, Y., Lubinevski, Y., Slabezki, Y. and Stern, Y. (1988). Varroa jacobsoni in Israel, 1984-1986. pp. 420-424.

Giurfa, M. (1991). Colour generalization and choice behaviour of the honeybee, *Apis mellifera* L. *Journal of Insect Physiology* 37: 41-44.

Giurfa, M., Backhaus, W. And Menzel, R. (1995). Color and angular orientation in the discrimination of bilateral symmetric patterns in the honeybee. *Naturwissenschaften* 82: 198-201.

Giurfa, M., Eichmann, B., and Menzel, R. (1996a). Symmetry perception in insect. *Nature* 382: 458-461.

Giurfa, M., Vorobyev, M., Kevan, P. and Menzel, R. (1996b). Detection of coloured stimuli by honeybees: minimum visual angles and receptor specific contrasts. *Journal of Comparative Physiology* 178: 699-709.

Gojmerac, W. L. (1980). *Beekeeping, Honey and pollination*. Westport: The AUI Publishing. pp. 263.

Goodman, G. (2007). Nosema disease in honey bee. Agriculture Notes, 1329-8062. Retrieved December 25, 2009, from http://www. dpi.vic.gov.au.

Goodwin, M., Ten Houten, *A.*, Perry, J., and Blackmann, R. (1990). Cost benefit analysis of using fumagillin to treat Nosema. *N. Z. Beekeeper* 208: 11-12.

Gould, J. L., and Gould, C. G. (1988). *The Honey Bee* (New York: W.H. Freeman) 231 pp.

Graham, J. M. (1992). *The Hive and The Honey bee*. Michigan: Dadant & Sons. pp. 1324.

Griffiths, D. *A.* (1988). Functional morphology of the mouthparts of Varroa jacobsoni and Tropilaelaps clareae as a basis for the interpretation of their life-styles, pp. 479-486.

Gupta, M. (1992). Scanning electron microscopic studies of antennal sensilla of adult workers of *Apis florea* F. (Hymenopter: Apidae). *Apidologie* 23: 47-56.

Hachiro, S. and Knox, D. *A.* (2000). *Diagnosis of Honey Bee Diseases*. United States: Department of Agriculture.

Handel, S. N. (1983). *Pollination ecology, plant population structure and gene flow. Pollination Biology*. Orlando: Academic. pp. 163-212.

Hansen, D. L., Brodsgaraard, H. F. and Enkegaard, *A.* (1999). Life table characteristics of Macrolophus callglnosus preying upon Tetranychus urticae. *Entomologia Experimentalis et Applicata* 93: 269-275.

Hansson, B. S. (1999). *Insect Olfaction*. New York: Springer-Verlag Berlin Heidelberg.

Harborne, J. B. (1993). *Ecological Biochemistry*. London: Academic Press Limited.

Hartwig, *A.* and Przelecka, *A.* (1971). Nucleic acids in the intestine of *Apis* meliffera infected with Nosema *Apis* and treated with fumagillin DCH; cytochemical and autoradiographic studies. *Journal of Invertebrate Pathology* 18: 331-336.

Hassanein, M. H. (1951).The influence of Nosema *Apis* on the larval honeybee. *Annual of Applied Biology* 38: 844-846.

Hassanein, M. H. (1952). The effects of infection with Nosema *Apis* on the pharyngeal salivary glands of the worker honey bee. *Proceedings of the Royal Entomological Society of London* 27: 22-27.

Hassanein, M. H. (1953). The influence of infection with Nosema *Apis* on the activities and longevity of the worker honeybee. *Annual of Applied Biology* 40: 418-423.

Heinrich, B. (1979). Keeping a cool head: honeybee thermoregulation. *Science* 205: 1269-1271.

Heinrich, B. and Esch, H. (1994). Thermoregulation in bees. *American Science* 82: 164-170.

Hepburn, H. R., Jones, G. E. and Kirby, R. (1994). Introgression between *Apis mellifera* capensis Escholtz and *A. mellifera* scutellata Lepeletier: The sting apparatus. *Apidologie* 25: 557-565.

Hepburn, H. R. and Radloff, S. E. (2011). *Honeybees of Asia*. Springer: Berlin.

Hepburn, H. R., Smith, D. R., Radloff, S. E. and Otis, G. W. (2001). Infraspecific categories of *Apis* cerana: morphometric, allozymal and mtDNA diversity. *Apidologies* 32: 3-23.

Higes, M., Hernandez, R. M., Bailon, E. G., Palencia, P. G. and Meana, *A.* (2007a). Detection of infective Nosema ceranae (Microsporidia) spores in corbicular pollen of forager honey bees. *Journal of Invertebrate Pathology* 97: 76-78.

Higes, M., Martin, H. R., Garrido, B. E., Botias, C., Gafcia, P. P., and Meana, *A.* (2008a). Detection of infective Nosema ceranae (Microsporidia) spores in corbicular pollen of forager honeybees. *Journal of Invertebrate Pathology* 97: 76-78.

Higes, M., Martin, H. R., Garrido, B. E., Botias, C., Gafcia, P. P. and Meana, *A.* (2008b). Regurgitated pellets of Merops apiaster as fomites of infective Nosema ceranae (Microsporidia) spores. *Environmental Microbiology* 10(5): 1374-1379.

Higes, M., Martin, R. and Meana, *A.* (2006). Nosema ceranae, a new microsporidian parasite in honeybees in Europe. *Journal of Invertebrate Pathology* 92: 93-95.

Higes, M., Palencia, G. P., Hemandez, M. R. and Meana, *A.* (2007b). Experrimental infection of *Apis mellifera* honey bees with Nosema ceranae (Microsporidial). *Journal of Invertebrate Pathology* 94: 211-217.

Hirai, Y., Yasuhara, T., Yoshida, H. and Nakajima, T. (1981). A new mast cell degranulating peptide, mastoparan-M, in the venom of the hornet Vespa mandarinia Biomed. Res. 2: 447-449.

Hirschfelder, H. and Sachs, H. (1952). Recent research on the acarine mite. *Bee World* 33: 201-209.

Hirst, S. (1921). On the mites (Acar*Apis* woodi (Rennie) associated with Isle of Wighth bee disease. *Annual Magazine Natural History* 7: 509-519.

Hoppe, H. and Ritter, W. (1989). The influence of the Nasonov pheromone on the recognition of house bees and foragers by Varroa jacobsoni. *Apidologie* 19: 165-72.

Hrassnigg, N. and Crailsheim, K. (1998). Adaptation of hypopharyngeal gland development to the brood status of honeybee (*Apis mellifera* L.) colonies. *Journal of Insect Physiology* 44: 929-939.

Huang, W. F., Jiang, J. H. and Wang, C. H. (2005). Nosema ceranae infection in *Apis mellifera. 38th Annual Meeting of Society for Invertebrate Pathology. Anchorage, Alaska.*

Huang, W. F., Jiang, J. H., Chen, Y. W., and Wang, C. H. (2007). A Nosema ceranae isolate from the honeybee *Apis mellifera. Apidologie* 38: 30-37.

Huang, W. F., Bocquet, M., Lee, K. C., Sung, I. H., Jiang, J. H., Chen, Y.W., and Wang, C. H. (2007). The comparison of rDNA spacer region of Nosema ceranae isolates from different hosts and locations. *Journal of Invertebrate Pathology* 97: 9-13.

Hughes, P. R. (1974). Myrcene: a precursor of pheromones in Ips beetles. Journal of Insect Physiology 20: 1271-75.

Hunt, J. H. (1982). *Trophaliaxis and the evolution of eusocial Hymenoptera*. The Biology of Social Insect. New York: Chapman and Hall. pp. 201-205.

Ifantidis, M. D. (1983). Ontogenesis of the mite Varroa jacobsoni in worker and drone honeybee brood cells. *Journal of Apicicultural Research* 22:200-206.

Insuan, S., Deowanish, S., Klinbunga, S. and Sittipraneed, S. (2007). Genetic differentiation of the giant honey bee (*Apis dorsata*) in Thailand analyzed by mitochondrial genes and microsatellites. *Biochemical Genetics* 45: 345-361.

Jones, J. C. and Oldroyd, B. P. (2007). Nest thermoregulation in social insects. *Advances in Insect Physiology* 33: 153-191.

Juven, B. J., Kanner, J., Schved, F. and Weisslowicz, H. (1994). Factors that interact with the antibacterial action of thyme essential oil and its active constituents. *Journal of Applied Bacteriology* 76: 626-631.

Kaissling, K. E. (1971). Insect olfaction. In: L.M. Beidler (Ed.), Handbook of Sensory Physiology IV (pp.341-431). Heidelberg: Springer Verlag.

Kaissling, K. E. (1972). Kinetic studies of transduction in olfactory receptors of Bombyx mori. In: D. Schneider (Ed.), Olfaction and Taste IV (pp. 207-213). Stuttgart: Wissenschaftliche Verlagsgesellschaft.

Kaissling, K. E. (1974). Sensory transduction in insect olfactory receptors. In: L. Jaenicke (Ed.), Biochemistry of Sensory Functions (pp. 243-273). Berlin: Springer Verlag.

Katznelson, H. and Jamieson, C. *A.* (1952). Control of Nosema disease of honey bees with fumagillin. *Science* 115: 70-71.

Katzav-Gozansky, T., Soroker, V. and Hefetz, *A.* (2002). *A.* Evolution of worker sterility in honey bees: egg-laying workers express queen-like secretion in Dufour's gland. *Behavioral Ecology and Sociobiology* 51 (6): 588-589.

Kerr, W. E., Blum, M. S. Pisani, J. F. and Stort, *A.* C. (1974). Correlation between amounts of 2-heptanone and isopentyl acetate in honeybees and their aggressive behaviour. *Journal of Apicultural Research* 13: 173-176.

King, G. E. (1993). *The larger gland in the worker honeybee – A correlation of activity with age and with physiological functioning*. Ph.D. Thesis University of Illinois. Urbana: Illinois.

Klee, J., Besana, *A.* M., Genersch, E., Gisder, S., Nanetti, *A.*, Tam, D. Q., Chinh, T. X., Puerta, F., Ruz, J. M., Kryger, P., Message, D., Hatjina, F., Korpela, S., Fries, I.,and Paxton, R. J. (2007). Widespread dispersal of the microsporidian Nosema ceranae, an emergent pathogen of the western honey bee, *Apis mellifera. Journal of Invertebrate Pathology* 96: 1-10.

Klee, J., Tay, W. T. and Paxton, R. (2006). Specific and sensitive detection of Nosema bombi (Microsporidia:Nosematidae)in bumble bees (Bombus spp;Hymenoptera: Apidae) by PCR of partial rRNA gene sequences. *Journal of Invertebrate Pathology* 91: 98–104.

Knecht, D. and Kaatz, H. H. (1990). Pattern of larval food production by hypopharyngeal glands in adult worker honey bees. *Apidologie* 21: 457-468.

Koeniger, N. (1969). Experiments concerning the ability of the queen (*Apis mellifera* L.) to distinguish between drone and worker cells. *XXII International Beekeeping Congress Summit*. p. 138.

Koeniger, N. (1970). Factors determining the laying of drone and worker wggs by the queen honeybee. *Bee World* 51: 166-169.

Koeniger, G., Koeniger, N., Anderson, D. L., Lekprayoon, C. and Tingek (2002). Mites from debris and sealed brood cells of *Apis dorsata* colonies in Sabah (Borneo) Malaysia, including a new haplotype of Varroa jacobsoni. *Apidologie* 33: 15-24.

Koeniger, N., Koeniger, G, de Guzman, L. L. and Lekprayoon, C. (1993). Survival of Euvarroa sinhai Delfinado and Baker (Acari: Varroidae) on workers of *Apis* cerana Fabr., *Apis florea* Fabr. and *Apis mellifera* L. in cages. *Apidologie* 24(4): 403-410.

Koeniger, N. and Veith, H.J. (1983). Glyceryl-1, 2-dioleate-3-palmitate, a brood pheromone of the honeybee (*Apis mellifera* L.). *Experientia* 39: 1051-1052.

Koeniger, N., Weiss, J. and Maschwitz, U. (1979). Alarm pheromones of the sting in the genus *Apis*. *Journal of Insect Physiology* 25: 467-476.

Koning, R. E. (1994). Honeybee Biology. Plant Physiology Website.
http:// plantphys.info/plants_human/bees/bees.html (your visit date).

Kronenberg F. and Heller, H.C. (1982).Colonial thermoregulation in honey bees (*Apis mellifera*). *Journal of Comparative Physiology* 148:65-76.

Kubo, T., Sasaki, M., Nakamura, J., Sasagawa, H., Ohashi, K., Takeushi, H. and Natori, S. (1996). Change in the expression of hypopharyngeal gland proteins of the worker honeybees (*Apis mellifera* L.) with age and/or role. *Journal of Biochemistry* 119(2): 291-295.

Laigo, F. M. and Morse, R. *A.* (1969). Control of the bee mite, Varroa jacobsoni Oudemans and Tropilaelapse clareae. Delfinado and Baker with Chorobenzilate. *Philippines Entomology* 1: 144-148.

Latif, *A.*, Qayyum, *A.* and Abbas, M. (1960). The role of *Apis* indica in the pollination of oil seeds Toria and Sarson (Brassica campestris Var), Toria and Dichotoma. *Bee World* 41: 283-286.

Leal, W. (2010). The treacherous scent of a human. *Nature* 464: 37.

Le Conte, Y., Arnold, G., Trouiller, J., Masson, C., Chappe, B. and Ourisson, G. (1989). Attraction of the parasitic mite Varroa to the drone larvae of honey bees by simple aliphatic esters. *Science* 245: 638-639.

Leadbeater, E and Chittka, L. (2007). The dynamics of social learning in an insect model, the bumblebee (Bombus terrestris). *Behaviour Ecology Sociobiology* Doi 10.1007/s00265-007-0412-4.

Lehnert, T., Shimanuki, H. and Knox, D. (1973). Transmission of Nosema disease from Infected workers of the honey bee to queens in mailing cages. *American Bee Journal* 113: 413-414.

Leiro, J., Cano. E., Ubeira, F. M., Orallo, F. and Sanmartìn, M. L. (2004). In vitro effects of resveratrol on the viability and infectivity of the Microsporidian Encephalitozoon Cuniculi. *Antimicrobial Agents Chemotherapy* 48: 2497-2501.

Lekprayoon, C. and Tangkanasing, P. (1991). Euvarroa wongririi, A new species of bee mite from Thailand. *Internaional Journal of Acarology* 17 (4): 255-258.

Lekprayoon, C. and Tangkanasing, P. (1993). Comparative morphology of Euvarroa sinhai and Euvarroa wongririi: Parasites of *Apis florea* and *Apis andreniformis*. Asian Apiculture. p. 427-433.

Li, J., Feng, M., Zhang, Z. and Pan, Y. 2008. Identification of the proteome complement of hypopharyngeal glands from two strains of honeybees (*Apis mellifera*). *Apidologie* 39: 199-214.

Lindauer, M. (1952). Ein Beitrag zur Frage der Arbeitsteilung im Bienenstaat. *Z. vergl. Physiol.* 34, 299-345. (Translation, *Bee world*, 34: 63-73, 85-90).

Lindauer, M. (1956). Uber Veestandingung bei indischen Bienen. *Z. vgl. Physiol.* 38: 521-557.

Lindauer, M. (1961). *Communication among social bees*. Cambridge: Harvard University. 346 pp.

Lindquist, E. E. (1986). The world genera ofTarsonemidae (Acari: Heterostigmata): a morphological, phylogenetic, and systematic revision, with a reclassification of Z family-group taxa in the Heterostigmata. *Mem. Entomol. Soc. Can.* 136:1-517.

Liu, T. P. (1973). Effects of fumadil B on the spores of Nosema *Apis* and on lipids of the host cell as reviealed by freeze-etching. *Journal of Invertebrate Pathology* 22: 364-368.

Maa, T. C. (1953). An inquiry into the systematics of the Tribus Apidini or honeybees (hymenoptera). *Treubia* 21: 525-640.

Mackensen, O. (1943). The occurrence of parthenogenetic females in some strains of honeybees. *Journal of Economic Entomology* 36: 465-467.

Mahmoud, *A.* L. E. (1999). Inhibition of growth and aflatoxin biosynthesis of Aspergillus flavus by extracts of some Egyptian plants. *Journal of Applied Microbiology* 29: 334-336.

Maistrello, L., Lodesani, M., Costa, C., Leonardi, F., Marani, G., Caldon, M., Mutinelli, F. and Granato, *A.* (2008). Screening of natural compounds for the control of Nosema disease in honeybees (*Apis mellifera*). *Apidologie* 39: 436-455.

Maksong, S. (2008). *Identification of bee flora from the midgut of honeybees of Thailand.* Burapha university. Chon Buri.

Malone, L. *A.* and Gatehouse, H. S. (1998). Effects of Nosema *Apis* infection on honey bee (*Apis mellifera*) digestive proteolytic enzyme activity. *Journal of Invertebrate Pathology* 71: 169-174.

Malone, L. *A.*, Gatehouse, H. S. and Tregidga, E. L. (2001). Effects of time, temperature and honey on Nosema *Apis* (Microsporidia: Nosematidae) a parasite of the honey bee *Apis mellifera* (Hymenoptera: Apidae). *Journal of Invertebrate Pathology* 77: 258-268.

Malone, L. *A.*, Giacon, H. *A.* and Newton, M. R. (1995). Comparison of the responses of some New Zealand and Australian honey bees (*Apis mellifera* L.) to Nosema *Apis* Z. *Apidologie* 26: 495–502.

Malone, L. *A.* and Stefanovic, D. (1999). Comparison of the response of two races of honeybees to infection with Nosema *Apis*. *Apidologie* 30: 375–382.

Marcucci, M. C. (1995). Propolis: chemical composition, biological properties and therapeutic activity. *Apidologie* 26: 83-99.

Matheson, *A.* (1993). World bee health report. *Bee World* 74: 176–212.

Matheson, *A.* (1996).World bee health update 1996. *Bee World* 77: 45–51.

Mattila, H. R. and Otis, G. W. (2006). Effects of pollen availability and Nosema infection during the spring on division of labour and survival of worker honey bees. *Environmental Entomology* 35(3): 708-717.

McGregor, S. E. (1976). *Insect pollination of cultivated crop plant.* Agricultural Handbook. Washington D.C.: USDA-ARS. New York. 496 pp.

Michener, C. D. (2000). *The bees of the world*. Johns Hopkins University Press, New York, New York.

Moffet, J. O., Lackett, J. J. and Hitchcock, J. D. (1969). Compounds tested for control of Nosema in honeybees. *Journal of Economic Entomology* 62: 886-889.

Morin, C.E. and Otis, G.W. (1993). Observations on the morphology and biologyo of Euvarroa wongsirii (Mesostigmata: Varroidae), a parasite of *Apis andreniformis* Hymenoptera: Apidae). *International Journal of Acarology* 19: 167-172.

Morse, R. *A*. (1978). *Arachnids: Acarina (mites and ticks)*, p. 197-209, In Morse RA, ed. Honey Bee Pests, Predators, and Diseases. Cornell University Press, Ithaca. 430 pp.

Morse, R. *A*., and Boch, R. (1971). Pheromone concert in swarming honeybees. *Annals of the Entomological Society of America* 64: 1414-1417.

Morse, R. *A*. and Shimanuki, H. (1990). *Summary of control methods*. In R. *A*. Morse and R. Nowogrodzki (eds.), Honey Bee Pests, Predators, and Diseases (pp. 341-354). Ithica and London: Cornell University Press.

Mossadegh, M. S. (1990). Development of Euvarroa sinhai (Acarina: Mesostigmata), a parasitic mite of *Apis florea*, on *A. mellifera* worker brood. *Experimental Applied Acarology* 9: 73-78.

Mossadegh, M.S. and Komeili, B.A. (1986). Euvarroa sinhai Delfinado and Baker (Acarina: Mesostigmata): a parasitic mite on *Apis florea* F. in Iran. *American Bee Journal* 126: 684-685.

Naik, D. G., Bhongle, *A*. S., Kapadi, *A*. H., Suryanyayana, M .C., Chawda, S. S. and Chauhavy, O. P. (1995). Antennal sensilla of adult *Apis* cerana indica. *Journal Apiculture Reseach* 34: 205-208.

Nair, R. K. K. (1985). *Melissopalynology. Essentials of Palynology*. New Delhi: Today and Tomorrow's. pp. 59-64.

Nakamura, L. K. (1996). Paenibacillus apiarius sp. nov./nt. *Journal of Systematic Bacteriology* 46: 688-693ю

Nakamura, J. and Seeley, T. D. (2006). The functional organization of resin work in honey bee colonies. *Behavioral Ecology and Sociobiology* 60:339-349.

Nieh, J. C. (1998). The role of scent beacon in the communication of food location by the stingless bee, Melipona panamica. *Behaviour Ecology Sociobiology* 43: 47-58.

Nieh, J. C., Contrera, F. *A*. L. and Nogueira-Neto, P. (2003). Pulsed mass recruitment by a stingless bee, Trigona hyalinata. *Proceedings of the Royal Society of London* 270: 2191-2196.

Nieh, J. C., and Roubik, D. W. (1998). Potential mechanisms for the communication of height and distance by a stingless bee, Melipona panamica. *Behaviour Ecology Sociobiology* 43: 387-399.

Ohashi, K., Natori, S. and Kubo, T. (1999). Expression of amylase and glucose oxidase in the hypopharyngeal gland with an age dependent role change of the worker honeybee (*Apis mellifera* L.) *European Journal of Biochemistry* 265(1): 127-133.

Ohashi, K., Sasaki, M., Sasagawa, H., Nakamura, J., Natori, S. and Kubo, T. (2000). Functional flexibility of the honey bee hypopharyngeal gland in a dequeened colony. *Zoology Science* 17: 1089-94.

Oldroyd, B. P., Somlenski, *A.* J., Cornuet, M., Wongsiri, S., Estoup, *A.*, Rinderer, T. E. and Crozier, R. H. (1996). Levels of polyandry, and intracolonial genetic relationship in *Apis dorsata* (Hymenoptera: Apidae). *Genetic* 276-283.

Oldroyd, B. P. and Wongsiri, S. (2006). *Asian Honey Bees. Biology, Conservation and Human Interactions.* Harvard University Press, Cambridge, Massachusetts.

Otani, H., Oyama, M. and Tokita, F. (1985). Polyacrylamide gel electrophoretic and immunochemical properties of proteins in royal jelly. *Japanese Journal Diary Food Science* 34: 21-25.

Otis, G. W. (1990). *Diversity of Apis in Southeast Asia.* In G. K. Vearesh, B. Malik, and H. Viraktanathan (Eds.), Social insects and Environment (pp. 725-726). Oxford: IBA.

Otis, G. W. (1991). *A review of the diversity of species within Apis*, Diversity of the Genus *Apis*. New Delhi: Oxford and IBH.

Park, O. W. (1949). *Activities of honeybees.* The Hive and Honeybee. Hamilton: Dadant & Sons. pp. 79-152.

Partap, T. (1992). *Honey plant sources in mountain areas.* Honeybee in Mountain in Agriculture. New Delhi: Oxford and IBH. pp. 91-112.

Partap, U. and Partap, T. (1997). *Managed crop pollination. The Missing Dimension of Mountain Agricultural Productivity.* Kathmandu: International centre for Integrated Mountain Development. pp. 95-102.

Partap, U. and Verma, L. R. (1994). Pollination of radish by *Apis* cerana. *Journal of Apicultural Research* 33: 237-241.

Partap, U. and Verma, L. R. (1998). Asian bees and bee keeping: Issues and initiatives, Asian bees and bee keeping progress of research and development. *Proceeding of Fourth Asian Apicultural Association International Conference*, Kathmandu, March 23-28, 1998. pp 3-14.

Paxton, R. J., Klee, J., Korpela, S. and Fries, I. (2007). Nosema ceranae has infection *Apis mellifera* in Europe since at least 1998 and may be more virulent than Nosema *Apis*. *Apidologie* 38: 558-565.

Payne, T. L. (1970). Electrophysiological investigations on response to pheromones in bark beetles. *Contrib. Boyce Thompson Znst* 24: 275-82.

Payne, T. L., Shorey, H. H. and Gaston, L. K. (1970). Sex pheromones of noctuid moths: Factors influencing antennal responsiveness in males of Trichoplusia ni. *Journal of Insect Physiology* 16: 1043-1055.

Pederson, K. (1976). Chalkbrood: Possible methods of control, and the effect of additional heat. *Birokteren* 92:18-22.

Phelan, L. P., Smith, *A.*W. and Needham, G.R. (1991). Mediation of host selection bycuticular hydrocarbons in the honey bee tracheal mite Acar*Apis* woodi (Rennie). *Journal of Chemical Ecology* 17: 463-73.

Plettner, E., Lazar, J., Prestwich, E. G. and Prestwich, G. D. (2000). Discrimination of pheromone enantiomers by two pheromone binding proteins from the gypsy moth Lymantria dispar. *Biochemistry* 39: 8953-8962.

Plettner, E., Slessor, K. N.,Winston, M. L. and Oliver, J. E. (1996). Caste-selective pheromone biosynthesis in honeybees. *Science* 271(5257): 1851-1853.

Plettner, E., Slesser, K. N., Winston, M. L., Robinson, G. E. and Page, R. E. (1993). Mandibular gland component and ovarian development as measures of caste differentiation in the honeybee (*Apis mellifera* L.). *Journal of Insect Physiology* 39: 235-240.

Punchihewa, R. W. K., Koeniger, N., Kevan, P. G. and Gadawski, R. (1985). Observations on the dance communication and the natural foraging ranges of *Apis* cerana, *Apis florea* and *Apis dorsata* in Sri Lanka. *Journal of Apicultural Research* 24(3): 168-175.

Pyramarn, K. and Wongsiri, S. (1986). Bee flora for four species of *Apis* in Thailand. *Journal Science Research Chulalongkorn University* 11(2): 95-103.

Raffiudin, R. (2002). *Honeybee behavioural evolution and ITPR gene structure studies*. Ph.D. Thesis James Cook University. Canada, 115 pp.

Raffiudin, R. and Crozier, R. H. (2007) Phylogenetic analysis of honey bee behavioral evolution. *Molecular Phylogenetics and Evolution* 43: 543-552.

Rasmidatta, *A*., Suwannapong, G. and Wongsiri, S. (1999). Ultrastructure of the compound eyes of *Apis dorsata. Asian Bee Journal* 1 (1): 60-64.

Rath, W., Boeking, O. and Drescher, W. (1995). The phenomena of simultaneous infestation of *Apis mellifera* in Asia with the parasitic mites Varroa jacobsoni Oud. and Tropilaelaps clareae Delfinado and Baker. *American Bee Journal* 135:125-27.

Ratnieks, F. L. W. (1993). Egg-laying, egg-removal, and ovary development by workers in queenright honey bee colonies. *Behaviour Ecology Sociobiology* 32: 191-198.

Rennie, J. (1921). Acarine disease in hive bees: its cause, nature and control. *N. Scotland Coll. Agricultural Bullentin* 33: 3-34.

Rice, R. N. (2001). *Nosema Disease in Honeybees Genetic Variation and Control*. Australian: Canprint.

Richards, *A*. J., (1953). The Social Insects: London: Mc Donald.

Richards, *A*. J. (2001). Does low biodiversity resulting from modern agricultural practice affect crop pollination and yield. *Annals of Botany* 88: 165-172.

Rijk, P. D., Gatehouse, H. S. and Wachter, R. D. (1998). The secondary structure of Nosema *Apis* large subunit ribosomal RNA. *Biochimca et Biophysica Acta* 1442: 326-328.

Rinderer, T. E., de Guzman, L. I., Lancaster, V. *A*., Delatte, G. T. and Stelzer, J. *A*. (1999). Varroa in the mating yard: I. The effects of Varroa jacobsoni and *Apis*tan on drone honey bees. *American Bee Journal* 139:134-39.

Rinderer, T. E., and Elliott, K. D. (1977). Worker honey bee response to infectionwith Nosema *Apis*: influence of diet. *Journal of Economic Entomology* 70: 431-433.

Rinderer, T. E. and Sylvester, H. *A*. (1978). Variation in response to Nosema *Apis*, longevity, and hoarding behavior in a free-mating population of the honey bee. *Annals of the Entomological Society of America* 71: 372-374.

Rinderer, T. E., Oldroyd, B. P., Wongsiri, S., Sylvester, H. *A.*, de Guzman, L. I., Stelzer, J. *A.* and Riggio, R. M. (1995). A morphological comparison of the dwarf honey bees of southeastern Thailand and Palawan, Philippines. *Apidologie* 26(5): 387-394.

Robinson, G. E. (1987). Regulation of honey bee age polyethism by juvenile hormone. *Behavioral Ecology and Sociobiology* 20: 329-338.

Robinson, G. E. (1994). *Regulation of division of laboring honeybee colonies: integrated hormonal, genetic and neural analyses of social behavior*. In *A.*Lenoir, G. Arnold, and M. Lepage (Eds). Les Insectes Sociaux, 12thCongress of the International Union for the Study of Social Insects IUSSI (pp. 5-8). Paris: Universite Paris Nord.

Robinson, G. E., Page, R. E., Strambi, C. and Strambi, *A.* (1989). Hormonal and genetic control of behavioral integration in honey bee colonies. *Science* 246: 109-112.

Roubik, D. W. (1995). *Pollination of Cultivated Plants in the Tropics*. Rome: Food and agriculture organization. pp. 122.

Royce, L. *A.* and Rossignol, P. *A.* (1991). Sex bias in tracheal mite [Acar*Apis* woodi (Rennie)] infestation of honey bees (*Apis mellifera* L.). *Bee Science* 1:159-61.

Ruttner, F. (1966). The life and the flight activity of drone. *Bee World* 47: 93-100.

Ruttner, F. (1988). *The Genus Apis, Biogeography and Taxonomy of Honeybees*. Berlin: Spriger-Verjag. pp. 3-11.

Sakagami, S. F. (1953). Untersunchungen über die arbeitsteilung in einem zwergvolk derhonigbiene. Beiträge zur biologie des bienenvolkes, *Apis Mellifera L. Japanese Journal of Zoology*. 11: 117-185.

Sakagami, S. F., Matsumura, T. and Ito, K. (1980). *Apis laboriosa* in Himalaya, the little known world's largest honey bee (Hymenoptera, Apidae). *Insecta Matsumurana* 19: 47-77.

Sammataro, D. (1997). Report on parasitic honey bee mites and disease associations. *American Bee Journal* 137: 301-302.

Sammataro, D., Cobey, S., Smith, B. H. and Needham, G. R. (1994). Controlling tracheal mites (Acari: Tarsonemidae) in honey bees (Hymenoptera: Apidae) with vegetable oil. *Journal of Economic Entomology* 87: 910-16.

Sammataro, D., Gerson, U. and Needham, G. R. (2000). Parasitic mites of honeybees: Life history, Implications, and impact. *Annual Review Entomology*, 45: 519-548.

Sammataro, D. and Needham, G. R. (1996). Host-seeking behaviour of tracheal mites (Acari: Tarsonemidae) on honey bees (Hymenoptera: Apidae). *Experimenatl Applied Acarology* 20:121-36.

Sanpa, S. and Chantawannakul, P. (2009). Survey of six bee viruses using RT-PCR in Northern Thailand. *Journal of Invertebrate Pathology* 100: 116- 119.

Sasagawa, Y., Matsuyama, H. S. and Peng, C. Y. S. (1999). Recognition of a parasite: hygienicallo-grooming behavior induced by parasitic Varroa mites in the Japanese honey bee, *Apis* cerana japonica RAD. p. 415.

Schmidt, J. O. and and Hurley, R. (1995). Selection of Nest Cavities by Africanized and European Honey Bees. *Apidologie* 26: 467-745.

Schneider, D. (1962). Electrophysiological investigations of the olfactory specificity of sexual attracting substances in different species of moths. *Journal of Insect Physiology* 8: 15-30.

Schneider, D. and Steinbrecht, R. *A.* (1968). Checklist of olfactory sensilla. *Symposium of the Zoological Society of London* 23: 279-297.

Seeley, T. D. (1982). How honey bee find a home. *Scientific American* 247, 158-168.

Seely, T. D. (1985). *Labour Specialization by Workers*. Honeybee Ecology. Prinecton: New Jersy. pp. 31-35.

Seely, T. D. (1996). *Wisdom of the hive*, Harvard university press. ISBN 978-044553765.

Sellier, N. and Cazaussus, *A.* (1991). Structure determination of sesquiterpenes in Chinese vetiver oil by gas chromatography-tandem mess spectrometry. *Journal of chromatography* 557: 451-458.

Shanbhag, S. R., Muller, B. and Steinbrecht, R. *A.* (1999). Atlas of olfactory organs of Drosophila melanogaster. Types, external organization, innervation and distribution of olfactory sensilla. *International Journal of Insect Morphology and Embryology* 28: 377-397.

Shanbhag, S. R., Muler, B. and Steinbrecht, R. *A.* (2000). Atlas of olfactory organs of Drosophila melanogaster. 2. Internal organization and cellular architecture of olfactory sensilla. *Arthropod Structure and Development* 29: 211-229.

Shearer, D. *A.* a nd Bosch, R. (1962). 2-Heptanone in the mandibular gland secretion of the honeybee. *Nature* 206: 530.

Shearer, D. and Boch, R. (1965). 2-Heptanone in the mandibular gland secretion of the honeybee. *Nature* 206: 530-532.

Shimanuki, H. (1981). Controlling the greater wax moth – a pest of honey combs. *United States Department of Agriculture Farmers' Bulletin* 2217, Washington, DC, 12 pp.

Shimanuki, H. (1990). *Bacteria in: Honey Bee Pests, Predators, and Diseases*, 2nd edition. R.A. Morse and R. Nowogrodzki (Eds). New York: Cornell University Press. pp. 28-47.

Shimanuki, H., Knox, D. *A.*, Furgala, B., Caron, D. M. and Williams, J. L. (1992). *The Hive and the Honey Bee*. Illinois: Hamilton.

Shuel, R. W. (1992). *The production of nectar and pollen by plants*. The Hive and The Honey bee. Hamilton, Illinois: Dadant & sons. pp. 345-455.

Sihag, R. C. (1988). Effect of pesticides and bee pollination on seed yield of some crops in India. *Journal of Apicultural Research* 27(1): 49-54.

Simpson, J. (1960). The functions of the salivary glands in *Apis mellifera. Journal of Insect Physiology* 4: 107-121.

Simpson, J. (1966). Repellency of mandibular gland scent of worker honeybees. *Nature* 209: 531-532.

Simpson, J., Inge, B. M. and Wilding, N. (1968). Invertase in hypopharyngeal glands of the honeybee. *Journal of Apicultural Research* 7(1): 29-36.

Singh, Y. (1975). Nosema in Indian honey bee (*Apis* cerana indica). *American Bee Journal* 115: 59.

Singh, Y.P. (1981). Studies on Pollen gathering capacity of lnd. Honey bee (*Apis* cerana irrdica F.) under Saharanpur conditions. *Prog. Horf.* 12: 31-38.

Sittipraneed, S., Sihanuntavong, D. and Klinbunga, S. (2001). Genetic differentiation of the honey bee (*Apis* cerana) in Thailand revealed by polymorphism of a large subunit of mitochondrial ribosomal DNA. *Insectes Sociaux* 48 (3): 266-272.

Slamovits, C. H., Williams, B. *A.* P. and Keeling, P. J. (2004). Transfer of Nosema locustae (Microsporidia) to Antonospora locustae n. comb. Based on molecular and ultrastructural data. *Journal of Eukaryotic Microbiology* 51(2): 207-213.

Slifer E. H., Prestage, J. J. and Beams, H. W. (1959). The chemoreceptors and other sense organs on the antennal flagellum of the grasshopper (Orthoptera: Acrididea). *Journal of Morphology* 105: 145-191.

Smith, *A.* W., Page, R. E., and Needham, G. R. (1991). Vegetable oil disrupts the dispersal of tracheal mites, Acar*Apis* woodi (Rennie), to young host bees. *American Bee Journal* 131: 44-46.

Smith, C. R., Toth, *A.* L., Suarez, *A.* V. and Robinson, G. E. (2008). Genetic and Genomic analyses of division of labour in insect societies. *Natural Reviews Genetics* 9: 735-748.

Smith, D. R., Villafuerte, L., Otis, G. and Palmer, M. R. (2000). Biogeography of *Apis* cerana F. and *A.* nigrocincta Smith: insights from mtDNA studies. *Apidologie* 31: 265-279.

Snodgrass, R. E. (1925). *Anatomy and physiology of the honeybee*, McGraw-Hill Book Company, New York.

Snodgrass, R. E. (1956). *Anatomy of the Honey Bee*. New York: Cornell University Press.

Snodgrass, R. E. and Erickson E. H. (1992). *The anatomy of the honeybee*, in: Graham J.M. (Ed.), The hive and the honey bee, Dadant and Sons (pp. 103-169). Hamilton: Illinois.

Steinbrecht, R. *A.* (1996). Are odorant-binding proteins involved in odorant discrimination? *Chemical Senses* 21: 718-725.

Steinbrecht, R. *A.* (1997). Pore structures in insect olfactory sensilla: a review of data and concepts. *International Journal of Insect Morphological Embryology* 26: 229-245.

Steinbrecht, R. *A.* (1998). Odorant-binding proteins: expression and function. *Annual New York Academic Science* 855: 323-332.

Steinbrecht, R. *A.*, Ozaki, M. and Ziegelberger, G. (1992). Immunocytochemical localization of pheromone-binding protein in moth antennae. *Cell Tissue Research* 270: 287-302.

Suppasat, T., Smith, D. B., Deowanish, S. and Wongsiri, S. (2007). Matrilineal origins of *Apis mellifera* in Thailand. *Apidologie* 38: 323-334.

Suwannaong, G. (2000). *Ultrastructure and Pheromones of the Mandibular Glands of Honeybee Foragers in Thailand*. Ph.D Thesis, Chulalongkorn university. pp. 177.

Suwannapong, G., Chaiwongwattanakul, S. and Benbow, M. E. (2010a). Histochemical comparision of the hypopharyngeal gland in *Apis* cerana Fabricius, 1793 and *Apis mellifera* Linneaus, 1758 Workers. Psyche: *A journal of Entomology* (article ID181025, 7 pages. Doi:10.1155/2010/101025.

Suwannapong, G., Maksong, S., Seanbualuang, P. and Benbow, M. E. (2010b). Experimentalinfection of dwarf honeybee with Nosema ceranae isolated from *Apis* florae Fabricius 1787: A new parasite in Thai honeybee. *Journal of Asia- Pacific Entomology* 13 (4): 361-364.

Suwannapong, G. and Wongsiri, S. (1999). Ultrastructure of the compound eyes of the giant honey bee queens, *Apis dorsata* Fabricius, 1793. *Journal STREC* 7(1-2): 60-68.

Suwannapong, G. and Wongsiri, S. (2005). Pheromonal activities of the mandibular gland pheromones on foraging activity of dwarf honeybees. Apimondia. *39th Apimondia International Apicultural Congress*, Dublin, Ireland. pp. 89-90.

Suwannapong, G., Seanbualuang, P. and Wongsiri, S. (2007). A histochemical study of the hypopharyngeal glands of the dwarf honey bees *Apis andreniformis* and *Apis florea. Journal of Apicultural Research* 46(4): 260-264.

Suwannapong, G., Seanbualuang, P., Gowda, S. V. and Benbow, E. M. (2010c). Detection of odor perception in Asiatic honeybee, *Apis* cerana Frabicius, 1793 workers by changing in membrane potential of the antennal sensilla. *Journal of Asia Pacific Entomology* 13 (3): 197-200.

Suwannapong, G., Yamor, T., Boonpakdee, C. and Benbow, M. E. (2011) Nosema ceranae, a new parasite in Thai honeybees. *Journal of Invertebrate Pathology* 106(2): 236-241.

Tangjingjai, W., Verakalasa, P., Sittipraneed, S., Klinbunga, S. and Lekprayoon, C. (2003). Genetic differentiation between Tropilaelaps claereae and Tropilaelaps koenigerum in Thailand based on ITS and RAPD analysis. *Apidologie* 34: 514-524.

Tew, J. E. (2006). *An overview of the usual swarm episode*. Bee culture: The Magazine of American Beekeeping.

Thibout, E. (1972). Male and female exocrine glands intervening in the courtship behaviour of Acrolepia assectella (Lepidoptera: Plutellidae). *Annales de la Societe Entomologique de France* 8(2): 475-480.

Tirgari, S. (1971). On the biology and manipulation of *Apis* (Micr*Apis*) *florea* F. in Iran. *23rd International Beekeeping Congress*, 330-332.

Tribe, G. D. and Fletcher, D. J. C. (1977). *Rate of development of the workers of Apis mellifera adansonii L.* In D. J. C. Fletcher (Ed.), African bees: their taxonomy, biology, and economic use (pp. 115-119). Pretoria, South Africa: Apimondia.

Underwood, B. *A.* (1991). Thermoregulation and energetic decision-making in the honeybees *Apis* cerana, *Apis* dorstata and *Apis laboriosa. Journal of Experimental Biology* 157: 19-34.

Vallet, *A.*, Cassier, P. and Lensky, Y. (1991). Ontogeny of the fine structure of the mandibular gland of honeybee *Apis mellifera* L. and pheromonal activity of 2-heptanone. *Journal of Insect Physiology* 37: 789-804.

Van der Meer, J. W., Tonjes, P. and de Waal, J. P. (1998). *A code for dike height design and examination.* In: N.W.H. Allsop (Ed.), Coastlines, Structures and Breakwaters (pp. 5-19). London: ICE. Thomas Telford.

Viollon, C. and Chaumont, J. P. (1994). Antifungal properties of essential oils e their main components upon Cryptococcus neoformans. *Mycopathologia* 123: 151-153.

Von Frisch, K. (1967). *The dance language and orientation of bees*, Harvard University Press, Cambridge, Mass.

Von Frisch, K. (1971). *Bees, Their Vision, Chemical Senses and Language, Ithaca*. New York: Cornell University Press.

Waite, R. J., Brown, M. *A.*, Thompson, H.M. and Bew, M. H. (2003). Controlling European foulbrood with the shook swarm methodand oxytetracycline in the UK. *Apidologie* 34 (6): 569-575.

Wang, der I. and Moeller, F. E. (1969). Histological comparison of the development of hypopharyngeal glands in healthy and Nosema-infected worker honeybee. *Journal of Invertebrate Pathology* 14: 135-142.

Wang, der I. and Moeller, F. E. (1970). The division of labour and queen attendance behavior of Nosema-infected worker honeybees. *Journal of Economical Entomology* 63: 1539-1541.

Wang, der I. and Moeller, F. E. (1971). Ultrastructural changes in the hypopharyngeal glands of worker honey bees infected by Nosema *Apis*. *Journal of Invertebrate Pathology* 17: 308-320.

Warhurst, P. and Goebel, R. (1995). *The bee book. Beekeeping in the Warmer Areas of Australia*. Australia: Copyright Act. pp. 244.

Webster, T. C. (1993). Nosema *Apis* spore transmission among honey bees. *American Bee Journal* 133: 869-870.

Webster, T. C., Pomper, K. W., Hunt, G., Thacker, E. M. and Jones, S. C. (2004). Nosema *Apis* infection in worker and queen *Apis mellifera*. *Apidologie* 35: 49-54.

White, G. F. (1919). *Nosema Disease*. United States: Department of Agriculture Bull.

White, J. W. (1975). *Honey*. The Hive and The Honey bee. Illinois: Dadant & sons. pp. 491-530.

Winston, M. L. (1979). Intra-colony demography and reproductive rate of the Africanized honeybee in South America. *Behavioral Ecology and Sociobiology* 4: 279-292.

Winston, M. L. (1982). *The Biology of the Honey Bee*. Cambridge: Harvard University Press.

Winston, M. L. (1987). *The biology of honeybee*. Cambridge: Harvard University Press.

Winston, M. L. (1992). *The honeybee colony: life history; The hive and the honeybee*. Michigan: Dadant & Sons. pp. 73-101.

Winston, M. L. and Fergusson, L. *A.* (1985). The effect of worker loss on temporal caste structure in colonies of the honeybee (A. *mellifera* L.). *Canada Journal Zoology* 63: 777-780.

Winston, M. L. and Katz, S. J. (1982). Foraging differences between cross-fostered honeybee workers (*Apis mellifera* L.) of European and Africanized races. *Behavioral Ecology and Sociobiology* 10: 125-129.

Wongsiri, S. and Tangkanasing. P (2000). *Biology of Honeybee (Thai version)*. Chualongkorn university press: Bangkok. 213 pp.

Wongsiri, S., Lekprayoon, C., Thapa, R., Thirakupt, K., Rinderer, T., Sylvester, H. and Oldroyd, B. (1996a). Comparitive biology of *Apis andreniformis* and *Apis florea* in Thailand. *Bee World* 77(4): 25-35.

Wongsiri, S., Limpipichai, K., Tangkanasing, P., Mardan, M., Rinderer, T., Sysvester, H. *A.*, Koeneiger, G. and Otis, G. W. (1990). Evidence of reproductive isolation confirms that

Apis andreniformis is a separate species from sympatric *Apis florea* (Fabricius, 1787). *Apidologie* 31: 3-7.

Wongsiri, S., Rinderer, T. E. and Sylvester, H. *A.* (1991). *Biodiversity of Honeybees in Thailand.* Bangkok: Prachachon. pp. 50-63.

Wongsiri, S. Thapa, R. Oldroyd, B. and Burgett, D. M. (1996b). A magic bee tree: home to *Apis dorsata. American Bee Journal* 136: 796-799.

Woods, W. *A.*, Jr, Heinrich, B. and Stevenson, R. D. (2005). Honeybee flight metabolic rate: does it depend upon air temperature. *Journal of Experimental Biology* 208: 1161-1173.

Woyciechowski, M. and Czekonska, K. (1999). The effect of temperature on Nosema *Apis* Zander (Microsporida, Nosematidae) infection in honey bees (*Apis mellifera*). *Parasite* 6: 185-187.

Woyke, J. (1963). What happens to diploid drone larvae in a honeybee colony? *Journal of Apicultural Research* 2: 73-75.

Woyke, J. (1969). A method of rearing diploid droned in a honeybee colony. *Journal of Apicultural Research* 8: 65-74.

Woyke, J. (1973). Repropductive organs of haploid and diploid drone honeybee. *Journal of Apicultural Research* 12: 35-51.

Woyke, J. (1977). *The heredity of colour patterns in the honeybee.* Genetic, Selection and Reproduction of the Honeybee. Bucharest: Bee Biology in Moscow. pp. 49-55.

Woyke, J. (1987). Length of stay of the parasitic mite Tropilaelaps clareae outsidesealed honeybee brood cells as a basis for its effective control. *Journal of Apicultural Research* 26:104-9.

Woyke, J. (1994a). Mating behavior of the parasitic honeybee mite Tropilaelaps clareae. *Experimental Applied Acarology* 18:723-33.

Woyke, J. (1994b). Tropilaelaps clareae females can survive for four weeks whengiven open bee brood of *Apis mellifera. Journal of Apicultural Research* 33:21-25.

Woyke, J. (1998). Difference in body color expression between European and Asian honeybees. *Proceeding of Fourth Asian Apicultural Association International Conference*, Kathmandu, March 23-28, 1998. pp 20-23.

Wu, Y. and Kuang, B. (1987). Two species of small honeybee- a study of the genus Micr*Apis*. *Bee World* 68: 153-155.

Yoder, J. *A.*, Sammataro, D. Peterson, J. *A.*, Needham, G. R. and Wa, B. (1999). Waterrequirements of adult females of the honey bee parasitic mite, Varroa jacobsoni (Acari: Varroidae) and implications for control. I. *International Journal of Acarology* (in press).

Yucel, B. and Dogaroglu, M. (2005). The impact of Nosema *Apis* Z. infestation of honey bee (*Apis mellifera* L.) colonies after using different treatment methods and their effects on the population levels of workers and honey production on consecutive years, Pak. *Journal of Biological Science* 8: 1142-1145.

Zander, E. and Weiss, K.(1964). *Das leben der biene*. Ulmer: Stuttgart. pp. 189.

In: Bees
Editor: Richard M. Florio

ISBN: 978-1-61324-825-6

Chapter 2

ASIAN HONEYBEES: BIOLOGY, THREATS AND THEIR CONSERVATION

Narayanappa Nagaraja[1]
UGC Academic Staff College, Bangalore University,
Central College Campus, Bangalore, India

Abstract

Asian continent is blessed with eight indigenous and one exotic honeybee species. Among these, five species are nesting in cavities and construct multiple parallel combs. In contrast, remaining four are open-nesting species which build single exposed combs. Asian honeybees exhibit a rich biodiversity in respect of distribution, nesting behavior, sex pheromone communication, defensive behavior etc. Honeybees have got an immense value for their various products and also for meticulous services they render in cross pollination of varieties of crops. Thai sac brood, a major viral disease is primarily responsible for mortality of millions of *Apis cerana* colonies. Frequent incidence and infestation of brood mites, *Varroa destructor* and *Tropilaelaps clareae,* predation by various species of ants, wasps and birds render a greater hindrance in establishment of *Apis mellifera* in most parts of Asia. Similarly, traditional methods of honey harvesting, habitat destruction, deforestation and pesticide poisoning are apparently threatening open-nesting and hive honeybee species. There is a greater scope to safeguard honeybees from pests, predators, parasites and diseases through economically viable and environmentally sustainable bee-friendly scientific methods. Obviously, creating awareness on anthropogenic impact on these bee fauna would play a tremendous role in conservation of honeybees in Asia.

Keywords: Asian honeybees, biology, conservation, deforestation, honey hunting, pesticide poisoning.

[1] E-mail address: nagaraja@bub.ernet.in.

1. Introduction

Honeybees live in highly organized societies that compete our own in complexity and succeed in dealing with varied challenges posed in their social life. They exhibit a few unique and interesting characteristics Honeybees always construct combs with wax secreted by them. The foragers communicate the direction and distance of their food in the field to their nest-mates through dance language. The 9-oxo-2-decenoic acid (9-ODA) is a chief component of the queen's substance and the isopentyl acetate is a major alarm pheromone. The new colonies are usually founded by swarms in which, an old queen is generally accompanied by a large group of worker bees. Honeybees maintain a constant nest temperature and exhibit division of labor which generally varies with the age of worker bees (Robinson, 1992).

Asian continent is rich with diversified bee fauna and harbors all honeybee species on introduction of western honeybee, *Apis mellifera*. Beekeeping is an age old practice in the continent with the native hive bee, *Apis cerana* in addition to exploiting honey from wild honeybee species. Asia is a major honey producer with huge number of bee colonies. The recent invasion and infection of Thai sac brood viral disease coupled with frequent swarming and absconding tendency in the colonies of *A. cerana* forced the faster introduction of *A. mellifera* for higher honey production and better pollination in Asia. China is a premier beekeeping country comprising one half of the total number of bee colonies of Asia. Out of existing more than 8.5 million colonies of hive honeybees, 70% are *Apis mellifera* and only remaining 30% are *Apis cerana* (Zhen *et al.*, 1992). Beekeeping in India is generally a mixed type as both honey hunting and traditional beekeeping co-exists along with the modern beekeeping with a native honeybee, *A. cerana* and the introduced honeybee *A. mellifera* (Reddy, 2003). India leads in beekeeping in south and south East Asia with millions of *A. cerana* and *A. mellifera* colonies. *A. mellifera* is well established in most parts of northern and in a few parts of South India. Equal numbers of open-nesting honeybees, *Apis florea* and *Apis dorsata* also exist in the country. Unfortunately, the population demography of feral colonies of both native hive and open-nesting species are yet to be ascertained. In Hindu Kush Himalayan region, beekeeping with *A. cerana* is being almost replaced by *A. mellifera* in a faster rate that, the populations of former is declining in a rate that is no longer viable (Verma, 2003). East Asia is the world centre for the production of royal jelly. In Japan, and South Korea, beekeeping with *A. cerana* has been more or less replaced by *A. mellifera* and a few beekeepers are raising *A. cerana* colonies (Choi, 1984; Sakai, 1992).

2. Diversity of Honeybee Species

Bees in general, belong to the family Apidae under the order Hymenoptera comprise of Euglossini (orchid bees), Bombini (bumble bees), Meliponini (stingless bees) and Apini (honeybees). The tribe Apini has only one genus, the *Apis* with nine different species. Asian continent is blessed with all nine species of the genus *Apis*. Among these, eight species are indigenous and remaining one species, *Apis mellifera* is exotic. Based on the nature of nesting behavior, honeybees can be broadly categorized into open-nesting and cavity nesting species.

The red dwarf honeybee, *Apis florea* Fabricius, black dwarf honeybee, *Apis andreniformis* Smith, common giant honeybee, *Apis dorsata* Fabricius and giant mountain honeybee, *Apis laboriosa* Smith exist under feral conditions in nature. These are open-nesting species and invariably build single exposed combs. The eastern honeybee, *Apis cerana* Fabricius, red honeybee, *Apis koschevnikovi* Enderlein, mountain honeybee, *Apis nuluensis* Tingek, Koeniger and Koeniger, Sulawesian honeybee, *Apis nigrocincta* Smith and western honeybee, *Apis mellifera* Linnaeus are cavity nesting hive honeybees species and construct multiple parallel combs.

Apis florea apparently occupies a primitive position within the genus *Apis* (Alexander, 1991, Makinson *et al.*, 2011). Its ecological niche is found in the stratum of dense bushes and small trees of tropical Asia. These colonies generally secure shade and occur in plains and rarely in places higher than 1500 meters above mean sea level. *A. florea* stretches from Oman through southern Iran and tropical mainland Asia, and to Palawan and the islands of Indonesia. These build combs, often suspended from branches of hedges, house chimneys, empty caves and piles of dried sticks. Ability of these species to survive in dry climates is one of the outstanding traits (Ruttner, 1987). These bees migrate periodically between plains and adjacent lower hills according to variation in availability of pollen and nectar by deserting old combs and rapidly building the new ones (Nagaraja and Rajagopal, 1999). *A. florea* shows striking differences in size among its castes. The small size of the worker bee could be an adaptation to the nesting site and foraging activity. These bees take along the pollen and honey stores and even the wax while migrating to new sites. Its foragers perform dances on a horizontal surface, pointing directly towards the source of the food, despite dances on a vertical surface of the combs have also been recorded (Gould *et al.*, 1985). Black dwarf honeybee, *A.andreniformis* is distributed from Palawan to China and Myanmar and both *A. florea* and *A. andreniformis* are found to overlap in Southeast Asia. These bee species are more or less similar to each other in many characteristics and were determined to be re-classified as separate species (Wu and Kuang, 1987). Virgin queens of these bees often produce piping sounds similar to those of *A. cerana* and *A.mellifera* and induce a cessation of activity in the worker bees (Wongsiri *et al.* 1996). Rinderer *et al.* (2002) reported that both the species of dwarf honeybees have a tendency to form colony aggregations in a spatial distribution.

Apis dorsata is well distributed in Southeast Asian regions both in plains and hilly regions of up to 1600 meters. The colonies are most frequently found on the branches of trees, rocks and ceilings of buildings, protected from direct sunrays and rainfall. Congregation of these colonies is very common and the number ranges from 50 to 200 on a tree or rock where the bee forage sources are quite abundant (Rajagopal and Nagaraja, 1999, Oldroyd *et al.*, 2000). There was been found more than 500 colonies of these species on a single banyan tree in Bangalore, India recently. The foragers of *A. dorsata* perform dances on its vertical comb without directing towards the food source. Koeniger and Koeniger (1980) observed that, dances of *A. dorsata* foragers were performed only on the brightest spot of the comb which moves with the position of the sun. These species are continued to dance in dark during nights also (Divan and Salvi, 1965). Colonies of *A.dorsata* are highly migratory and normally move to hilly regions in dry season and return back to plains during monsoon and post monsoon seasons where, the food is quite abundant (Nagaraja and Rajagopal, 2000; Woyke *et al.*, 2005). Migrated *A. dorsata* colonies are able return back to their original nest sites even after years and is possibly for the ability of returning swarms in identifying these sites by recognizing the chemical signatures left by the departing swarm (Paar *et al*, 2000; Paar *et al.* 2004; Neumann *et al.*, 2000). *Apis laboriosa* is found to occur exclusively in mountainous regions up to 1200 meters, where night temperatures are most frequently fall below the freezing level (Roubik *et al.,* 1985). It is well distributed in Uttar Pradesh in India through Nepal, Bhutan, China and Northern Laos and builds nests on the cliffs and rock faces, more frequently in dense aggregations. These are well adapted to high altitudes of Himalaya and have a drone flight between 12.00 to 14.00 hours (Underwood, 1990). Two subspecies of *A. dorsata* viz. *Apis dorsata breviligula* Maa and *Apis dorsata binghami* Cockerell have been reported. The former is found in the northwest of the Merrill line in Luzon, the Philippines whereas the latter in the east of the Wallace line in Sulawesi and Butung of Indonesia.

Eastern hive bee, *Apis cerana* is an important indigenous hive honeybee of Asia. It is also distributed in the east of Iran and south of the great mountain regions and central deserts. These species are gentle in temperament, easy to handle and industrious in food collection. However, it exhibits frequent swarming and absconding of colonies. Its feral colonies are generally found in hollows of trees, clefts in rocks, termite mounds, crevices of walls, telephone poles etc. *A. cerana* foragers perform dances on vertical comb and usually fan the hive entrance by facing outwards the hive entrance. The taxonomic studies on cavity-nesting honeybees in Asia have confirmed three more distinct species recently. They are *A. koschevnikovi* (Tingek *et al.*, 1988), *A. nigrocincta* (Hadisoesilo *et al.*, 1996) and *A. nuluensis* (Tingek *et al.*, 1996). *A. koschevnikovi* is well distributed in the regions of Java, Sumatra, peninsular Malaysia and southern Thailand. It is relatively larger than *A. cerana* and is reddish yellow in color. These bee species differ each other in a few morphometric characteristics and also in mitochondrial and nuclear DNA composition. However, their phylogenetic relationships are yet to be established (Arias and Sheppard, 2005). *A. nigrocincta* is distributed in Sulawesi and Mindanao regions. Drones of these species perform mating flights about 2 hours later than that of *A. cerana* with partial overlapping. *A.nuluensis* is confined to mountain regions at 1500 m and has been known to occur at Sabah in Borneo. Its drone flight period was found between 10.45 to 13.15 hours. (Koeniger *et al.*, 1996). Phylogenetic studies based on the DNA analysis indicate that *A. nuluensis* and *A.cerana* are distantly related to each other (Arias *et al.*, 1996; Tanaka *et al.*, 2001). However, no reports are available on colony aggregations in Asian hive bees.

The western hive honeybee, *Apis mellifera*, is widely distributed all over the world including Asia mostly on its introduction from parts of Europe and Africa as a good honey producer and efficient pollinator. *A. mellifera* has more than 26 subspecies established in different geographical regions of Europe, western Asia and Africa (Engel, 1999, Hepburn and Radloff, 1998). These bees are gentle in nature and industrious in food collection with less swarming and absconding tendency. Among these subspecies, *A.mellifera ligustica* has been well popularized in many parts across the world. It is well established in south East Asian regions including northern parts of India and produces considerably larger quantity of honey and royal jelly in China and Japan. Foragers of *A. mellifera* perform dances on the vertical combs of their hive and they fan at the entrance of their hive, by facing towards the hive entrance. It has been found to form nest aggregations in feral conditions (Oldroyd *et al.*, 1995).

3. Honeybee Biology

3.1. Development

Honeybee colony normally consists of a fertile queen and tens of thousands of worker bees. However, during flow season, the colonies may rear handful of drones. Queen is the mother of the colony, which lays eggs and coordinates the activities of colony through its glandular secretions, primarily by the queen substance. Honeybee during its course of development passes through egg, larva, pupa and the adult stages Queens are reared in conical cells constructed at the lower edge of the comb in the colonies. Queen develops from fertilized egg laid by the queen to the stage of an adult in about 16 days and was fed with royal jelly secreted by the nurse bees. Occasionally, traces of pollen have been recorded in the food of queen larvae (Ribbands, 1953). After a few days of emergence, the matured queen would mates with 10-20 drones only once in her lifetime. It stores the sperms in the spermatheca and lays millions of eggs during her life span of about 3 to 5 years. The queen has a reproductive monopoly since the ovaries of worker bees are normally inactive and the eggs laid by them are eaten by their fellow workers. But, if any colony becomes queenless, the workers start to lay eggs and stop policing (Miller and Ratneiks, 2001). However, the workers of *A. florea* from other colonies exploit queenless period as an opportunity to move in and lay their own eggs while no policing is in force. Such parasitism of queenless colonies does not occur in other honeybee species and may be facilitated by the accessibility of *A. florea* nests (Nanork *et al.*, 2005).

Drones are reared in larger hexagonal cells constructed on the combs. They develop normally from the unfertilized eggs laid by the queen bee and are fed with drone jelly initially for one to three days followed by a modified drone jelly added with honey and pollen. The developmental period of drones from the egg to an adult stage is about 24 days which is comparatively longer than that of worker and queen bees. The chief function of the drone bees is to mate with virgin queen which on maturation flies out generally in the afternoon hours of the day to drone congregation areas where mating occurs. Worker bee develops in normal hexagonal cells from the fertilized egg laid by the queen. Worker larvae are fed initially with worker jelly for two days followed by modified worker jelly added with pollen and honey. Larval feeding ceases on attaining the age of 5-6 days when the cells are sealed by wax capping. After maturation, the pupa emerges as an adult. The developmental period of worker bees is about 21 days, despite the period varies among different species of honeybees.

Most of the worker activities within the hive are social, which in turn depends on the stage of development of various glandular systems in relation to their age. Generally, the younger bees tend to stay inside the hive for a period of first two to three weeks. They feed a mixture of honey and pollen to larvae older than three days. They perform orientation flights also called as 'play flights' at the hive entrance. The worker bees involve in duties such as feeding young larvae, removing debris and dead bees, comb building, packing pollen in the cells, cell capping, nectar ripening and guarding the colony. The house bee on attaining the age of about three weeks, switch over to out-door duties such as collection of nectar, pollen, water and propolis depending on needs of the colony. The foraging bees gather pollen and nectar from the flowers and carried to the hive in specially modified structures known as 'pollen basket' and 'honey sac' respectively. They transfer the nectar to house bees on reaching the hive, which later spread the nectar and convert it into honey. They raise brood generally in the central parts of the comb cells. Honey is stored in the upper portions of the combs and pollen is packed in parts of the comb cells situated between the brood and honey stores. Honeybees repeatedly reuse the combs and even accept the combs of other honeybee colonies.

3.2. Drone Congregation Areas (DCAs)

Polyandry, mating males with only one female but females mating with several males is rare among insects. Interestingly, virgin queens of honeybee species mate with several drones possibly from different colonies. Drones of honeybees fly to drone congregation areas (DCAs), the identified mating yards to mate with visiting queens (Koeniger *et al.*, 2005). The location of DCAs are known to be remains constant for several years. The time of the day, the drones perform mating flights and heights at which they congregate would vary among honeybee species (Buawangpang *et al.*, 2009). The drones of *A. mellifera* congregate in an open air at a height between 5-40 m above the ground level during afternoon hours. In most Asian honeybee species, the drones congregate close to the vegetation and tree canopies (Otis *et al.*, 2000). *A. cerana* drones fly in an open space within or near the canopies of the trees. Similarly, the drones of *A. koschevnikovi* assemble under the canopy of lower trees and they fly for a long period of nearly 2 hours (Koeniger *et al.*, 1994). Drones of *A. dorsata* assemble under the canopy of tall emergent trees. Evidence for the existence of DCAs of *A. florea* is not yet determined despite they are known assemble at the DCAs located close to small trees with dense leafage (Nagaraja and Brockmmann unpubl.). The drone flight period of *A. florea* was between 13.45 and 15.30 hours and ceased within an hour (Koeniger *et al.*, 1996). However, no information is available on DCAs of *A. nuluensis* , *A. nigrocincta* , *A. laboriosa* and *A. andreniformis*. The mechanism followed by drones and queens to locate the congregation areas is very interesting. When the queen visits a congregation site, the drones are attracted for queen's sex pheromones. Drones chase the queen in the air by forming a comet-like swarm around her. Several drones copulate successively in flight with the queen (Gries and Koeniger 1996).

3.3. Sex Pheromone Communication

Honeybees have evolved an intricate system of chemical communication to regulate their complex social interactions (Slessor *et al.*, 2005). Sex pheromone communication in honeybee colonies is closely associated with mandibular gland secretions of virgin queens (Butler *et al.*, 1967). In contrast, mated queens use these secretions to signal their presence to workers in the hive (Brockmann *et al.*, 1998). The sex pheromone communication was thought to be similar with the general idea that queens of all honeybee species use 9- Oxo-2-decenoic acid (9-ODA), a queen mandibular gland component, as the major sex pheromone (Koeniger and Koeniger, 2000). However, chemical analyses of queen mandibular gland secretions have shown that, 9-ODA is not a major component in virgin queens of all honeybee species (Keeling *et al.*, 2000). Queens of *A. florea* as well as unmated queens of *A. mellifera* synthesize higher amounts of 10-hydroxy 2-decenoic acid (10-HDA) than 9-ODA

(Plettner *et al.*, 1997). It was found that, the drones of *A.mellifera* were attracted towards 9-ODA, however, 10-HDA increases its attractiveness (Brockmann *et al.*, 2006). Sannasi *et al* (1971) reported a evidence that drones of *A. florea* are attracted to 9-ODA, whereas, attempts to attract these drones towards 9-ODA had failed (reviewed in Oldroyd and Wongsiri, 2006). In queens of these species, 10-HDA is a major component in the mandibular gland secretion. *A. florea* drones were attracted towards both 9-ODA as well as 10-HDA. However, 10-HDA attracted higher numbers of drones.

3.4. Defensive Behavior

Colony defense is one of the survival strategies to defend their colonies against pests and diseases in honeybees. The sting is a primary defensive organ and is well developed in worker bees. Open-nesting bee species normally construct nests on the sites almost free from pests, predators and natural hazards such as direct sunlight, rainfall, high wind velocity etc. (Reddy and Reddy, 1993). *A. florea* tends to select shaded sites especially dense bushes to protect the colonies against birds and mammals and elicit a hissing sound against any disturbance. If an intruder tries to land on its nest forcibly, is aggressively attacked by a group of workers by clumping around it. On continuous disturbances, the colony would rather abscond than defending itself. However, if an intruder approaches the nests of *A.andreniformis*, a tail of workers hangs beneath the nest by loosening the nest curtain and exhibit shimmering, hissing and roaring behaviours. *A.dorsata* colonies could be easily get alerted on exposed to predators and they warn them by visual and auditory signals. Worker bees of these species twist their bodies in a vertical direction by raising the abdomen and lowering the head whenever disturbance occurs on their nest curtain (Woyke *et al.*, 2008). However, on least disturbance, the workers perform a quick runs over the nest curtain by forming clusters at the lower edge of the colony. While noting this behavior, the workers from nearby giant honeybee colonies also emit defense waves on the surface of the nest (Seeley *et al.*, 1982). This behavior was also found in the colonies of *A. laboriosa* (Batra, 1996).

Feral colonies of hive honeybees have a narrow deep nest entrance. They maintain hive sanitation by cementing the cracks and crevices with propolis, a resinous substance collected from several species of plants. *A. mellifera* uses propolis in greater amounts to prevent the entry of pests, predators and parasites by tightly packing the cracks and crevices of the hive. *A. dorsata* uses propolis occasionally to strengthen the site of comb attachment on a branch, while, *A. cerana* colonies are do not thought to use these resins at all (Crane, 1990). The defensive mechanism in *A.cerana* consists of shimmering, aggressive and absconding behaviors. Butler (1974) called shimmering as a threatening behavior in which the bees perform abdominal, side-to-side and left-to-right movements towards an intruder at the hive entrance. *A. cerana* also exhibits response to mechanical stimuli. On a slight knock at the hive, the bees emit a conspicuous hissing sound which is transmitted among the nest-mates either by body contact or by a movement produced by the wings (Koeniger, 1995). In addition, a spectacular defense behavior was reported in the colonies of *A. cerana* (Ono *et al.*, 1995). Upon approach of flying hornets, the guard bees form groups and stay in tight contact with each other. When the hornet attacks, the bees form a ball around it and raise the temperature to lethal level by killing the hornet. The tactile contact stimulates the shimmering behavior and visual cues trigger the body shaking behavior in *A. nuluensis* (Koeniger *et al.*, 1996).

4. Economic Importance

4.1. Hive Products

Honeybees produce honey, the *'nature's golden wonder'* from the nectar collected from various flowers of plants. Honey is most accepted general sweetener in preparation of varieties of food products. It has been used by human beings to treat varieties of ailments through topical application. Honey is used in treatment of chronic and infected wounds (Armon, 1980, Dumronglert, 1983), dental infection (Elbagoury and Fayed, 1985) and gastrointestinal disorders (Salem, 1981). Beeswax is secreted by the young worker bees and is used in pharmaceuticals, cosmetics, waterproofing materials, polishes and also in manufacture of candles. Royal jelly is a secretion of the hypopharyngeal glands of young worker honeybees. It has been used widely as a healthy food in many parts especially in Thailand, China and Japan. Royal jelly is found to be hyposensitive, anti-tumor; anti-inflammatory, anti-fatigue and anti-allergic in human beings (reviewed in Baitala *et al.*, 2010). Bees collected pollen is rich source of proteins and has better antibiotic properties. Propolis is a hive product with variable composition and it has various pharmacological and nutritional applications (Marcucci, 1995). It has found varieties of medicinal properties (reviewed in Simone-Finstrom and Spivak, 2010). Honeybees use propolis to reinforce their combs and keep the hive environment aseptic. Bee venom is used in treatment of arthritis and other related ailments, despite it is allergic to a few children and adults.

4.2. Pollination of Crops

Pollinators play an important role in maintaining natural ecosystem since they facilitate gene flow among pollinated plants (Kevan, 1999; Kevan and Phillips, 2001). Insects are chief pollinators of angiosperms and among these, bees are one of the specialized groups (Danforth *et al.*, 2006). About two third of human food crops are dependent on insect pollination especially by honeybees (Richards, 2001; Klein *et al.*, 2007). Honeybees have become increasingly important as field sizes have increased and native bees have decreased, due to intensive land use and application of pesticides (Imperatriz-Fonseca *et al*, 2006). The branched hairs on bees' body safely accommodate the pollen grains for long distances during their flight. The hind legs have a specially modified structure, the 'pollen basket' to carry pollen loads safely to the hive. Honeybees involve in pollen and nectar collection continuously throughout the day in large numbers. In addition, they exhibit floral constancy and restrict to flowers of one species on several trips if sufficient pollen and nectar are available (Beekman *et al.*, 2008). Bee colonies could also be shifted to crop plots where cross-pollination is required. Foraging bees in the act of food collection, visit huge numbers of flowers and transfer the pollen from male to female flowers belonging to same species leading to cross-pollination in the plants. Honeybees have been well recognized throughout the world as the best pollinators of many crops, if not all the crops (Mc Gregor, 1976; Suryanarayana *et al.*, 1990). Bee pollination is required for seed production of vegetables and found higher seed set and seed weight in bee pollinated cauliflower plots (Kumar *et al.*, 1988). Honeybee pollination can be equivocally proved to be increase seed or fruit yield (Free, 1993; Mahanta and Rahman, 1997), seed yield and oil content in oil seed crops (Rajagopal *et al.*, 1999; Nagaraja and Reddy, 2002, Rajagopal and Nagaraja, 2003). In addition, bees are also essential in pollination of numerous species of wild flora (Rinderer *et al.* 1996). In absence of these pollinators, there is a possibility of tremendous decline in number of wild floral species. Therefore, obviously, there is lots of scope to utilize the services of honeybee pollinators for increased food production and conservation of the environment.

5. Major Threats

5.1. Diseases, Parasites, Pests and Predators

a) Diseases

Asian honeybees are found to be most tolerant to pests and diseases. However, a few honeybee species are known prone to a few viral diseases and parasites recently. Thai sac brood virus disease which originated during early 1980s has had caused irreparable damage to beekeeping industry by killing about 80 to 90 % of *A. cerana* colonies. The viral pathogen infects the young larvae of 2-3 days old. The body color of the infected larva changes gradually from creamy white to pale yellow. It is compressed to become scale like and attaches to one side of the cell at the bottom. The sealed brood cells are irregularly scattered on the combs with perforations on the capping. Adult bees of the diseased colonies become sluggish and a small portion of bees goes out for foraging and carry a small quantity of pollen and nectar. During severe infection, the worker bees fail to cope up with the removal of infected brood and colonies would abscond gradually (Rajagopal and Kencharaddi, 2000). The absconded colonies could settle in new sites, if the adult population is large with sufficient number of young workers along with a young queen. The colonies with low bee population would die due to starvation within a few days. This disease is responsible for death of millions of *A. cerana* colonies for the past three decades in Asia. It spreads through contaminated food, overcrowding of bee colonies, swarming, robbing and also by *Varroa* mites (Nagaraja and Rajagopal, 2009).

Kashmir bee virus is responsible for death of thousands of *A. cerana* colonies in Kashmir during 1970s (Bailey and Woods, 1977). The virus which was infecting the original host, *A. cerana* also made a species jump and began to infect a new host, *A. mellifera* on its introduction into south Asia. The virus infects adult bees and bees are partly or completely hairless, with a dark upper thoracic surface and exhibit trembling uncoordinated movements. Kashmir bee virus declines the adult bee population initially and finally leading to death of bee colonies in a faster rate. Similarly, the Iridovirus is a pathogen of *A. cerana* and the virus infected worker bees become sluggish and form clusters at the hive entrance. It was also recorded in the colonies of *A. mellifera* from northern states of India (Mishra *et al.*, 1980).

American foulbrood is a most destructive bacterial brood disease killing millions of *A. mellifera* colonies throughout the world. Its occurrence was also reported on *A.cerana* brood in Uttar Pradesh, India. The diseased brood irregularly intermingles with healthy brood with uncapped, punctured or sunken capping in the form of a pepper box. European foulbrood is another contagious bacterial disease of *A.mellifera.* It also infects the young larvae of *A. cerana* , where *A. mellifera* colonies are kept in some parts of central and southern states of India (Kshirsagar and Godbole, 1974). This disease was also found in the colonies of *A. laboriosa* (Bailey and Ball, 1991). The virulence of this pathogen is most common during peak brood rearing seasons.

b) Parasitic Mites

Honeybee colonies are most frequently exploited by many species of parasitic mites. Among these, the tracheal mite, *Acarapis woodi,* and the brood mites, *Varroa destructor* and *Tropilaelaps clareae* are very important. *A. woodi* is an endoparasitic mite inhabiting the tracheae and air sacs of adult honeybees. It invades into the tracheae of young bees through first pair of thoracic spiracles. Mites feed on the tracheal haemolymph by piercing their sharply pointed chelicerae and form a wound. The infested bees become sluggish, paralytic and form scattered clusters on the brood. Liu (1989) found a secondary coating on the interior part of the trachea of infested bees. These secondary coatings lead to insufficient supply of oxygen causing breathlessness in adult bees. The punctures caused by the mites become melanized resulting in dark brown spots on the tracheal wall. The infested bees also show bacterial infection in their haemolymph occasionally and probably this infection may cause more damage than the mites alone. *A. woodi* also parasitizes young bees of *A. cerana* and *A. mellifera* in India (Singh, 1957), Nepal (Abrol, 2000) and also on *A.dorsata* in Pakistan (Delfinado-Baker *et al.,* 1989). Transmission *A.woodi* between *A.cerana* and *A.mellifera* has been observed (Adlakha, 1976).

Varroa destructor and *Tropilaelaps clareae* are the most destructive ectoparasitic mites parasitizing the brood of honeybee colonies. *Varroa destructor* is widely distributed in the colonies of *A.mellifera* and *A.cerana.* Anderson and Trueman (2000) identified the haplotypes concealed with the complex of mites infesting *A.cerana*, which have become pests of *A.mellifera* worldwide. The Korean haplotype being a parasite of *A.cerana* in Korea became pests of *A.mellifera* in Europe, New Zealand, the Middle East, Africa, Asia, Canada, North and South America. The Japan and Thailand haplotype, being a parasite of *A.cerana,* became a pest of *A.mellifera* in Japan, Thailand and North America. The Korean haplotype of *V.destructor* appears to be more parasitic on *A.mellifera* than Japan and Thailand type.

Varroa destructor is a most serious parasite of *A.mellifera* (Johnson *et al.*, 2009). The mites are used to pierce the soft intersegmental tissues of the larvae and feed on the haemolymph. The larvae die in the pre-pupal stage with characteristically raised heads (Garg and Kashyap, 1998). They also feed on the young bees (Hoppe and Ritter, 1989). Mite infested adult bees show a stunted growth with deformed legs and wings and on severe infestation they usually become crippled or die. The mite infested bee colonies generally become weak with declined adult bee population. *Varroa* mites transmit various pathogens which cause diseases in honeybees (Ball and Allen, 1988). *Varroa* spp. which infests and reproduces on the drone brood of *A.cerana* other than mainland of Asia is *V. jacobsoni*. It also parasitizes drone brood of *A. nigrocincta* (Anderson and Trueman, 2000). In absence of drone brood, these mites may invade worker brood but they do not lay eggs due to non-activation of oogenesis. They feed on the developing pupae of worker bees. In some parts of Asia, genotypes of *V.jacobsoni* may invade *A.mellifera* colonies at low levels but they do not reproduce either on worker or drone brood (Anderson, 1994). Similarly, *Varroa underwoodi* is a parasite of *A. cerana* in Nepal (Delfinado-Baker and Aggarwal, 1987). *Varroa rindereri* infests the brood of *A.nigrocincta* and *A.nuluensis* (Otis and Kralj, 2001) and *A. koschevnikovi* (de Guzman and Delifinado-Baker, 1996). *Euvarroa sinhai* is an ectoparasite of *A. florea* and these mites feed on the brood and adult bees. However, *Euvarroa wongsirii* infests only the drone brood of *A.andreniformis*. *Euvarroa haryanensis* was found to be associated with *A.florea* and subsequently on *A.mellifera* (Kapil *et al.*, 1985).

Tropilaelaps clareae is an ectoparasitic mite, which exploits the brood of honeybees possibly by surviving during broodless periods. *T. clareae* has been evolved away from its presumed predatory ancestry to become a parasite on the brood of honeybees. It is distributed widely in Asia, from Iran in the North West to Papua New Guinea in south east (Ellis and Munn, 2005). *T. clareae* is a original parasite of giant honeybee, *A.dorsata* (Bhardwaj, 1968) and is considered to be most important limiting factor on establishment of beekeeping with exotic honeybee, *A.mellifera* in Asia (Nagaraja, 2000, Nagaraja and Rajagopal, 2001). The shortened life cycle coupled with a brief stay on adult bees trigger the faster development of its population. Bee colonies infested with *T.clareae* show a scattered brood pattern with sunken capping. It infests both worker and drone brood of *A.mellifera* and the intensity of infestation vary according to seasons (Nagaraja, 1998). The infested bee pupae have mutilated wings with stunted abdomen. Woyke (1994) recorded about 55% mite infestation in the brood of *A.mellifera. T. clareae* also infests the brood of *A. dorsata* which are found to migrate after brood rearing season by leaving the mite infested brood in the deserted combs. In addition, these species do not open the mite infested pupae in the capped cells. Therefore, the parasite which kills the pupae does not reach the adult bees or other brood cells (Woyke *et al.*, 2004). *Tropilaelaps koenigerum* is a new species of ectoparasite on *A.dorsata* in Sri Lanka (Delfinado-Baker and Baker, 1982). It was also found on the brood of *A.cerana* and *A.mellifera* (Abrol and Putatunda, 1995) and *A. laboriosa* (Delfinado-Baker *et al.*, 1985). *Tropilaelaps mercedesae* parasitizes *A. dorsata* and *A.mellifera* in mainland of Asia and Indonesia except Sulawesi Island. *Tropilaelaps thaii* parasitizes *A.laboriosa* in Himalayan mountain regions (Anderson and Morgan, 2007).

c) Pests and Predators

Honeybee colonies are most frequently attacked by varieties of insect pests during dearth season for bees. The greater wax moth, *Galleria mellonella* is a serious pest of honeybees causing severe damage especially in weaker bee colonies in tropical Asia. Combs of all the species of *Apis* are freely damaged by wax moth larvae. The extent of its damage varies in bee colonies through geographical regions. During dearth season for bees, the bee colonies become weak and are attacked by wax moth. The wax moths lay eggs most frequently in cracks and crevices between supers, bottom board and on the frames when combs are fully occupied by bees. The larvae are cylindrical in shape and create small tunnels between the cells and midribs of the comb through silken strands of web. They prefer to live in darker combs and confine mostly to the midribs and base of the cells in empty combs. The combs are infested in store, as well as in hives unoccupied by bees. The larvae live in the silken tunnels and feed on the propolis, pollen and beeswax in the combs. The matured larvae spin dense silken cocoons and they usually attach firmly to the hive parts. During severe infestation, the combs are seen covered with silken web containing numerous black faecal particles by destroying the combs. This leads to greater damage of bee combs and the colonies would abscond finally (Nagaraja and Rajagopal, 2003). Mahindre (1983) found about 90% wax moth infestation in the combs of *A. dorsata.* However, the combs of *A. laboriosa* do not appear to be attacked by wax moths as these moths are unable to survive in the cool climate and freezing winter conditions, where these bee colonies exist (Underwood, 1986). The lesser wax moth, *Achroia grisella* Fab. also damages the bee combs and is less destructive.

Ants are the most destructive predators of honeybees in tropical and sub-tropical regions. The weaver ant, *Oecophylla smaragdina*, black ants, *Camponotus compressus*, *Camponotus rufoglaucus* and small brown ant, *Monomorium* spp. are major predators of bee colonies. *O. smaragdina* attacks bee colonies and eat or carry away the comb contents such as brood, adult bees and honey. However, *A. florea* has developed a special defensive mechanism against these ants. It protects the nests with two rings of sticky bands one on either side around the substratum. These sticky bands are roughened most frequently through bees salivary secretions which act as repellent against ants. In addition, if these ants try to attack on the bee colony, the worker bees' counter attack on them by clumping together (Pirk *et al.* 2002). Ants generally forage in groups and kill the weaker bee colonies. These are known to be one of the severe impediments in establishment of *A.mellifera* in parts of Asia. The ants cause greater nuisance to bee colonies during dearth season. Bee colonies are most frequently fed with sugar syrup during dearth season. This sugar syrup in turn attracts the ants towards the colonies initially to feed on sugar syrup and subsequently prey upon the brood and adult bees.

Many species of wasps are widely distributed predators of honeybees in many parts of Asia. They attack on bees at the hive entrance and also in the field. They catch the bees, macerate and even feed them to their young ones. The wasp causing greater damage to bee colonies is giant hornet, *Vespa mandarina* in the Middle East and Asia. It attacks on bee colonies by taking away both brood and adult bees. The yellow banded wasp, *Vespa tropica* causes greater menace during monsoon period (Hanumanthaswamy, 2000). *V. orientalis* is very common on the sweet meat shops and occasionally predate on honeybees. The brown wasp, *Vespa velutina* is a notorious predator and troublesome to beekeeping in hilly regions during July to September (Shah and Shah, 1991).

Birds are the major predators of honeybees in tropical Asia. The common green bee-eater, *Merops orientalis* prey upon a wide variety of insects in Indian sub-continent. These birds catch up the bees, beaten them against the perch until death and eaten up later. The blue bearded bee-eater, *Nyctyornis athertoni* disturbs *A. dorsata* nests by making a quick fly past and flies away by pursuing bees to a sheltered spot and feed upon them (Kastberger and Sharma, 2000). The drango, *Dicrurus adsimilis*, visits apiaries especially on cloudy days, and prey upon bees. *Dicrurus leucophaeus* is a highly migratory bird and catches bees at the hive entrance. The oriental honey buzzard, *Pernis ptilorhyncus* is found throughout Asia and predate on *A. dorsata* colonies (Thapa and Wongsiri, 2003). Swifts prey upon honeybees at flight and easily catch up drone bees. The Philippine purple needle tailed swift, *Hirundapus celebensis* is a major constraint to beekeeping in the Philippines.

The widely distributed mammalian predators of honeybees are the bears. Bears usually dismantle the hives to feed upon the honey, pollen, brood and adult bees. They tear the hives into pieces and quickly carry off honey combs to escape from mass stinging of bees. Rhesus monkey, *Macaca mulatta*, has been found damaging the brood of *A. florea*. Problem of monkeys is very common in south India, where troops of monkeys attack on the colonies of *A. cerana* and *A. florea.* The monkeys in troops jump down on the bee hives, open the top cover and shake the combs for the bees to fly away. Later, they carry off combs with honey and brood. Seeley *et al* (1982) observed a troop of macaque monkeys which shook an *A. florea* nest until bees deserted the comb and made off with the brood.

5.2. Deforestation

There is a tremendous increase in human population in South and Southeast Asian countries for the past century. Huge population explosion coupled with new policies and programmes on development, intensive agriculture, industrialization and growing urbanization are primarily responsible for deforestation in Asia. South East Asia recorded highest rates of deforestation in the world. However, the information on impact of deforestation on honeybee fauna is meager in these regions. Removal of trees often cause dearth of pollen and nectar to honeybees. (Kevan and Viana, 2003, Sodhi *et al.*, 2004). For example, University of Agricultural Sciences, Bangalore, India has had been found more than 250 *A. dorsata* colonies on a banyan tree and 150 colonies on the buildings especially during winter for past two decades primarily for rich source of nectar from eucalyptus plantations. However, on uprooting of most of these eucalyptus trees by the university a few years ago has made the university campus almost free from *A. dorsata* colonies. Similar tendency was also observed in case of feral colonies of *A. cerana* and *A. florea* whose numbers were declined in a rapid way. Removal of trees for widening of roads is also known responsible for declined number of giant honeybee colonies in rural areas. Similarly, deforestation was considered to be the main threat to most indigenous bee species in neotropics (Freitas *et al.*, 2009). In addition, decreasing areas of forested land increase the hunting pressure on the remaining forested pockets (Nath *et al.*, 1994). Oldroyd and Nanork (2009) have also observed that, the deforestation and excessive honey hunting pressure are the primary threats to honeybees in East Asia.

5.3. Honey Hunting

Hunting honeybees for their honey and other beneficial products has been found to be practiced from time immemorial (Crane, 1999). Colonies of *A. florea* are being easily hunted for their tasty honey by merely shaking the bees off by holding the branch of the nest located. Hunters in the act of honey collection destroy the whole nest by killing brood and occasionally adult bee population. In rural areas, even school children could able to harvest honey from these species for their relatively less ferocious nature. Harvesting honey from these colonies is generally performed on day time, such that, the bees would move away with the queen and may re-establish their new colonies. Honey hunting is widely practiced from the nests of *A. dorsata* and *A. laboriosa*. *A. dorsata* is a major honey producer in parts of Asia. It has been estimated that, about 14000 tones of honey have been collected from these colonies annually in India. Honey from *A. dorsata* and *A. laboriosa* colonies is being harvested by unscientific traditional methods. The skilled hunters always prefer to hunt these colonies during night times to escape from their mass stinging. In the process of honey harvesting, the hunters generate a huge amount of smoke especially from wet leaves and cloths available in the locality rarely added with pesticides in and around the nest. They even expose the flames of the fire directly to the comb by killing the brood and adult bee population. This type of nocturnal hunting destroys a huge number of giant honey bee colonies. Rarely, some of the harvested nests may re-establish their colonies if, the queens survive with sufficient number of young worker bees. Nowadays, a few skilled honey hunters could able to harvest honey from giant honeybee colonies during day time itself. Though, the combs are being destroyed in the act of honey collection, the bee population may flee the spot and re-establish their colonies. Honey is also harvested traditionally from the feral colonies of native hive honeybees. However, as honey hunting takes place in light on day times, the queens along with adult bee population may establish their colonies on new sites.

5.4. Habitat Loss

In most parts of the Asia, the native hive bees also exist in feral conditions by nesting in darker cavities like termite mounds, trunk cavities of older trees, electric poles etc. Removal of older trees for fuel and timber wood destroys the natural habitats of bees. Similarly, destruction of termite mounds, as termites are the pests of various agricultural and horticultural crops and conversion of wasteland, where bee colonies exist, into agricultural land gradually reduces the nesting habitats of *A. cerana.* In addition, clearing of bushes and huge trees in the forest as well as on road sides usually destroys the habitats of *A. florea* and *A. dorsata* respectively. Forest fire is a common problem in tropics, wherein the wild fire spreads through thousands of kilometers by destroying the nesting habitats such as termite mounds, tree cavities, bushes and even damaging the huge trees. It leads to death of millions of feral honeybee colonies inhabiting in the region (Taylor *et al.*, 1999). For example, *A. cerana* which normally nests in darker cavities, has initiated constructing its nests in large scale on window roofs, unused switch boards, car sheds etc as a suitable nest sites are limited in metropolitan cities like Bangalore, India.

5.5. Pesticide Poisoning

Most of the agricultural and horticultural crops in tropics are being sprayed by varieties of pesticides in pest control practices. Despite a few pesticides are no or least toxic to honeybees, most of them are moderately or highly toxic to honeybees. These pesticides kill the brood as well as adult population and contaminate the hive products (Desneux *et al.*, 2007, De La Rua *et al.*, 2009). Honeybees are most frequently exposed to pesticide poisoning on visit to flowers duly sprayed by the pesticides. They also carry poisoned nectar and pollen to the hive and feed on to the brood and young bees (Tasei *et al.,* 1987). Some pesticides may even persist on the pollen and nectar and cause long-term mortality. For example, mango plantations are more numerous in tropical countries like India. These plantations produce enormous quantity of nectar and are being attracted by bee foragers in large numbers. However, these plantations are most frequently sprayed by highly toxic pesticides in control of varieties of insect and non-insect pests during blooming stage of the flowers. Visitation of bee foragers to these pesticide sprayed plantations, leads to large scale mortality of adult bees initially and death of entire colonies on feeding upon poisonous nectar to the brood as well as young bees in the hive. Varieties of acaricides are used in control of various bee mites in honeybee colonies. They are generally toxic and cause poisoning to brood and contaminate the hive products. Similarly, fumigation of paradichlorobenzene in the hives against the wax moth larvae also contaminates honey. Mortality of *A. dorsata* colonies on direct application of insecticides is one of the serious problems in a few metropolitan cities in Asia. For example, *A.dorsata* colonies rarely attack on school children and public in metropolitan cities like Bangalore by constructing nests on the public and school buildings. It attracts owners of the buildings to destroy these colonies through pest control agencies in the locality. The pest control agents usually destroy *A. dorsata* nests by spraying highly toxic pesticides directly. This mode of brutal killings is responsible for large scale mortality of *A.dorsata* colonies in urban areas.

5.6. Introduction of Exotic Species

Western hive honeybee, *A. mellifera* has been successfully introduced to most parts of the Asia during early twentieth century. These species are capable of well adapting to diverse environmental conditions. It is a good honey and royal jelly producer in parts of Asia. Despite evidences on negative impact of these exotic bees on native honeybees is meager, these bees are known to compete with indiginous honeybees (Yang, 2005; Pratap, 1998). The foraging time of *A.mellifera* generally overlaps with *A.cerana* causing food competition between each other. Similarly, Moritz *et al.* (2005) pointed out that, native honeybees have suffered a lots of impediments from *A.mellifera* in respect of robbing, transfer of diseases and mating interferences. During mating season, queens of both *A. cerana* and *A. mellifera* attract hetero-specific drones (Wang *et al.*, 2003). It is obvious that, *A. mellifera* is gradually replacing the native hive bee, *A. cerana* in most parts of Asia.

6. Conservation

Asian honeybees are well adapted to their local conditions with high efficiency in tolerating various pests and diseases. However, these require timely protection against a few threats faced by them in their social life. Thai sac brood virus disease has become catastrophic to beekeeping industry with *A.cerana*. The disease which made its origin during 1976 in Thailand is continued to be threatening *A.cerana* colonies. It was opined that, these bees would get natural resistance against the disease within few years. But, so far such disease resistant colonies if any have been not found. This is primarily responsible for declined beekeeping with *A. cerana* in Asia. Apparently, there is a possibility of spread of this disease to its sister honeybee species in near future. This would lead to greater damage to Asian beekeeping despite a few management practices are quite effective in combating the disease. Most of the ports and airports of Asia give priority to the free flow of goods for varied reasons. This is responsible for rapid spread of Thai sac brood disease in the continent.

Therefore, it is advocated to strictly follow the quarantine measures on viral pathogens while importing or exporting of bee colonies. More research is needed on the systematic breeding of Thai sacbrood disease tolerant stocks of *A. cerana.*

Varroa destructor has become a global problem to beekeeping industry with *A. mellifera.* In addition, these mites are also threatening the nests of *A.dorsata* without much damage. Both nymph and adult mites parasitize on the bee brood and adult bees and leads to gradual dwindling of adult bee population in bee colonies. Similarly, *T. clareae* has become a serious threat in the colonies of *A.mellifera* in parts of Asia for its infestation with faster development. It may also expand its parasitization throughout the world in near future similar to *V.destructor.* Suitable steps should be taken to prevent the spread of these parasites by following strict quarantine measures while exporting or importing bee colonies. The acaricides used against mites during brood rearing seasons may contaminate honey and other hive products. The long-term measures required to reduce the mites problem is breeding and multiplication of mite tolerant bee stocks. Huge numbers of *Varroa* tolerant bee stocks have been developed in western countries. Such mite tolerant *A. mellifera* stocks have to be recommended for beekeeping in Asia. Similar to *Varroa* tolerant stocks, research should be focused on breeding of *Tropilaelaps* tolerant *A. mellifera* stocks in near future in Asia. Alternatively, screening and multiplication of some bio-control agents which are efficient in reducing mite population is required for effective management of bee mites. The improved technologies especially on the application of suitable management practices against the pests and diseases have to be implemented and such technology should be more effective, safer and easily accessible to the local beekeepers.

It is obvious that, the growing human population coupled with more demand for timber in the market is responsible for increased rates of deforestation in Asia. Removal of trees suitable for nesting often reduces the floral resources for honeybees. Similarly, clearing of bushes and uprooting of trees in a rapid rate declines the nesting habitats of open-nesting honeybees (Kremen *et al.*, 2004). Encouraging afforestation, watershed management, and community forests slows down the rates of deforestation. Similarly, allowing local communities and civil society to participate in natural resource management is very much essential in management of forests.

Traditional methods of honey hunting decline the population of open-nesting in a large scale. The nocturnal crude ways of honey hunting are responsible for death of millions of honeybee colonies. Therefore, awareness should be created among the honey hunters on eco-friendly scientific methods of honey harvesting especially from *A. dorsata* colonies. Honey may be harvested even during day time with protective clothing and blowing a few puffs of smoke on the colonies if required. It prevents the brutal killing of colonies during night times. Attraction of *A.dorsata* swarms by constructing rafters on specific locations is very much encouraged. This facilitates harvested of honey from lower heights without destroying the nests. In some parts of Asia, efforts are being made to recommend harvesting of honey from *A. dorsata* nests in a non-destructive manner (Tan *et al*, 1997, Waring and Tump, 2004). Some of such conservative methods have been initiated in certain parts of the world (Tan and Ha, 2002, Waring and Tump, 2004). However, as the giant bees is major honey and wax producers, more research is needed on developing economically viable and bee-friendly methods of honey harvesting.

In addition, killing of giant honeybee colonies directly by exposing to highly toxic pesticides should be discouraged. In most parts of Asia, strict control on use of many pesticides, including several banned is lacking. However, in some parts, there are restrictions and regulations to the use of pesticides but with little enforcement. More emphases should to be given on investigating the impact of honey hunting and pesticide poisoning on population demography of open-nesting and feral hive honeybee species. If we adopt a few if not all, the Asia may become safer continent for honeybees with greater honey production and better pollination.

CONCLUSION

Asian continent is homeland of all species of honeybees with rich biodiversity. *A. cerana* beekeeping is being well practiced with surplus honey production in many parts of the continent. However, there is immediate need to combat Thai sac brood viral disease, which caused irreparable damage to beekeeping industry through selective modern breeding techniques. Most of the native Asian honeybees are no or least damaged by the brood mites. However, recent invasion and infestation of the brood mites, *T.clareae* and *V. destructor* are endemic to *A.mellifera* colonies in some parts of Asia. There is urgent need to develop respective mite tolerant stocks for better beekeeping with *A. mellifera* in Asia. Asian open-nesting honeybees are more or less free from most of the pests and diseases for their own defensive mechanisms. Surprisingly, the honey obtained from these species is most commonly referred to as "organic honey" as foragers of these species produce honey mainly from the nectar collected from wild and forest flora which are no or least exposed to pesticides. Harvesting of honey from *A.dorsata* colonies without damaging the nests in a sustainable manner on day time would safeguard millions of these colonies. A thorough training is pre-requisite for honey hunters on bee-friendly non-destructive methods of honey harvesting. Apparently, creating awareness on impact of anthropogenic activities *viz* deforestation, habitat destruction, bee poisoning etc on honeybee species among the public is need of the hour in conservation of Asian honeybees.

REFERENCES

Abrol, D. P. (2000). Beekeeping with *Apis cerana* in Jammu and Kashmir: present status and future perspectives. *Bee Wld.* 81: 149-152.

Abrol, D. P. and Putatunda, B. N. (1995). Discovery of an ectoparasitic mite *Tropilaelaps koenigerum* Delfinado-Baker and Baker (Acari: Laelapidae) associated with honeybees, *Apis dorsata, Apis mellifera* and *Apis cerana* in Jammu, India. *Curr. Sci.* 68: 90.

Adlakha, R. L. (1976). Acarine disease of adult honey bees in India. *American Bee J.* 116: 324, 344.

Alexander, B. A. (1991). Phylogenetic analysis of the genus *Apis* (Hymenoptera: Apidae). *Ann. Entomol. Soc. Am.* 84: 137-149.

Anderson, D. L. (1994). Non-reproduction of *Varroa jacobsoni* in *Apis mellifera* colonies in Papua New Guinea and Indonesia. *Apidologie* 25: 412-421.

Anderson, D. L. and Morgan, M. J. (2007). Genetic and morphological variations of bee parasitic *Tropilaelaps* mites (Acari: Laelapidae): new and re-classified species. *Exp. Appl. Acarol.* 43: 1-24.

Anderson, D. L. and Trueman, J. W. H. (2000). *Varroa jacobsoni* (Acari: Varroidae) is more than one species. *Exp. Appl. Acarol.* 24: 165-189.

Arias, M. C. and Sheppard, W. S. (2005). Phylogenetic relationships of honeybees (Hymenoptera: Apinae) inferred from nuclear and mitochondrial DNA sequence data. *Mol. Phylogenetic. Evol.* 37: 25-33.

Arias, M. S., Tingek S., Kelitu, A. and Sheppard, W. S. (1996). *Apis nuluensis* Tingek, Koeniger and Koeniger and its genetic relationship with sympatric species inferred from DNA sequences. *Apidologie* 27: 415-422.

Armon, P. J. (1980). Use of honey in treatment of infected wounds. *Trop. Doct.* 10(2): 91.

Bailey, L. and Ball, B. V. (1991). *Honeybee Pathology.* Academic Press, London, 193 p.

Bailey, L. and Woods, R. D. (1977). Two more small RNA viruses from honeybees and further observations on sac brood and acute bee paralysis viruses. *J. Gen. Virol.* 37: 175-182.

Baitala, T. V., Faquinello, P., de Toledo, V. A. R., Mangolin, C. A., Martins, E. N. and Ruvolo-Takasusuki, M. C. C. (2010). Potential use of major royal jelly proteins (MRJPs) as molecular markers for royal jelly production in Africanized honeybee colonies. *Apidologie* 41: 160-168.

Ball, B. V. and Allen, M. F. (1988). The prevalence of pathogens in honeybee (*Apis mellifera*) colonies infested with the parasitic mite, *Varroa jacobsoni. Annl. Appl. Biol.* 113: 237-244.

Batra, S. W. (1996). Biology of *Apis laboriosa*, a pollinator of apple at high altitudes in the greater Himalaya range of Garhwal, India (Hymenoptera: Apidae). *J. Kansas Entomol. Soc.* London. B 265: 2009-2014.

Beekman, M., Gloag, R. S., Even, N., Wattanachaiyingcharoen, W. and Oldroyd, B. P. (2008). Dance precision of *Apis florea*-clues to the evolution of the honeybee dance language? *Behav. Ecol. Sociobiol.* 62: 1259-1265.

Bhardwaj, R. K. (1968). A new record of the mite *Tropilaelaps clareae;* a parasite of honeybee in south east Asia. *Bee Wld.* 49: 115.

Brockmann, A., Bruckner, D. and Crewe, R. M. (1998). The EAG response spectra of workers and drones to queen honey bee mandibular gland components: the evolution of a social signal. *Naturwissenschaften* 85: 283-285.

Brockmann, A., Dietz, D., Spaethe J. and Tautz, J. (2006). Beyond 9-ODA: sex pheromone communication in the European honey bee *Apis mellifera* L. *J. Chem. Ecol.* 32: 657-667.

Buawangpang, N., Sukumalanand, P. and Burgett, M. (2009). *Apis florea* drone flight: longevity and flight performance. *Apidologie* 40: 20-25.

Butler, C. G. (1974). *The World of Honeybee,* Collins, London, 226 p.

Butler C. G., Calam D. H. and Callow R. K. (1967). Attraction of *Apis mellifera* drones by the odors of the queens of two species of honeybees. *Nature* 213: 423-424.

Choi, S. Y. (1984). Brief report on the status of Korean beekeeping. *Proc. Expert Consult*, FAO, pp. 170-190.

Crane, E. (1990). *Bees and Beekeeping: Science, practice and World Resources*. Heinemann Newnes, Oxford, 614 p.

Crane, E. (1999). *The world history of beekeeping and honey hunting*. Routledge, New York, 720 p.

Danforth, B. N., Sipes, S., Fang, J. and Brady, S. G. (2006). The history of early bee diversification based on five genes plus morphology. *PNAS* 103(41): 15118-15123.

de Guzman, L. I. and Delfinado-Baker, M. (1996). A new species of *Varroa* (Acari: Varroidae) associated with *Apis koschevnikovi* (Apidae: Hymenoptera) in Borneo. *Intern. J. Acarol.* 22: 23-27.

De La Rua, P., Jaffe, R., Dall'Olio, R., Munozi, I. and Serrano, J. (2009). Biodiversity, conservation and current threats of European honeybees. *Apidologie* 40: 263-284.

Delfinado-Baker, M. and Aggarwal, K. (1987). A new *Varroa* (Acari: Varroidae) from the nest of *Apis cerana* (Apidae). *Intern. J. Acarol.* 13: 233-237.

Delfinado-Baker, M. and Baker, E. W. (1982). A new species of *Tropilaelaps* parasitic on honeybees. *Am. Bee J.* 122: 416-417.

Delfinado-Baker, M., Baker, E. W. and Phoon, A. C. G. (1989). Mites (Acari) associated with bees (Apidae) in Asia, with description of a new species. *Am. Bee J.* 129: 609-613.

Delfinado-Baker, M., Underwood, B. A. and Baker, E. W. (1985). The occurrence of *Tropilaelaps* mites in brood nests of *Apis dorsata* and *Apis laboriosa* in Nepal with descriptions of the nymphal stages. *Am. Bee J.* 125: 703-706.

Desneux, N., Decourtye, A. and Delpuech, J. M. (2007). The sub lethal effects of pesticides on beneficial arthropods. *Ann. Rev. Entomol.* 52: 81 106.

Divan, V. V. and Salvi, S. R. (1965). Some interesting behavioral features of *Apis dorsata* F. *Indian Bee J.* 27: 52.

Dumronglert, E. (1983). A follow-up study and chronic wound healing dressing with pure natural honey. *J. Nat.Res. Counc. Thail.* 15(2): 39-66.

Elbagoury, E. F. and Fayed, N. A. (1985). Application of natural honey after surgical removal of impacted third molar. *Egypt Dent. J.* 31: 203-211.

Ellis, J. D. and Munn, P. A. (2005). The worldwide health status of honeybees. *Bee Wld.* 86: 88-101.

Engel, M. S. (1999). The taxonomy of recent and fossil honeybees (Hymenoptera: Apidae). *Am. Museum Novitates*, 3296: 1-11.

Free, J. B. (1993). *Insect pollination of crops*. Academic Press, London, UK.

Freitas B. M., Imperatriz-Fonseca V. L., Medina L. M., Kleinter A. M. P., Galetto L., Nates-Parra, G. and Quezada-Euán J. J. G. (2009). Diversity, threats and conservation of native bees in the Neotropics. *Apidologie* 40: 332–346

Garg, R. and Kashyap, N.P. 1998. Mites and other enemies of honeybees in India. In: *Perspectives in Indian Apiculture* (Eds. Mishra, R.C. and Garg, R.). pp. 264-303, *Agro Botanica,* New Delhi, India.

Gould, J. L., Dyer, F. C. and Towne, W. F. (1985). Recent progress in the study of the dance language. *Fortschr. Zool.* 31: 140-161.

Gries, M. and Koeniger, N. (1996). Straight forward to the queen: pursuing honeybee drones (*Apis mellifera* L.) adjust their body axis to the direction of the queen. *J. Comp. Physiol.* 179: 539-544.

Hadisoesilo, S., Otis, G. W. and Meixner, M. (1996). Two distinct populations of cavity nesting honeybees (Hymenoptera: Apidae) in south Sulawesi, Indonesia. *J. Kans. Entomol.* Soc. 68: 399-407.

Hanumanthaswamy, B. C. (2000). Natural enemies of honeybees with special reference to Bioecology and Management of greater wax moth, *Galleria mellonella* Linnaeus (Lepidoptera: Pyralidae). *Ph.D thesis,* University of Agricultural Sciences, Bangalore, India, 257 p.

Hepburn, H. R. and Radloff, S. E. (1998). *Honeybees in Africa*. Springer, Berlin.

Hoppe, H. and Ritter, W. (1989). The influence of the Nasonov pheromone on the recognition of house bees and foragers by *Varroa jacobsoni. Apidologie*. 19: 165-172.

Imperatriz-Fonseca, V. L., Saravia, A. M. and De Jong, D. (2006). *Bees as pollinators in Brazil: assessing the status and suggesting best practices.* Ribeirao Preto, Holos, Editora, 112 p.

Johnson, R. M., Evans, J. D., Robinson, G. E. and Berenbaum, M. R. (2009). Changes in transcript abundance relating to colony collapse disorder in honey bees (*Apis mellifera*). *Proc. Natl. Acad. Sci.* 106: 14790-14795.

Kapil, R. P., Putatunda, B. N. and Aggarwal, K. (1985). Studies on ectoparasitic mites of *Apis* species. *Tech. Rep.* Dept. of Zoology, Haryana Agric. University. Hissar, India,191 p.

Kastberger, G. and Sharma, D. K. (2000). The predator-prey interaction between blue bearded bee eaters (*Nyctyornis athertoni* Jardine and Selby) and giant honeybees (*Apis dorsata* Fabricius). *Apidologie* 31: 727-736.

Keeling, C. I., Slessor, K. N., Koeniger N., Koeniger, G. and Punchihewa, R. W. K. (2000). Quantitative analysis of the mandibular gland components of the dwarf honey bee (*Apis florea* Fabricius) *Apidologie* 31: 293-299.

Keven, P. G. (1999). Pollinators as bioindicators of the state of the environment: species, activity and diversity. *Agric. Ecosyst. Environ*, 74: 373-393.

Kevan, P. G. and Phillips, T. P. (2001). The economic impacts of pollinator decline: an approach to assessing the consequences. *Conserv. Ecol*. 5: 8.

Kevan, P. G. and Viana, B. F. (2003). The global decline of pollination services, *Biodiversity* 4: 3-8.

Klein, A. M., Vaissiere, B. E., Cane, J. H., Steffan- Dewenter, I., Cunningham, S. A., Kremen, C. and Tscharntke, T. (2007). Importance of pollinators in changing landscapes for world crops. *Proc. R. Soc.B-Biol. Sci*. 274: 303-313.

Koeniger, N. (1995). Biology of the eastern honeybee, *Apis cerana* (Fabricius 1773). In: *The Asiatic hive bee: Apiculture, biology and role in sustainable development in tropical and subtropical Asia*. (Ed. Kevan, P.G.) pp. 29-30, Enviroquest Ltd., Canada.

Koeniger, N. and Koeniger, G. (1980). Observations and experiments on migration and dance communication of *Apis dorsata* in Sri Lanka. *J. Apic. Res.* 19: 21-34.

Koeniger, N. and Koeniger, G. (2000). Reproductive isolation among species of the genus, *Apis. Apidologie* 31:313-339.

Koeniger G., Koeniger, N.and Tingek S. (1994). Cross fostered drones of *Apis cerana* Fabricius, 1793 and *Apis koschevnikovi* v. Buttel-Reepen, 1906 fly at their species specific mating times. *Insectes Soc.* 41: 73-78.

Koeniger, G., Koeniger, N., Tingek, S. and Phiancharoen, M. (2005). Variance in spermatozoa number among *Apis dorsata* drones and among *Apis mellifera* drones. *Apidologie* 36: 279-284.

Koeniger, N., Koeniger, G., Gries, M., Tingek, S. and Kelitu, A. (1996). Observations on colony defense of *Apis nuluensis* Tingek, Koeniger and Koeniger 1996 and predatory behaviour of the hornet, *Vespa multimaculata* Perez. *Apidologie* 27: 341-352.

Kremen, C., Williams, N. M., Brigg, R. L., Fay, J. P. and Thorp, R. W. (2004). The area requirements of an ecosystem service: crop pollination by native bee communities in California. *Ecol. Lett.* 7: 1109-1119.

Kshirsagar, K. K. and Godbole, S. H. (1974). Indian *Streptococcus pluton. Indian Bee J.* 36: 39.

Kumar, J., Gupta, J. K., Mishra, R. C. and Dogra, G. S. (1988). Pollination studies in some cultivars of Cauliflower (*Brassica oleraceae* var. *botrysis* L.) *Indian Bee J.* 50(4): 93-95.

Liu, T. P., Mobus, B. and Braybrook, G. (1989). A scanning electron microscope study on the prothoracic tracheae of the honeybee, *Apis mellifera* L. infested by the mite, *Acarapis woodi* (Rennie). *J. Apic. Res.* 28: 81-84.

Mahanta, M. and Rahman, A. (1997). Studies on pollination of Litchi (*Litchi chinensis*) by honeybees (*Apis cerana* F). *Indian Bee J.* 59(2): 74-75.

Mahindre, D. B. (1983). Handling rock bee colonies. *Indian Bee J.* 45: 72-73.

Makinson, J.C., Oldroyd, B.P., Schaerf, T.M.,Wattanachaiyingcharoen, W. and Beekman, M. 2001. Moving home: nest site selection in the red dwarf honeybee, *Apis florea*. *Behav. Ecol. Sociobiol.* 65: 945-958.

Marcucci, M.C. (1995). Propolis: chemical composition, biological properties and therapeutic activity. *Apidologie* 26: 83-99.

Mc Gregor, S. E. (1976). *Insect pollination of cultivated crop plants*. USDA handbook No. 496, Washington DC, 411 p.

Miller, D. G. and Ratneiks, F. L. W. (2001). The timing of worker reproduction and breakdown of policing behavior in queen honeybee (*Apis mellifera* L.) societies. *Insects Soc.* 48: 178-184.

Mishra, R. C., Dogra, G. S. and Gupta, P. R. (1980). Some observations on iridovirus of bees. *Indian Bee J.* 42: 9-10.

Moritz, R. F. A., Hartel, S. and Neumann, P. (2005). Global invasions of the western honeybee (*Apis mellifera*) and the consequences for biodiversity. *Ecoscience* 12: 289-301.

Nagaraja, N. (1998). Rate of invasion and infestation of *Tropilaelaps clareae* in sulphur treated and untreated colonies of *Apis mellifera. Him. J. Env.Zool.* 12: 131-136.

Nagaraja, N. (2000). Impact of parasitic brood mites, *Tropilaelaps clareae* and *Varroa jacobsoni* on *Apis mellifera* beekeeping in Karnataka. *J. Ecotoxicol. Environ. Monit.* 10: 205-210.

Nagaraja, N. and Brockmann, A. (2009). Drones of dwarf honeybee, *Apis florea* are attracted to (2E)-9-oxodecenoic acid and (2E)-10-hydroxydecenoic acid. *J. Chem. Ecol.* 35: 653-655.

Nagaraja, N. and Rajagopal, D. (1999). Colony establishment, nesting and foraging activity of little honeybee, *Apis florea* F. (Hymenoptera: Apidae). *J. Ent. Res.* 23: 331-338.

Nagaraja, N. and Rajagopal, D. (2000). Foraging and brood rearing activity of rockbee, *Apis dorsata* F. (Hymenoptera: Apidae). *J. Ent. Res.* 24: 243-248.

Nagaraja, N. and Rajagopal, D. (2001). Biology of brood mite, *Tropilaelaps clareae* in the brood of *Apis mellifera* (Hymenoptera: Apidae). *Asian Bee J.* 3: 20-26.

Nagaraja, N. and Rajagopal, D. (2003). Pests, predators and parasites of European honeybee, *Apis mellifera ligustica* Spin. in Karnataka, India. *Indian Bee J.* 65: 120-127.

Nagaraja, N. and Rajagopal, D. (2009). *Honeybees: Diseases, parasites, pests, predators and their management.* MJP Publishers, Chennai, India 210 p.

Nagaraja, N. and Reddy, C. C. (2002). Influence of bee pollinators on the quality of seed production in sunflower (*Helianthus annus*) and Coconut (*Cocas nucifera*). *Asian Bee J.* 4(1-2): 1-5.

Nanork, P., Paar, J., Chapman, N. C., Wongsiri, S. and Oldroyd, B. P. (2005). Asian honeybees parasitize the future dead. *Nature* 437: 829.

Nath, S., Roy, P., Leo, R. and John, M. (1994). Honey hunters and beekeepers of Tamil Nadu, A Survey document, Keystone, Pondicherry, 86 p.

Neumann, P., Koeniger, N., Koeniger, G., Tingek, S., Kryger, P. and Moritz, R. F. A. (2000). Home site fidelity of migratory honeybees. *Nature* 406: 474-475.

Oldroyd, B. P. and Nanork, P. (2009). Conservation of Asian honeybees. *Apidologie* 40: 296-312.

Oldroyd, B. P. and Wongsiri, S. (2006). *Asian honeybees: Biology, Conservation and Human interactions.* Harvard University Press, USA, 340 p.

Oldroyd, B. P., Osborne, K. E. and Mardan, M. (2000). Colony relatedness in aggregations of *Apis dorsata* F.(Hymenoptera: Apidae). *Insects Soc.* 47: 94-95.

Oldroyd, B. P., Smolenski, A. I. J., Cornuet, J. M., Wongsiri, S., Estoup, A., Rinderer, T. E. and Crozier, R. H. (1995). Levels of polyandry and intracolonial genetic relationship in *Apis florea. Behav. Ecol. Sociobiol.* 37: 329-335.

Ono, M., Igarashi, T. Ohno, E. and Sasaki, M. (1995). Unusual thermal defense by a honeybee against mass attack by hornets. *Nature* 377: 334-336.

Otis, G. W., Koeniger, N., Rinderer, T. E., Hadisoesilo, S., Yoshida, T., Tingek, S., Wongsiri, S. and Mardan, M. (2000). Comparative mating flight times of Asian honeybees. *Proc. 7th Intl. Conf. Tropical bees and 5th AAA,* IBRA, pp 137-141.

Otis, J. and Kralj, J. (2001). Parasitic mites not present in North America. In: *Mites of the honeybee*, (Eds.: Webster, T.C. and Delaplane, K.S.) pp. 251-272, Dadant and Sons, Hamilton, USA.

Paar, J., Oldroyd, B. P. and Kastberger, G. (2000). Giant honeybees return to their nest sites. *Nature* 406: 475.

Paar, J., Oldroyd, B. P., Huettinger, E. and Kastberger, G. (2004). Genetic structure of migration of an *Apis dorsata* population: the significance of migration and colony aggregation. *J. Heredity*, 95: 119-126.

Pirk, C. W. W., Hepburn, R., Radloff, S. E. and Erlandsson, J. (2002). Defense posture in the dwarf honeybee, *Apis florea. Apidologie* 33: 289-294.

Plettner, E., Otis, G. W., Wimalarntnee, P. D. C., Winston, M. L., Slessor, K. N., Pankiw, T. and Punchihewa, P. W. K. (1997). Species and caste determined mandibular gland signals in honeybees (*Apis*). *J. Chem. Ecol.* 23: 363-375.

Pratap, U. (1998). Foraging competition between *Apis cerana* and *Apis mellifera* and its impact on crop pollination. Asian bees and beekeeping: progress of research and development. *Proc. Fourth Asian Apic. Assoc. Intl. Conf.* Kathmandu, Nepal, pp. 206-208.

Rajagopal, D. and Kencharaddi, R. N. (2000). *Thai sac brood virus disease of Indian honeybee.* University of Agricultural Sciences, Bangalore, India,117p.

Rajagopal, D. and Nagaraja, N. (1999). Beekeeping status in Karnataka, India. *Asian Bee J.* 1: 50-59.

Rajagopal, D. and Nagaraja, N. (2003). Potentiality of honeybees in increasing crop production. In: *Potentials of Living Resources* (Eds. Tripathi, G. and Kumar, A.), pp. 105-158, Discovery Pub. House, New Delhi, India.

Rajagopal, D., Veeresh, G. K., Chikkadevaiah., Nagaraja, N. and Kencharaddi, R. N. (1999). Potentiality of honeybees in hybrid seed production of sunflower (*Helianthus annuus* L.) *Indian J. Agric. Sci.* 69: 40-43.

Reddy, C. C. (2003). Beekeeping policies and programmes in India. *Asian Bee J.* 5(1-2): 18-22.

Reddy, M. S. and Reddy, C. C. (1993). Studies on the distribution of nests of giant honeybee (*Apis dorsata* F.). *Indian Bee J.* 55: 36-39.

Ribbands, C. R. (1953). *The behaviour and Social life of Honeybees. Intern.* Bee Res. Assoc. London, 352 p.

Richards, A. J. (2001). Does low biodiversity resulting from modern agricultural practice affect crop pollination and yield? *Ann. Bot.* 88: 165-172.

Rinderer, T. E., Marx, B. D., Gries, M. and Tingek, S. (1996). A scientific note on stratified foraging by Sabahan bees on the yellow flame tree (*Peltophoram pterocarpum*). *Apidologie* 27: 423-425.

Rinderer, T. E., Oldroyd, B. P., de Guzman, L. I., Wattanachaiyingcharoen, W., and Wongsiri, S. (2002). Spatial distribution of the dwarf honeybees in an agro-ecosystem in southern Thailand. *Apidologie* 33: 539-543.

Robinson, G. E. (1992). Regulation of division of labor in insect societies. *Ann. Rev. Entomol.* 37: 637-665.

Roubik, D. W., Sakagami, S. F. and Kudo, I. (1985). A note on distribution and nesting of the Himalayan honeybee, *Apis laboriosa* Smith (Hymenoptera: Apidae). *J. Kansas Entomol. Soc.* 58: 746-749.

Ruttner, F. (1987). *Biogeography and Taxonomy of Honeybees.* Springer-Verlag, Berlin, 284 p.

Sakai, T. (1992). *Apis cerana* beekeeping in Japan, In: *Honeybees in mountain agriculture* Verma, L.R. (ed), Oxford and IBH Publishing Co. New Delhi, India.

Salem, S. N. (1981). Treatment of gastroenteritis by the use of honey. *Islamic Med.* 1: 358-622.

Sannasi, A., Rajulu, G. S. and Sundara, G. (1971). 9-Oxodec-trans- 2-enoic acid in the Indian honeybees. *Life Sci.* 10: 195-201.

Seeley, T. D., Seeley, R. H. and Akratanakul, P. (1982). Colony defense strategies of the honeybees in Thailand. *Ecol. Monographs* 52: 43-63.

Shah, F. A. and Shah, T. A. (1991). *Vespa velutina,* a serious pest of honeybees in Kashmir. *Bee Wld.* 72: 161-164.

Simone-Finstrom and Spivak, M. (2010). Propolis and bee health: the natural history and significance of resin use by honeybees. *Apidologie* 41: 295-311.

Singh, S. (1957). Acarine disease in the Indian bee, *Apis cerana* F. *Indian Bee J.* 19: 27-28.

Slessor, K. N., Winston, M. L. and Conte, Y. (2005). Pheromone communication in the honeybee (*Apis mellifera* L). *J. Chem. Ecol.* 31: 2731-2745.

Sodhi, N. S., Koh, L. P., Brook, B. W. and Ng, P. K. L. (2004). Southeast Asian biodiversity: an amending disaster. *Trends Ecol. Evol.* 19: 654-660.

Suryanarayana, M. C., Rao, M. G., Sihag, R. C. and Singh, T. S. M. S. (1990). Coconut palm, a pollen and nectar source to honeybees. *Indian Bee J.* 52: 41-43.

Tan, N. Q. and Ha, D. H. (2002). Socio-economic factors in traditional rafter beekeeping with *Apis dorsata* in Vietnam. *Bee World*, 83: 165-170.

Tan, N. Q., Chinh, P. H., Thai, P. H. and Mulder, V. (1997). Rafter beekeeping with *Apis dorsata*: some factors affecting occupation of rafters by bees. *J. Apic. Res.* 36: 49-54.

Tanaka, H., Roubik, D. W., Kato, M., Liew, F. and Gunsalam, G. (2001). Phylogenetic position of *Apis nuluensis* of northern Borneo and phylogeography of *Apis cerana* as inferred from mitochondrial DNA sequences. *Insects Soc.* 84: 44-51.

Tasei, J. N., Carre, S., Bosio, P. G., Debrey, P. and Hariot, J. (1987). Effects of the pyrethroid insecticide, WL 85871 and phosalone on adults and progeny of the leaf cutting bee, *Megachile rotundata* F. pollinators of Lucerne. *Pesticide Sci.* 21: 119-128.

Taylor, D., Saksena, P., Sanderson, P. G. and Kucera, K. (1999). Environmental change and rain forests on the Sunda shelf of Southeast Asia: drought, fire and the biological cooling of biodiversity by hotspots. *Biodiv. Conserv.* 8: 1159-1177.

Thapa, R. and Wongsiri, S. (2003). Flying predators of the giant honeybees, *Apis dorsata* and *Apis laboriosa* in Nepal. *Am. Bee J.* 143: 540-542.

Tingek, S., Koeniger, G. and Koeniger, N. (1996). Description of a new cavity dwelling species of *Apis* (*Apis nuluensis*) from Sabah, Borneo with notes on its occurrence and reproductive biology (Hymenoptera: Apoidea: Apini). *Sencken Bergiana Biol.* 76: 115-119.

Tingek, S., Mardan, M., Rinderer, G., Koeniger, N. and Koeniger, G. (1988). Re-discovery of *Apis vetchi* (Maa, 1953), the Sabah honeybee. *Apidologie* 19: 97-102.

Underwood, B. A. (1986). The natural history of *Apis laboriosa* in Nepal. *M.S. thesis,* Cornell University. Ithaca, USA.

Underwood, B. A. (1990). Seasonal nesting cycle and migration patterns of the Humalayan honeybee, *Apis laboriosa. Nat. Geogr. Res.* 6: 276-290.

Verma, L. R. (2003). Conservation and management of Asian hive bee, *Apis cerana. Asian Bee J.* 5(1-2): 1-10.

Wang, Q. F., Li, W. S., Zhang, Q. M. and Wu, S. S. (2003). The natural mating interference between *A. cerana* and *A.mellifera. Entomological Knowledge* 40: 164-167 (in Chinese).

Waring, C. and Tump, D. R. (2004). Rafter beekeeping in Cambodia with *Apis dorsata. Bee World,* 84: 14-18.

Wongsiri, S., Lekprayoon, C., Thapa, R., Thirakupt, K., Rinder, T. E., Sylvester, H. A., Oldroyd, B. P. and Booncham, U. (1996). Comparative biology of *Apis andreniformis* and *Apis florea* in Thailand. *Bee World*, 77: 23-35.

Woyke, J. 1994. Mating behaviour of the parasitic honeybee mite *Tropilaelaps clareae. Exp. Appl. Acarol.* 18: 723-733.

Woyke, J., Wilde, J. and Reddy, C. C. (2004). Open-air-nesting honey bees *Apis dorsata* and *Apis laboriosa* differ from the cavity-nesting *Apis mellifera* and *Apis cerana* in brood hygiene behaviour. *J. Inv. Pathol.* 86 : 1-6.

Woyke, J. Wilde, J. Reddy, C. C. and Nagaraja, N. (2005). Periodic mass flights of giant honeybee, *Apis dorsata* performed in successive days at two nesting sites in different environmental conditions. *J. Apic. Res.* 44: 180-189.

Woyke, J. Wilde, J. Wilde, M. Reddy, C. C. Cervancia, C. and Nagaraja, N. (2008). Comparison of defense body movements of *Apis laboriosa, Apis dorsata dorsata* and *Apis dorsata breviligula* honey bees. *J. Insect Behav.* 21: 481-494.

Wu, Y. and Kuang, B. (1987). Two species of small honeybee: a study of the genus *Micrapis. Bee Wld.* 68: 153-155.

Yang, G. H. (2005). Harm of introducing the western honeybee, *Apis mellifera* L. to the Chines honeybee, *Apis cerana* F. and its ecological impact. *Acta Entomologica Sinica* 48: 401-406 (in Chinese).

Zhen, M. J., Yang, G., Huang, S., Lians, S. and Ren, Z. (1992). The advancement of beekeeping science and technology in China. In: *Honeybees in mountain agriculture* (ed. Verma, L.R.). Oxford and IBH Publishing Co. New Delhi, India.

In: Bees
Editor: Richard M. Florio

ISBN: 978-1-61324-825-6

Chapter 3

ODOR, LEARNING AND BEHAVIOR

***Charles I. Abramson*[1], *Patrick H. Wells*[2], *Adrian M. Wenner*[3] *and Harrington Wells*[4]**

[1]Psychology, Oklahoma State University, Stillwater, OK, US
[2]Biology, Occidental College, Los Angeles, CA, US
[3]Ecology, Evolution & Marine Biology, University of California, Santa Barbara, CA, US
[4]Biology, University of Tulsa, Tulsa, OK, US

Abstract

This chapter presents an overview of the literature on the importance of odor in learning and behavior of the honey bee. We begin by illustrating the importance of odor in complex odor environments, and move on to describing the behavior of crop-attached and naïve recruit foragers. Information is also provided on the neural mechanisms of odor detection, limitations of cognitive interpretations of odor learning, and the role of odor perception in repellents and pesticides.

INTRODUCTION

For students of animal behavior honey bees are an intriguing organism, interacting in a complex eusocial colony setting as well as with the environment as they forage over wide areas. Much of that behavior is moderated by odors, which honey bees can detect at extremely low concentrations. Well known is the role of pheromones in regulating activities within the hive. Environmental odors as well are extremely important in determining the behavior of honey bees in and out of the hive. Honey bees visit a great variety of flowering plant species, and their ability to find 'nectar' extends beyond plants. They readily consume soft drinks in trash cans and even forage on the contents of discarded wine bottles from sporting events (Abramson *et al.* 2000, 2007), making them at times a nuisance. They are in many respects the ultimate generalist forager, which has been a boon for modern agriculture.

Odors turn out to be a key component in understanding these associations, not only for honey bees but also apparently for all species of social bees (Molet *et al.* 2009). Much has been learned about honey bee use of odor information from recent studies with harnessed bees in laboratory situations, and in field studies with bees foraging naturally at feeders and artificial flower patches. That body of work has implications for repellants, attractants, pesticide-use and agricultural pollination problems. Here we review that body of work, focusing on honey bee learning, implications for the cognitive architecture of the honey bee, and practical applications.

Complex Odor Environments

Honey bees deal with very complex odor environments. Floral scents alone typically contain over 20 different compounds (Knudsen *et al.* 2006). Further, a floral bouquet will vary within a species due to genetics, microhabitat effects, time of day and age of flower (Reinhard & Srinivasan 2009). Despite this complexity honey bees readily learn, discriminate, and recognize floral odors.

Almost a century ago von Frisch and other pioneers of bee-science began to explore the role of flower scents and colors in the foraging of honey bees. Honey bees were trained to visit a row of feeder boxes. Boxes differed in both color and scent. Foragers, finding that nectar was associated with a particular scent and color box, would repeatedly visit that particular box, ignoring all others. When the color and scent associated with nectar were then disassociated, the bees first flew to the box showing the color cue – but few entered the box. Instead, most foragers moved from box to box, and entered only the one marked with the odor originally associated with the nectar. In this instance, vision appeared to mediate the longer-distance perception of a nectar reward; sight of the color cue led the bee close enough to smell the odor associated with the nectar (Frisch 1919, 1950).

Honey bee foragers indeed learn all the individual elements of the floral bouquet, but not to the same extent. That is, some of the odor compounds when subsequently presented alone elicit a much stronger response than do others. This suggests that some scents are 'key odors' (Reinhart *et al.* 2010).

What is particularly interesting is a key odor for one mixture is not necessarily a key odor for another mixture. Functional groups (e.g. ketones, alcohols etc.) have been shown to influence acquisition rate and final level of learning (Guerrieri *et al.* 2005). Nevertheless, the functional group does not determine whether a molecule will be a key odor of a floral bouquet. Neither does the volatility of the odor molecule nor being chemically unique influence whether it will be a key odor of a bouquet (Reinhart 2010). Not surprising, selection as a key odor is linked to concentration relative to the other components of a floral bouquet, particularly when scents were weak.

Bees trained to complex mixtures responded better to a subset of key odorants than to single elements of the mixture. On the other hand, given mixtures composed of only three odor molecules, bees responded to each molecule equally (Laloi *et al.* 1997, 2000; Le Metayer *et al.* 1997). This may reflect the same complexity issues experienced when bees are given difficult energetic foraging problems involving both costs and rewards (Cnaani *et al* 2006; Sanderson *et al* 2006; Cakmak *et al.* 2009). Bees have been shown to focus on 'key' elements of the complex problem to reach decisions in these foraging problems. That is, they use only a portion of the available information, and this can lead to a decision that is uniform across foragers resulting in choosing a flower type that significantly reduces net energy gain (Sanderson *et al* 2006). This response appears to be very similar to the 'key odor' response seen by Reinhart *et al.* (2010). As the difficulty increases in foraging problems the response to the same problem becomes less uniform across individuals, but each individual is very consistent in her behavior (Cnaani *et al.* 2006; Cakmak *et al.* 2009). Fidelity in these cases can be based on rewards, energetic costs, or even a conditioned stimulus (Cakmak *et al.* 2009). Thus, it seems plausible that bees use all information available when presented with floral bouquet composed of a small number of compounds, but focus on a few key scents when presented with more complex mixtures.

Crop Attached Foragers

Once foragers find a particular location offering nectar those foragers will repeatedly return to harvest nectar from that source. These foragers fly in a 'beeline' (directly) each time to the location at a cruising speed of about 7.5 m/sec (25 feet/second), which makes the duration of flight from or to the hive a simple function of distance (Wenner 1963). Thus, it takes less than half a minute to travel to a nectar source 200 meters from the hive. There is little variation in measured flight times from hive to nectar source among bees returning to that nectar location. However, round trip times vary considerably due to the time it takes foragers to unload within the hive and to a lesser extent because bees may "pause" while drinking at the nectar source. These bees faithful to a nectar location are called 'crop-attached' foragers since they know where the nectar source is located.

Once crop-attached bees reach the foraging location, they are still faced with a recognition problem: What cues define the nectar source? The answer, at least partially, can be ascertained using artificial flowers since reward cues and nectar rewards can be easily manipulated. For example, when an artificial patch consists of blue flowers offering clove scented nectar and yellow flowers with cinnamon-scented nectar, any given bee will visit many flowers but of just one of those color-scent combinations. Thus, there is a set of bees only visiting blue flowers trip-after-trip from the hive, and another set of bees from the same hive showing extreme fidelity to yellow flowers each visit to the flower patch. If scents associated with the colors are switched, some individuals remain constant to the original color while others are faithful to the original odor, shifting their attention to the new color (Wells & Wells 1985).

The floral landscape for the crop-attached honey bee forager changes many times during a single day due to the endogenous patterns of nectar secretion and pollen presentation of angiosperms (Linnaeus 1756; Percival 1965; Endress 1994), as well as due to the effects of weather and the activity of other pollinators (Willmer & Stone 2004). When a flower patch (or feeding dish) stops yielding a nectar reward, bees crop-attached to it stop visiting. Although a few of the crop-attached individuals returned to the nectar location to inspect it from time to time, these bees are few in number. The full work force returns only when it again yields a reward. Re-recruitment of the work force is triggered when one of the few bees inspecting the foraging location again finds a nectar reward at the site and carries the food scent back into the hive. In fact, indeed, just re-introducing the scent associated with the nectar into the hive (Johnson & Wenner 1966; Wells & Rathore 1994) induces what Reinhard calls "scent-triggered navigation" (Reinhard *et al.* 2004). Not only do crop-attached bees learn to associate scent with food but also when again exposed to the scent they will fly directly and quickly to a place where they remember obtaining nectar with the same odor. They return to that familiar place regardless of conflicting in-hive scents or contacts with waggle dances of foragers harvesting other nectar sources (Gruter *et al.* 2008).

NAÏVE RECRUIT FORAGERS

Addition of naïve bees to the crop-attached population of foragers visiting a nectar location is an important element of honey bee sociality. In fact, the rapid recruitment of naïve bees to a foraging task provides honey bees an advantage over its competitors. Any such recruit bee is a 'naïve recruit' only on its first visit to the food source. On all subsequent visits to the flower patch that bee will harvest nectar as a crop-attached forager since it knows where the nectar resource is located, and will behave as described above (Hill *et al.* 1997).

Crop attached foragers approach a flower patch directly from the hive in a 'beeline' flight path; naïve recruit bees do not. Naïve recruits can be seen at some distance downwind of a scented nectar goal approaching in a zigzag flight pattern (Rosin 1991, 1999). This behavior resembles that of other insects engaged in odor-search behavior (Kennedy 1983). Here, scent is the long-distance attractant that may be supplemented by visual cues near the food source. Thus, odor is as important to naïve bees searching for a new food source as it is to their crop-attached companions, but in additional ways.

Potential naïve-recruits associate food reward with floral odors of nectar brought into the hive by returning crop-attached foragers (Farina *et al.* 2005, 2007). Although naïve-recruits may learn to associate an odor with nectar from contact with a returning crop-attached forager, they may also learn the association in an indirect manner. In-hive propagation of nectar-associated olfactory information by serial mouth-to-mouth contacts (trophallaxis) allows many bees to pre-learn scents of crops to which they later may be recruited (Gruter *et al.* 2007). Thus, although odor is necessary, it need not always enter the hive concurrently with recruitment.

Odor is critical for recruitment of naïve bees to a nectar location. Just how important odor is to recruitment is revealed by two observations. First, recruitment ceases with substitution of truly unscented nectar at the foraging location even though the crop-attached bees continue to visit that nectar source (Wenner *et al.* 1969; Wells & Wenner 1971; Friesen 1973). Second, if the scent crop-attached foragers have been bringing into the hive is offered at a place not visited by any bee and removed from the nectar at the crop-attached bee foraging location, searching naïve-recruit foragers will arrive at that new location (Wenner *et al.* 1969). Naïve-recruits arrived at the scented feeder in such a situation the day after their in-hive exposure to the conditioning scent (Wenner *et al.* 1969; Ohtani 2008).

Even though the ability of honey bees to recruit naïve bees to a nectar source is amazing, searching recruits take much longer than expected to locate food sources and there is tremendous variation in the flight durations of these recruit bees. For instance, when Esch &

Bastian (1970) measured success rate for 34 marked foragers recruited by trained foragers to a feeder 200m from the hive, only 14 found it — with 10 of those requiring more than one try. Also, the average naïve searching bee's flight duration was 8.5 minutes, although a direct 200m flight requires less than 30 seconds, and "the newcomers approach the food site from a considerable distance... in a zigzag flight" (Esch & Bastian 1970). Others have obtained similar measurements (Gould *et al.* 1970). Recruitment success rate is surprisingly low; and those that do succeed do not fly directly to the location so that the duration of a new recruits' flight is unpredictable in comparison to crop-attached foragers.

A crop-attached forager work force generates a substantial population of searching bees that are scattered across an odor landscape that is influenced by scents in prevailing air currents. This complexity means that 1) many of the naïve searching bees do not find the nectar source being visited by the crop-attached bees, 2) unlike crop attached bees, naïve recruits cannot not fly directly and quickly to the food source via odor cues, and 3) average search time of recruits is always many times greater than direct flight time to the destination (Friesen 1973). Nevertheless, some aspects of naïve-recruit searching efficiency are very predictable. Searching naive-recruits find nectar sources faster and farther form the hive when the nectar location is upwind, or even crosswind, of the hive in comparison to downwind nectar locations (Friesen 1973).

SENSING AND NEURONAL TRANSMISSION OF ODORS

Initially, to von Frisch and others of the early era of insect odor research, olfactory acuity of honey bees seemed to mirror that of human beings (Frisch 1919). However, bee olfaction was soon shown to be considerably better than that of man (Ribbands 1953). Just how much better was not totally clear until recently when bees were trained to replace dogs to find explosives around airports and to locate unexploded land mines (Bromenshenk *et al.* 1985; Bromenshenk pers. comm.). Honey bees have an exceptionally keen sense of smell, and they can be trained to use it to human advantage.

Although olfactory receptors exist in the forelimb tarsi, the major center for olfactory signal detection in honey bees is the antennae (de Brito Sanchez *et al.* 2008). In part this is a result of the number of sensilla located in each of these appendages, with at least 20 times more in the antennae (Whitehead & Larsen 1976). Even with the receptor density found on the antennae, classical conditioning studies show gustatory sensation is significantly more pronounced in the proboscis. The conditioned stimulus (odor) is detected primarily by the antennae, while the unconditioned stimulus (sucrose) is detected more effectively by the proboscis (Bitterman *et al.* 1983). Similar to scent detection, the forelimb tarsi are capable of unconditioned-stimulus gustatory-sensation but in classical conditioning tests elicit far inferior acquisition levels than occur with the proboscis receiving the unconditioned stimulus (de Brito Sanchez *et al.* 2008). What is of importance here is that the degree of learning seen in foraging honey bees, both in terms of speed and fidelity obtained, requires linking information from distinct sensory organs – usually the antennae for odor and proboscis for taste.

Returning to our focus, odor perception, the antennal lobes receive sensory input from the antennae olfactory receptor neurons. Within the antennal lobes are functional centers, glomeruli, which process information and send it on to the mushroom bodies and lateral protocerebrum via projection neurons (Flanagan & Mercer 1989; Abel *et al.* 2001). There are upwards of 150 glomeruli, and stimulation of the antennae with a specific odor produces both spatial and temporal activity in a subset of these functional centers. The activity pattern elicited by a specific odor is conserved between individuals (Galizia *et al.* 1999; Galizia & Menzel 2000). Thus, a specific odor produces activity in a specific subset of glomeruli, so that in effect the glomeruli are encoding information and all individuals are using the same neuronal code. Further, similar glomerular activity patterns are produced by different chemicals with similar molecular structures (Sachse *et al.* 1999).

At this point odor perception seems to be a rather 'mechanical process' for honey bees in that everyone has the same olfactory information in the decision process, and thus we should expect honey bee response to be quite uniform across foragers. In fact, in many cases it is, as is evidenced by the high acquisition levels reached very rapidly in conditioning studies (*e.g.* Abramson *et al.* 2010a). However, that is not the complete story. When either pollen or nectar foraging problems become difficult, then forager individuality becomes apparent in honey bees (Cnaani *et al.* 2006; Cakmak *et al.* 2009). Diverse response behaviors arise from the same information. Still, learning for each bee is quite apparent although bees are not reaching the same decision, and responses fall into a limited number of what might be termed behavior syndromes (Sih *et al.* 2004; Cakmak *et al.* 2009).

Honey bee olfactory sensitivity is broad. Foragers can recognize and discriminate among a wide range of scent molecules, scent mixtures, or individual components of scent mixtures (Smith & Abramson 2003); indeed, they can be trained to any natural or experimental olfactory signature of a food source. The magnitude of genetic adaptation to odor recognition was recently revealed by the analysis of the honey bee's DNA genome (Honeybee Genome Sequencing Committee 2006). At present, 170 odor-receptor genes have been identified, which is more than twice the number known for any other insect (Honeybee Genome Sequencing Committee 2006). While the homology of vertebrate/insect olfactory genes is clear, the honey bee genome apparently lacks an equivalent to the human language gene complex (Honeybee Genome Sequencing Committee 2006; Ohtani 2008).

ODOR PERCEPTION

Honey bees are known to detect odor molecules at extremely low concentrations and to be able to differentiate among a wide variety of scents (see: Wenner & Wells 1990). However, understanding the perceptual similarities of differing odors for bees has been a much more daunting challenge (Chittka & Brickmann 2005). The picture that has emerged is that within chemical groups (*e.g.* alcohol, aldehyde, ketone) the size of odor molecules (i.e. carbon chain length) determined perceptual similarity for bees (Guerrieri *et al.* 2005). That is particularly true for molecules with aldehyde groups (Getz & Smith 1990; Laska *et al.* 1990). In general, the greater the difference in the number of carbons, the more likely are bees to differentiate between the molecules. This relation appears to be also true in vertebrates (*e.g.* Kent *et al.* 2003; Xu *et al.* 2003) and other insects (*e.g.* Daly *et al.* 2004).

In honey bees, a linear relation does not appear to exist overall between molecular size and perceptual similarity. There is a grouping into small molecules versus large molecules within chemicals having the same functional group. Nevertheless, a perceptual continuum seems to exist within these two size classes of molecules in any of the functional groups studied (Guerrieri *et al.* 2005). Molecules with different functional groups are perceived by bees as categorically quite distinct. Here it is useful to think of perception more as a nominal data set. However, there do appear to be exceptions. Bees clearly see primary and secondary alcohols as similar odors (Guerrieri *et al.* 2005).

Coding of information by the antennal lobe glomeruli is quite intriguing, although still very incomplete due to the number of glomeruli that can be studied through optical imaging. The picture that is emerging is that there is a set of glomeruli that is involved in coding response to long chain molecules and another set that is involved in responding to shorter chain molecules. Further, these sets respond to molecules with different functional groups, albeit not with equal intensity (Sachse *et al.* 1999; Sachse & Galizia 2003). In this respect, odor neuronal coding has some properties like color vision (Chittka & Brockmann 2005). Vision is based on only three color receptor types each sensitive to a broad range of wavelengths, but each with different peak wavelength sensitivity. The visual system determines color by comparing the signals from receptors differing in spectral sensitivity (Lotto & Purves 2002; Yang *et al.* 2004). Using such an opponent mechanism, the central nervous system can interpret whether there is a stronger signal from the short wavelength receptor or the long wavelength receptor and in doing so can extract information about stimulus spectral quality (Chittka & Wells 2004). However, the situation is much more complex for honey bee odor perception because 1) there are about 170 receptor protein types (vs. 3 for color vision), 2) bees can perceive the components of a mixture (rather than perceiving an intermediate as in vision), 3) bees can attach unique properties to the mixture, and 4) there are asymmetries with respect to perceived similarities (Galizia & Menzel 2000; Deisig *et al.* 2003; Chittka & Brockmann 2005; Guerrieri *et al.* 2005). These asymmetries mean that the perceived similarity between two odors depends on which scent was presented first.

It is tempting to relate vertebrate to honey bee odor perception. There are basic functional and anatomical similarities in the way in which olfactory signals are detected and processed in the central nervous systems. This is especially true at the level of their first olfactory centers, the olfactory bulb of vertebrates and the antennal lobe of insects (Hildebrand et al. 1997). Nevertheless there are striking differences. Mammals have about a thousand odor receptor proteins in comparison to the honey bee's 170 (Buck & Axel 1991). Yet with the relatively poor compliment of receptors honey bees appear to do as well at perceiving odors as mammals like dogs (Bromenshenk et al. 1985). Partial similarities do not mean the corresponding systems have anywhere close to the same abilities. For example, about 40 receptor cells with approximately a combined 70 kHz range exist in bushcrickets, they converge on 3 interneurons ascending to the brain. Thus, the frequency resolution of the receptor cells is largely lost at the interneuron stage (Schul 1997; Höbel & Schul 2007; Triblehorn & Schul 2009).

The Importance of Nectar Odors for Foragers

Odor-based searching was seen in the early 1940s as important for naïve forager recruitment (Wenner 1993). Subsequent work implies that odor is not only important but also *essential* for naïve forager recruitment.

Recruitment rate of naïve bees is not altered by methods that disorient bee-dances if the nectar has a scent. However, the reverse is not true. When a nectar source is unscented, even though crop-attached bees may dance upon returning, there is no recruitment (Wells & Wenner 1973). Also, recruitment increased with increasing scent levels in nectar, while crop attached bee dancing upon returning to the hive decreases as does exposure of the Nasanov gland at the nectar location (Wells & Wenner 1971). This and additional work shows that Nasanov gland pheromone does not function as a forager attractant per se. However, bees can learn to use the Nasanov scent mixture as a food cue if it is linked to a reward in a short period of time as in proboscis conditioning experiments (Wenner & Wells 1990 excursus NG; Wells *et al.* 1993). Further, when odor-search and dance-language hypotheses were tested in experiments where results supportive of one hypothesis directly refute the other (*i.e.* crucial design), only the odor-search hypothesis was favored (Wenner *et al.* 1969).

The early 'step' (distance) and 'fan' (direction) studies that were once thought to demonstrate recruitment via a 'dance-language' (Frisch 1950) have now been shown to suffer from inadequate experimental designs (Johnson 1967; Wenner 1967). When more rigorous controls are added to those experiments of those designs, naïve-recruit arrival at an array of nectar reward sites distributed around the dance-specified location were distributed in a lognormal (random) distance pattern for "step" experiments and were inversely proportional to distance from the odor-field center for "fan" experiments (Wenner & Wells 1990 excursus NG; Wenner *et al.* 1991). Several subsequent attempts to demonstrate that bees use only language have at times drawn criticism as well (Wells & Wenner 1973; Rosin 1980, 1991, 1999; Veldink 1989; Wenner & Wells 1990; Kak 1991; Wenner *et al.* 1991; Vadas 1994; Wenner 1997, 2002, 2007). Several of these studies and additional discussion can be found at the beesource.com web site (*i.e.* beesource.com/pov/wenner/).

Technology has led to the development of a mechanical 'crop attached' bee to examine factors affecting naïve forager recruitment to nectar sources (Wenner 2007). Of surprise where the facts that recruitment occurred only if the "robot" bee provided a sugar-water reward *and* exposed potential recruits to *scent* of the food goal (Michelsen *et al.* 1989). Most of the naïve-recruits in step-experiments did not arrive at the distances signaled by the robot, but the pattern of recruitment to the reward-station array was consistent with a lognormal (random) distribution. Fan-experiment results fit a distance-from-odor-center model (Wenner *et al.* 1991; Michelsen *et al.* 1989). Experiments with the mechanical bee have yielded a wealth of data supportive of an in-hive sent-conditioning, odor-search hypothesis of forager recruitment.

The necessity of odor for recruitment of naïve foragers to a nectar source has led some to doubt whether the "dance" information is ever used (Wenner & Wells 1990) and others to significantly alter their view of recruitment of naïve bees. For example, direct in-hive observation of recruitment and other activities of a specific crop-attached forager and attending potential recruits led Ohtani to conclude that "the dance performances of honeybees possess physiological aspects which are inconsistent with the 'dance language' hypothesis"

(Ohtani 2008). Dance language advocates no longer expect dance attendants to fly quickly and directly to the target as originally proposed (Frisch 1950). To Riley, *et al* (2005), for example, "the honey bee (language) does not instantly specify a food location . . . (nor) with pinpoint accuracy . . . (and may require) several iterations of dance sessions and resultant search flight, and some never find the food at all." These authors champion non-specificity, not because it is of evolutionary benefit to bees, but because search inefficiency "would neatly account for the fact that the arrival of recruits at the source is often very much later than would be expected..." (Riley *et al.* 2005). Thus, it was not surprising in the 'radar tracking' study when none of the transponder toting recruits found the location offering unscented nectar reward, since in reality neither the odor-search nor dance language hypothesis predicted that recruits should find the unscented goal. But, even burdened with transponders, bees did fly off into the field and insect flight could be observed by radar, and with technological improvements in the future this avenue of research offers exciting possibilities for future research.

LIMITATIONS FOR ODOR COGNITION

Honey bee olfactory sensitivity is broad. Foragers can recognize and discriminate among a wide range of scent molecules, scent mixtures, or individual components of scent mixtures (Smith & Abramson 2003), which parallels the ability of those vertebrates with a keen sense of smell. However, the results on aversive conditioning open a larger and very fundamental question about the cognitive architecture of honey bees specifically and invertebrates in general. In avoidance experiments, it is not the absence of the aversive event that is reinforcing for invertebrates but rather the fact that the stimulus signaling the onset of the aversive event is paired with the actual negative event (Abramson 1986; Abramson *et al.* 1988; Abramson & Buckbee 1995). For example, in the honey bee shock avoidance study by Abramson (1986) extinction quickly followed removal of the shock since it is not the escape from shock (absence of shock) that is reinforcing a behavior. This questions whether the removal of a stimulus can serve as a signal predicting any type of event for an invertebrate?

Experimental comparison of addition and deletion tasks for any animal has been notably rare (reviewed in: Hearst & Wolff 1989), and this problem continues to be understudied (Miranda *et al.* 1992; Abramson & Buckbee 1995). Although honey bees can associate the onset of an odor with a reward they cannot do the same with the removal of the stimulus (*i.e.* the gap in a stimulus). This has been demonstrated using both a discrimination paradigm and a simple conditioning situation where the gap in a background stimulus served as the cue signaling a feeding (Abramson *et al.* 2010b). In contrast, several vertebrate groups have been shown to be able to associate the removal of a stimulus with an impending event, although it is much more difficult for all of the vertebrate species than associating the presentation of a stimulus with a reward or punishment (*e.g.* pigeons, Jenkins & Sainsbury 1969; rats, Crowell & Bernhardt 1979; monkeys, Pace *et al.* 1980; humans, Newman *et al.* 1980; Healy 1981). This inability to use the removal of a cue in learning either to expect a reward or punishment may represent a fundamental aspect of the cognitive architecture of bees that differs from vertebrates. Although the end result may be the same in terms of behavior, bees seem to reach this point by very different cognitive processes which maybe widespread among the invertebrates.

Thus, in considering the evolution of cognition (Dukas 1998, 2004), the end product may tell us less than knowing how the organism made the decision. This is certainly true for the honey bee where both signal gap experiments (Abramson *et al.* 2010b) and signaled avoidance studies (Abramson 1986, 1988; Abramson & Buckbee 1995; Suddendorf & Corballis 2007) suggest that honey bees cannot utilize the removal of an environmental cue in learning.

Odors as Repellants: Perspectives from Pesticide Studies

There has been substantial interest in insect repellants, including compounds that will repel bees and wasp. Inadvertently, studies of pesticide effects on honey bees have provided interesting insights about odors as repellents for honey bees.

Honey bees contribute substantially to the pollination of various wild plants and food crops. The annual value of agricultural crops benefiting from honey bee pollination is estimated at as much as $20 billion/year in the United States alone. Studying the influence of agrochemicals on honey bee behavior is important for the survival of honey bees, public policy issues, honey bee population regulation, environmental degradation, and the use of biological controls.

The use of toxic chemicals to control insect pests has a long history. Chemicals such as DDT, sevin, rotenon, diazionon, methoxychlor and imidacloprid have been used to control such pests as Colorado potato beetle, cabbageworm, and the gypsy moth. What has not always been known is how these chemicals affect honey bee behavior. Data generated over the past 50 years have shown that pesticides disrupt the functioning of the central nervous system, metabolic processes and some physiological processes such as molting and reproduction. Pesticides which are specially formulated to kill target insects usually do so by influencing receptor molecules in central nervous system mechanical, photo, and/or chemical receptors. Pesticides have also been developed that are synthetic analogs of enzyme substrates that interfere with metabolic pathways.

As a case study consider the Africanized honey bee in Brazil. The Africanized bee is important to the economy of Brazil in two main ways. Aside from the production of honey as a major agricultural product, bees serve as pollinators of cotton as well as many others cash crops in the Brazilian economy.

Cotton is an important crop for the agrarian sector and textile industry in Brazil. Cotton production was adversely affected by the appearance of the cotton boll weevil in 1983 and has led, for example, to unemployment, depreciated land value, and the closing of cotton gins and oil mills. The major strategy to combat the boll weevil is the use of pesticides such as baytroid, decis, endosulfan, and sevin. These pesticides, however, have had adverse effects on the honey bee population. When harnessed bees were exposed to field doses of baytroid and sevin, and subsequently tested using the classical conditioning of the proboscis extension reflex, death quickly resulted. Of more interest was the fact that bees exposed to endosulfan could acquire a learned response, but over the course of training the learning became unstable and soon disappeared. Those bees treated with decis showed a pattern of learning indistinguishable from untreated controls (Abramson *et al.*, 1999).

In an attempt to find alternatives to traditional pesticides, Bioganic Safety Brands, USA developed the first essential oil based pesticide. Exposing honey bees to Bioganic® Lawn and Garden Spray Multi-Insect Killer (Bioganic Safety Brands, USA) indicated that the pesticide was not lethal to bees whether consumed directly or applied to the abdomen. The effects of different concentrations of Bioganic on both Pavlovian conditioning and discrimination learning were also examined using harnessed foragers in proboscis extension studies. The results suggested that the pesticide affected learning; however, this conclusion is suspect because the bees would not feed on the pesticide, thus making it impossible to properly assess either Pavlovian conditioning or discrimination learning. Consequently, the effects of Bioganic on discrimination learning were examined in free flying bees trained to land on targets. The results of these experiments indicated that bees will not avoid a target associated with the smell of the pesticide but will avoid the target if they have to drink the pesticide as a reward (Abramson *et al.*, 2006a).

The study of toxic chemicals on honey bee behavior has extended to the area of sub-lethal effects. When a toxic chemical is released into the environment it is degraded, for example, by rain or ultraviolet rays from the sun. The result is that honey bees can be exposed to sub-lethal levels of agrochemicals even though the chemical was applied as a lethal dose. Evidence exists that sub-lethal doses of pesticides may be decreasing the number of honey bee colonies available for pollination and reducing the effectiveness of honey bees as pollinators. Sub-lethal doses of deltamethrin, for example, disrupt the homing flight of honey bees, while parathion disrupts in-hive behaviors of returning crop-attached foragers (Schricker & Stephen, 1970; Vandame *et al.*, 1995). In addition to the disruption of natural behavior, it is known that sub-lethal exposure to permethrin, coumaphos, and diazinon retards various aspects of learning (Taylor *et al.*, 1987; Mamood & Waller, 1990; Weick & Thon, 2002).

Recently, a new line of investigation has begun on agrochemical considered harmless to honey bees. These compounds include new generation pyrethroids, insect growth regulators, and metabolite by-products. Many of the new products containing these compounds are considered by the Environmental Protection Agency, and other regulatory bodies, as environmentally safe. However, little is known about their effects on honey bee behavior. In order to use these chemicals effectively, and without injuring important pollinators, it is necessary to know what effects agrochemicals have on honey bee behavior.

The first experiments on the study of chemicals considered "not harmful" to honey bees was an investigation of dicofol. Dicofol is a chlorinated hydrocarbon pesticide. It is considered nontoxic to most insects and is used primarily to control mites. Honey bees pretreated with dicofol, however, exhibited significantly lower levels of learning than honey bees not pretreated (Stone *et al.*, 1997).

Experiments have also been conducted using the insect growth regulators tebufenozide and diflubenzuron. The results of these experiments were similar to those with dicofol and equally unexpected. The learning ability of honey bees was again disrupted by agrochemicals once thought to be harmless (Abramson *et al.*, 2004).

As another example let us consider imidacloprid. Imidacloprid is a novel insecticide that mimics nicotine. It is applied to the seeds of crops, and as the plant develops, is transported to the stem and leafs of the plant. Aphids and other pests such as the Colorado potato beetle will die if they ingest imidacloprid. Imidacloprid is also used on sunflower seeds. Sunflowers are an excellent source of nectar for honey bees and sunflowers depend upon bees for pollination. Although it is toxic to honey bees, honey bees are not in direct contact with imidacloprid. It is known from the plant data that the average values of imidacloprid contained in the pollen of sunflowers, and of corn, is around 3 parts per billion, which is one fifth of the dose known to cause disorientation of dances in honey bees. Nevertheless, the French government decided to prohibit use of imidaclopride on sunflower seeds because of its effect on honey bees (Devillers & Pham-Delègue 2002). Imidacloprid is also known to influence learning, homing, and foraging activity (Medrzycki *et al.*, 2003; Decourtye *et al.*, 2005)

In addition to insecticides, effort has been directed at finding olfactory and gustatory insect repellents (Isman 2006). Until recently, and unlike pesticides, few behavioral mechanisms for identifying how repellents affect honey bee learning have been developed. Proboscis conditioning and conditioned suppression protocols in honey bees may provide the mechanistic model for identifying repellents.

In the case of the honey bee, putative olfactory and gustatory repellents are used and investigated for several reasons. These reasons include public safety issues (Abramson *et al.* 1997), reducing the effects of harmful agrochemicals (Atkins Jr. *et al.* 1975a, 1975b), separating bees from honey for apicultural purposes (Graham 1992), and studying comparative aspects of behavior across taxa (Abramson, 1994). Several studies suggest, for example, that *N*-octyl-, benzyl acetate, isopentil-acetate, and 2-heptanone are olfactory repellents for honey bees (Blum *et al.* 1978; Free 1987; Free *et al.* 1989). The majority of repellent studies base their conclusions on field tests (Schreck 1977).

If a stimulus is an olfactory repellent, not only should it repel honey bees in field tests but also should, in our view, be ineffective as a conditioned stimulus signaling a feeding opportunity. We also believe that an application to the antenna of the odor of a putative olfactory repellent to an already extended proboscis should produce a retraction of the proboscis as an avoidance reaction. Behavioral suppression to stimuli paired with aversive events is known in the psychological literature as conditioned suppression (Estes & Skinner 1941).

The classical conditioning of proboscis extension was used in a study investigating the repellent action of citronella to Africanized honey bees in Brazil. A previously published field test suggested that applying citronella to a diaper suspended above a feeding station reduced the number of bees visiting that station (Malerbo-Souza & Nogueira-Couto 2004). However, when proboscis conditioning procedures were employed to confirm the repellency of citronella it was found that Africanized bees readily associated the odor of citronella with feeding and that the application of citronella to the antennae did not disrupt feeding (Abramson *et al.* 2006b). A similar result was found for both DEET and Butyric acid (Abramson *et al.*, 2010a). These results suggested that the strongest evidence for testing olfactory repellency in the honey bee, and probably in other insects, is the use of both field tests and conditioning protocols.

Repellents also interfere with feeding and distribution of target organisms (Baker *et al.* 2005; Isman 2006; Werner *et al.* 2007; Carroll *et al.* 2008). The evidence for repellent action is often just the modification in some target behavior, and may lead to misidentification of mechanisms of action (Xue *et al.* 2001). For example, target insects leaving an area where a repellent has been applied may lead one to conclude that the repellent is effective. In actuality, what is causing the target pests to leave is that the environment has changed – that is, stimulus novelty. If this is indeed the case, the response to novelty may later lead to attraction once novelty is reduced. The importance of stimulus novelty was revealed in a recent study of the popular repellent DEET. DEET is effective because of odor masking (with evidence that a familiar attractant resulted in less responses in the presence of DEET, Ditzen *et al.* 2008) or through avoidance (with evidence that DEET is perceived and avoided, Syed & Leal 2008). The effect of butyric acid in practical use has been suggested to be an olfactory repellent for honey bees. In fact, in the Abramson *et al.* (2010b) experiment and those by Ditzen *et al.* (2008) and Syed & Leal (2008), direct contact with undiluted putative repellents (direct contact or mixed in food) led to inhibitory responses. However, in the Abramson *et al.* experiment, (2010b) as odors alone these chemicals were not repellent (similar to inference by Ditzen *et al.* 2008 for DEET).

The use of conditioning paradigms such as proboscis conditioning and conditioned suppression has much to recommend for the study of olfactory and gustatory repellents. The experiments are easy to perform and much data can be obtained not only on learned behavior to a conditioned stimulus, but on reflexive behavior as represented by the unconditioned response as well.

CONCLUSION

Scent is a crucial factor in food source recognition by crop-attached foragers. Great diversity and high sensitivity of olfactory receptors facilitates the honey bee's '*ultimate generalist pollinator'* role in the ecological/agricultural economy. Antennae are the primary organ involved in sensing odors, and the antennae-lobe glomeruli seem to be the central nervous systems coding region for odor stimuli. The chemical functional group appears to be very important in perception of similarity between scent chemicals, and to a lesser extent molecular size.

Crop-attached foragers are re-recruited and resume harvesting a food source when that crop's scent is again brought into the hive. Crop fidelity, the sequential visitation of only one type of flower, is largely based on scent. Recruitment of new workers to a forager force is mediated by in-hive learning of food scent (conditioning), followed by odor-search behavior. Culminating this odor-driven process, successful recruits are immediately added to a work force of crop-attached foragers as they harvest an available food resource. This odor-driven foraging process has many practical applications for agriculture (*e.g.* Rathore & Wells 1995).

There are limitations to learning involving scents as conditioning cues. First, when presented complex mixtures of scents in a floral bouquet bees although they can remember all the scents some are 'key odors' and elicit a stronger response. Second, bees cannot use the gap in an odor stimulus as a conditioning cue. Finally, pesticide studies suggest that there probably are no odors that are truly repellants. That is, as odors they do not elicit aversion behavior, and in fact can be used to elicit feeding responses. However, these odor-chemicals do produce aversion behavior if a bee is forced to consume the compound.

Acknowledgments

Writing of this chapter was supported in parts by National Science Foundation grants OIS –1043057 and DBI-0851651.

References

Abel, R, Rybak, J. & Menzel, R. (2001). Structure and response patterns of olfactory interneurons in the honeybee, *Apis mellifera*. *Journal of Comparative Neurology* **437**, 363-383.

Abramson, C. I. (1986). Aversive conditioning in honey bees (*Apis mellifera*). *Journal of. Comparative Psychology* **100**, 108-116.

Abramson, C. I. (1994). A primer of invertebrate learning: the behavioral perspective. *American Psychological Association*, Washington, D.C.

Abramson, C. I. & Buckbee, D. A. (1995). Pseudoconditioning in earthworms (*Lumbricus terrestris*): Support for nonassociative explanations of classical conditioning phenomena through an olfactory paradigm. *Journal Comparative Psychology* **109**, 390-397.

Abramson, C. I., Wells, H. & Bozic, J. (2007). A social insect model for the study of ethanol induced behavior: The honey bee. *In*: R. Yoshida (*ed.*), *Trends in Alcohol Abuse and Alcoholism Research*. Nova Science Publishers, Hauppauge, N.Y.

Abramson, C. I., Aquino, I. S., Azeredo, G. A. & Price, J. M. (1997). The attraction of Africanized honey bees (*Apis mellifera* L.) to soft drinks and perfumes. *Journal of General Psychology* **124**, 166-181.

Abramson, C. I., Aquino, I. S., Ramalho, F. S. & Price, J. M. (1999). Effect of insecticides on learning in the Africanized honey bee (*Apis mellifera* L.). Archives of Environmental Contamination and Toxicology **37**, 529-535.

Abramson, C. I., Armstrong, P. M., Feinman, R. A. & Feinman, R. D. (1988). Signaled avoidance in the eye withdrawal reflex in the green crab. *Journal of the Experimental Analysis of Behavior* **50**, 483-492.

Abramson, C. I., Nolf, S. L., Mixon, T. A. & Wells, H. (2010b). Can honey bees learn the removal of a stimulus as a conditioning cue? *Ethology* **116**, 843-854.

Abramson, C. I., Squire, J., Sheridan, A., &, Mulder, Jr., P. G. (2004). The effect of insecticides considered harmless to honey bees (*Apis mellifera* L.): Proboscis conditioning studies using the insect growth regulators Confirm®2F (Tebufenozide) and Dimilin®2L (Diflubenzuron). *Environmental Entomology* **33**, 378-388.

Abramson, C. I., Singleton, J. B., Wilson, M. K., Wanderley, P. A., Ramalho, F. S., & Michaluk, L. M. (2006a). The effect of an organic pesticide on mortality and learning in Africanized honey bees (*Apis mellifera* L.) in Brasil. *American Journal of Environmental Science* **2**, 37-44.

Abramson, C. I., Wilson, M. K., Singleton, J. B., Wanderley, P. A., Wanderley, M. J. A., & Michaluk, L. M. (2006b). Citronella is not a repellent to Africanized honey bees *Apis mellifera* L. (Hymenoptera: Apidae). *BioAssay*, **1,** 1-7 (13).

Abramson, C. I., Giray, T, Mixon, T.A., Nolf, S. L., Wells, H., Kence, A. & Kence, M. (2010a). Proboscis conditioning experiments with honey bees (*Apis mellifera caucasica*) with butyric acid and DEET mixture as conditioned and unconditioned stimuli. *Insect Science* **10**, 1-22.

Abramson, C. I., Stone, S. M., Ortez, R. A., Luccardi, A., Vann, K. L., Hanig, K. D. & Rice, J. (2000). The Development of an Ethanol Model Using Social Insects I: Behavior Studies of the Honey Bee (*Apis mellifera* L.). *Alcoholism: Clinical and Experimental Research* **24**, 1153-1166.

Atkins Jr., E. L., McDonald, R. L. & Greywood-Hale, E. A. (1975a). Repellent additives to reduce pesticide hazards to honey bees: field tests. *Jouranl of Environmenal Entomology* **4**, 207-210.

Atkins Jr., E. L., McDonald, R. L., McGouern, T. D., Beroza, M. & Greywood-Hale, E. (1975b). Repellent additives to reduce pesticide hazards to honey bees: laboratory testing. *Journal of Apicultural Research* **14,** 85-97.

Baker, S. E, Elwood, S. A., Watkins, R. & MacDonald, D. W. (2005). Non-lethal control of wildlife: using chemical repellents as feeding deterrents for the European badger *Meles meles*. *Journal of Applied Ecology* **42**, 921-931.

Barnea, G, O'Donnell S, Mancia, F, Sun, X, Nemes A, et al. (2004). Odorant receptors on axon termini in the brain. *Science* **304**, 1468-1468.

Bitterman, M. E., Menzel, R., Fietz, A. & Schäfer, S. (1983). Classical conditioning of proboscis extension in honeybees (*Apis mellifera*). *Journal of Comparative Psychology* **97**,107-119.

Blum, M. S., Fales, H. M., Tucker, K. W., & Collins, A. M. (1978). Chemistry of the sting apparatus of the worker honey bee. *Journal of Apicultural Research* **17**, 218-221.

Bortolotti, L., Montanari, R., Marcelino, J., Medrzycki, P., Maini, S., & Porrini, C. (2003). Effects of sub-lethal imidacloprid doses on the homing rate and foraging activity of honey bees. *Bulletin of Insectology* **56**, 63-67.

Bromenshenk, J. J, Carlson, S. R., Simpson, J. C. & Thomas, J. M. (1985). Pollution monitoring on Puget Sound with honey bees. *Science* **277**, 632-634.

Buck, L. & Axel, R. (1991). A novel multigene family may encode odorant receptors: A molecular basis for odor recognition. *Cell* **65**, 175-187.

Cakmak, I., Sanderson, C., Blocker, T. D., Pham, L. L., Checotah, S., Norman, A. A., Harader-Pate, B. K., Reidenbaugh, R. T., Nenchev, P., Barthell, J. F. & Wells, H. (2009). Different solutions by bees to a foraging problem. *Animal Behaviour* **77**, 1273-1280.

Carroll, J. F., Benante, J. P., Klun, J. A., White, C. E., Debboun, M., Pound, J. M., et al. (2008). Twelve-hour duration testing of cream formulations of three repellents against *Ambylomma americanum*. *Medical and Veterinary Entomology* **22**, 144-151.

Chittka, L., & Brockmann, A. (2005). Perception space – The final frontier. PLoS Biol **3**(4): e137.

Chittka, L. & Wells, H. (2004). Color vision in bees: Mechanisms, ecology and evolution. *In*: F. Prete (*ed.*) *Complex Worlds from Simpler Nervous Systems*. MIT Press. Boston, MA. pp. 165-191.

Cnaani, J., Thompson, J. D. & Papaj, D. R. (2006). Flower choice and learning in foraging bumblebees: effect of variation in nectar volume and concentration. *Ethology* **112**, 278-285.

Crowell, C. R. & Bernhardt, T. P. (1979). The feature-positive effect and sign-tracking behavior during discrimination learning in the rat. *Animal Learning and Behavior* **7**, 313-317.

Daly, K.C., Wright, G.A. & Smith, B. H. (2004). Molecular features of odorants systematically influence slow temporal responses across clusters of coordinated antennal lobe units in the moth *Manduca sexta*. *Journal of Neurophysiology* **92**, 236-254.

de Brito Sanchez, M. G., Chen, C., Li, J., Liu, F, Gauthier, M. & Giurfa, M. (2008). Behavioral studies on tarsal gustation in honeybees: sucrose responsiveness and sucrose-mediated olfactory conditioning. *Journal of Comparative Physiology* [A] **194**, 861-869.

Decourtye, A., Devillers, J., Genecque, E., Le Menach, K., Budzinski, H., Cluzeau, S. & Pham-Delègue, M. H. (2005). Comparative sublethal toxicity of nine pesticides on olfactory learning performance of the honeybee *Apis mellifera*. *Archives of Environmental Contamination and Toxicology* **48**, 242-250

Deisig, N., Lachnit, H., Sandoz, J. L., Lober, K. & Giurfa, M. (2003). A modified version of the unique cue theory accounts for olfactory compound processing in honeybees. *Learning & Memory* **10**, 199-208.

Devillers, J. & Pham-Delègue, M. H. (2002). (Eds.), *Honey Bees: Estimating the Environmental Impact of Chemicals*. Taylor & Francis, London.

Ditzen M., Pellegrino M., & Vosshall L. B. (2008). Insect odorant receptors are molecular targets of the insect repellent DEET. *Science* **319**, 1838-1842.

Dukas, R. (1998). *Cognitive Ecology: the Evolutionary Ecology of Information Processing and Decision Making*. University of Chicago Press, Chicago.

Dukas, R. (2004). Evolutionary biology of animal cognition. *Annual Review of Ecology and Evolution* S. **35**, 347-374.

Endress, P. K. (1994). *Diversity and Evolutionary Biology of Tropical Flowers*. Cambridge University Press, Cambridge.

Esch, H. & Bastian, J. A. (1970). How do newly recruited honey bees approach a food site? *Zeitschrift fuer vergleichende Physiologie* **68**,175-181.

Farina, W. M., Gruter, C. & Acosta, L. (2007). Honeybees learn floral odors while receiving nectar from foragers within the hive. *Naturwissenschaften* **94**, 55-60.

Farina, W. M., Gruter, C. & Diaz, P. C. (2005). Social learning of floral odours inside the honeybee hive. *Proceedings of the Royal Society London* B **272**, 1923-1928.

Flanagan, D. & Mercer, A. (1989). Morphology and response characteristics of neurons in the deutocerebrum of the brain in the honeybee *Apis mellifera*. *Journal of Comparative Physiology* [A] **164**, 483-494.

Free, J. B. (1987). *Pheremones of Social Bees*. Chapman Hall, N.Y.

Free, J. B., Ferguson, A. W. & Simpkins, J. R. (1989). Honey bee responses to chemical components from the worker sting apparatus mandibular glands in field tests. *Journal of Apicultural Research* **28**, 7-21.

Friesen, L. J. (1973). The search dynamics of recruited honey bees, *Apis millifera ligustica* Spinosa. *Biological Bulletin* **144**, 107-131.

Frisch, K. von. (1919). Uber den Geruchsinn der Bienen und seine blutenbiologische Bedeutung. *Zoologische Jahrbucher, Abt.* 3, **37**, 1-238.

Frisch, K. von. (1950). *Bees: Their Vision, Chemical Senses, and Language*. Cornell University Press, Ithaca, N.Y.

Galizia, C. G. & Menzel, R. (2000). Odour perception in honeybees: Coding information in glomerular patterns. *Current Opinion in Neurobiology* **10**, 504-510.

Galizia, C. G., Menzel, R. & Hőlldobler, B. (1999). Optical imaging of odour evoked glomerular activity patterns in the antennal lobes of the ant *Camponotus rufipes*. *Naturwissenschaften* **86**, 533-537.

Getz, W. & Smith, K. (1990). Odorant and odor mixture perception in the free-flying honey bees (*Apis mellifera*). *Chemical Senses* **15**, 111-128.

Gould, J. L., Henerey, M. & MacLeod, M. C. (1970). Communication of direction by the honey bee. *Science* **169**, 544-554.

Graham, J. M. (ed.). (1992). *The hive and the honey bee*. Dadant and Sons, Hamilton, IL.

Gruter, C., Acosta, L. E. & Farina, W. M. (2007) Propagation of olfactory information within the honeybee hive. *Behavioral Ecology and Sociobiology* **60**, 707-715.

Gruter, C., Balbuena, M. S. & Farina, W. M. (2008). Informational conflicts created by the waggle dance. *Proceedings of the Royal Society* B **275**,1321-1327 .

Guerrieri, F., M. Schubert, J.-C. Sandoz, & M. Giurfa. (2005). Perceptual and neural olfactory similarity in honeybees. *PLoS Biology* **3,** e60.

Healy, A. F. (1981). The effects of visual similarity for misspelling. *Memory and Cognition* **9**, 453-460.

Hearst, E. & Wolff, W. T. (1989). Addition versus deletion as a signal. *Animal Learning and Behavior* **17**, 120-133.

Hildebrand J. & Shepherd G. (1997). Mechanisms of olfactory discrimination: Converging evidence for common principles across phyla. *Annual Review of Neuroscience* **20**, 595-631.

Hill, P. S. M., Wells, P. H. & Wells, H. (1997). Spontaneous flower constancy and learning in honey bees as a function of colour. *Animal Behaviour* **54**, 615-627.

Höbel, G. & Schul, J. (2007). Listening for males and bats: Spectral processing in the hearing organ of *Neoconocephalus bivocatus* (Orthoptera: Tettigoniidae). *Journal of Comparative Physiology* [A] **193**, 917-925.

Honeybee Genome Sequencing Committee. (2006). Insights into social insects from the genome of the honeybee *Apis mellifera. Nature* **443**, 931-949.

Isman, M. B. (2006). Botanical insecticides, deterrents, and repellents in modern agriculture in an increasingly regulated world. *Annual Review of Entomology* **51**, 45-66.

Jenkins, H. M. & Sainsbury, R. S. (1969). The development of stimulus control through differential reinforcement. *In*: N. J. Mackintosh & W. K. Honig, (*eds.*). *Fundamental Issues in Associative Learning*. Dalhousie University Press, Halifax, pp. 123-161.

Johnson, D. L. (1967). Honeybees: Do they use the direction information contained in their dance maneuver? *Science* **155**, 847-849.

Johnson, D. L. & Wenner, A. M. (1966). A relationship between conditioning and communication in honey bees. *Animal Behaviour* **14**, 261-265.

Kak, S. C. (1991). The honeybee dance language controversy. *Mankind* [Quarterly] **31**, 357-365.

Kennedy, J. S. (1983). Zigzagging and casting as a programmed response to wind-boune odor: A review. *Physiological Entomology* **8**,109-120.

Kent, P. F. Mozell, M. M., Youngentob, S. L. & Yurco, P. (2003). Mucosal activity patterns as a basis for olfactory discrimination: Comparing behavior and optical recordings. *Brain Research* **981**, 1-11.

Knudsen, J., Eriksson, R., Gershenzon, J. & Stahl, B. (2006). Diversity and distribution of floral scent. *Botanical Review* **72**, 1-120.

Laloi, D., Bailez, O., Blight, M., Roger, B., Pham-Delegue, M-H., *et al.* (2000). Recognition of complex odors by restrained and free-flying honeybees, *Apis mellifera. Journal of Chemical Ecology* **26**, 2307-2319.

Laloi, D., Roger, B., Blight, M. M., Wadhams, L. J. & Pham-Delegue, M.-H. (1999). Individual learning ability and complex odor recognition in the honey bee, *Apis mellifera* L. *Journal of Insect Behavior* **12,** 585-597.

Laska, M., Galizia, C. G., Giurfa, M. & Menzel, R. (1999). Olfactory discrimination ability and odour structure-activity relationships in honeybees. *Chemical Senses* **24**, 429-438.

Le Metayer, M., Marion-Poll, F., Sandoz, J., Pham-Delegue, M., Blight M, *et al.* (1997). Effect of conditioning on discrimination of oilseed rape volatiles by the honeybee: use of a combined gas chromatography-proboscis extension behavioural assay. *Chemical Senses* **22**, 391-398.

Linnaeus, C. (1756). *Calendarium Florae*. Hojer, Uppsala, Sweden.

Lotto, R. B. & Purves, D. (2002). A rationale for the structure of color space. *Trends in Neuroscience* **25**, 84-88.

Malerbo-Souza, D. T. & Nogueira-Couto, R. H.. (2004). Efficiency of n-Octyl-Acetate, 2-Heptanone and citronellal in repelling bees from Brasil (*Ocimum sellowii* - Labiatae). *Brazilian Archives of Biology and Technology* **47**, 121-125.

Mamood, A. N. & Waller, G. D. (1990). Recovery of learning responses by honeybees following a sublethal exposure to permethrin. *Physiological Entomology* **15**, 55-60.

Michelsen, A., Anderson, B. B., Kirchner, W. H. & Lindauer, M. (1989). Honeybees can be recruited by a mechanical model of a dancing bee. *Naturwissenshaften* **76**, 277-280.

Miranda, N., Jackson, L. S., Bentley, D. M., Gash, G. H. & Nallan, G. B. (1992). Children discover addition more easily and faster than deletion. *Psychological Record* **42**, 117-129.

Molet, M., Chittka, L. & Raine, N. E. (2009). How floral odors are learned inside the bumblebee (*Bombus terrestris*) nest. *Naturwissenschaften* **96**, 213-219.

Newman, J., Wolff, W. T. & Hearst, E. (1980). The feature-positive effect in adult human subjects. *Journal of Experimental Psychology Human Learning* **6**, 630-650.

Ohtani, T. (2008). Recruitment efficiency of waggle dances performed by a worker honeybee and the influence on her dance by recruited foragers. *Humans and Nature* **19**, 71-88.

Pace, G. M., McCoy, D. F. & Nallan, G. B. (1980). Feature-positive and feature-negative learning in the Rhesus monkey and pigeon. *American Journal of Psychology* **93**, 409-427.

Percival, M. (1965). *Floral Biology*. Pergamon Press, Oxford.

Rathore, R. R. S. & Wells, H. (1995). Training Asian rock bees (*Apis dorsata*) to forage at specific locations. *Indian Bee Journal* **57**, 8-9.

Reinhard, J., Sinclair, M., Srinivasan, M.V. & Claudianos, C. (2010). Honeybees Learn Odour Mixtures via a Selection of Key Odorants. *PLoS ONE* **5**(2): e9110.

Reinhard, J. & Srinivasan, M. (2009). The role of scents in honey bee foraging and recruitment. *In*: M. Hrncir & S. Jarau, (*eds.*) *Food exploitation by social insects*. CRC Press. N.Y. pp. 165-182.

Reinhard, J., Srinivasan, M. V. & Zhang, S. (2004). Scent-triggered navigation in honeybees. *Nature* **427**, 411.

Ribbands, C. R. (1953). *The Behaviour and Social Life of Honeybees*. London: Bee Research Association.

Riley, J. R., Greggers, U., Smith, A. D., Reynolds, D. R. & Menzel, R. (2005). The flight paths of honeybees recruited by the waggle dance. *Nature* **435**, 295-207.

Rosin, R. (1980). The honey-bee "language" hypothesis and the foundations of biology and behavior. *Journal of Theoretical Biology* **87**, 457-481.

Rosin, R. (1991). Much abuzz about nothing! (The honeybee "dance language."). *American Bee Journal* **131**, 525-526. beesource.com/pov/wenner/abjaug1999.htm

Rosin, R. (1999). Do honey bees still have a "dance language"? *American Bee Journal* **139**, 577-578.

Sachse, S. & Galizia, C.G. (2003). The coding of odour-intensity in the honeybee antennal lobe: Local computation optimizes odour representation. *European Journal of Neuroscience* **18**, 2119-2132.

Sachse, S., Rappert, A. & Galizia, C. G. (1999). The spatial representation of chemical structures in the antennal lobe of honeybees: Steps towards the olfactory code. *European Journal of Neuroscience* **11**, 3970-3982.

Sanderson, C. E., Orozco, B. S., Hill, P. S. M. & Wells, H. (2006). Honey bee (*Apis mellifera ligustica*) response to differences in handling time, rewards and colors. *Ethology* **112,** 937-946.

Schreck, C. E. (1977). Techniques for the evaluation of insect repellents: a critical review. *Annual Review of Entomology* **22**, 101-119

Schricker, B., & Stephen, W. P. (1970). The effect of sublethal doses of parathion on honeybee behaviour. I. Oral administration and the communication dance. *Journal of Apicultural Research* **9**, 141-153.

Schul, J. (1997). Neuronal basis of phonotactic behaviour in *Tettigonia viridissima*: Processing of behaviourally relevant signals by auditory afferents and thoracic interneurons. *Journal of Comparative Physiology* [A] **180**, 573-583.

Sih, A., Bell, A. M., Johnson, J. C. & Ziembq, R. E. (2004). Behavioral syndromes: an integrated overview. *Quarterly Review of Biology* **79**, 241-277.

Smith, B. H. & Abramson, C. I. (2003). Case studies in insect behavior. *In*: J. H. Byrne, H. Ruediger & R. F. Thompson (*eds*). *Encyclopedia of Learning and Memory*, 2nd ed. Macmillian, N.Y., pp. 258-263.

Stone, J. C., Abramson, C. I., & Price, J. M. (1997). Task dependent effects of dicofol

(Kelthane) on learning in the honey bee (*Apis mellifera*). *Bulletin of Environmental Contamination and Toxicology* **58**, 177-183.

Suddendorf, T. & Corballis, M. C. (2007). The evolution of foresight: What is mental time travel, and is it unique to humans? *Behavioral and Brain Sciences* **30**, 299-351.

Syed Z. & Leal W. S. (2008). Mosquitoes smell and avoid the insect repellent DEET. *Procedings of the National Academy of Sciences*, USA **105**, 13598-13603.

Taylor, K. S., Waller, G. D. & Crowder, L. A. (1987). Impairment of a classical conditioning response of the honey bee (*Apis mellifera* L.) by sublethal doses of synthetic pyrethroid insecticides. *Apidologie* **18,** 243-252.

Triblehorn, J. D. & Schul, J. (2009). Sensory encoding differences contribute to species-specific call recognition mechanisms. *Journal of Neurophysiology* **102**, 1348-1357.

Vadas, R. L., Jr. (1994). The anatomy of an ecological controversy: Honey-bee searching behavior. *Oikos* **69**, 158-166.

Vandame, R., Meled, M., Colin, M. & Belzunces, L. P. (1995). Alteration of the homing-flight in the honey bee *Apis mellifera* L. exposed to sublethal doses of deltamethrin. *Environmental Toxicology and Chemistry* **14**, 855-860.

Veldink, C. (1989). The honey-bee language controversy. *Interdisciplinary Science Reviews* **14**, 166-175.

Weick, J. & Thorn, R. S. (2002). Effects of acute sublethal exposure to coumaphos or diazinon on acquisition and discrimination of odor stimuli in the honey bee (Hymenoptera: Apidae). *Journal of Economic Entomology* **95**, *227-236.*

Wells, H. & Rathore, R. R. S. (1994). Discriminant conditioning of foragers in Asian honey bees, *Apis cerana* and *Apis Dorsata. Ecological Entomology* **20**, 374-379.

Wells, P. H. & Wells, H. (1985). Ethological isolation of plants 2. Odour selection by honeybees. *Journal of Apicultural Research* **24**, 86-92.

Wells, P. H. & Wenner, A. M. (1971). The influence of food scent on behavior of foraging honey bees. *Physiological Zoology* **44**, 191-209.

Wells, P. H. & Wenner, A. M. (1973). Do bees have a language? *Nature* **241**, 171-174.

Wells, P. H., Wells, H., Vu, C., Vadehra, N., Lee, C., Han, R., Han, K. & Chang, L. (1993). Does honey bee nasanov pheromone attract foragers? *Bulletin of the Southern California Academy of Sciences* **92**, 70-77.

Wenner A. M. [with K. v. Frisch] (1993). The language of bees. *Bee World* **74**, 90-98.

Wenner, A. M. (1963). The flight speed of honeybees: A quantitative approach. *Journal of Apicultural Research* **2**, 23-32.

Wenner, A. M. (1967). Honeybees: Do they use the distance information contained in their dance maneuver? *Science* **155**, 847-849; beesource.com/pov/wenner/scifeb1967b.htm

Wenner, A. M. (1997). The role of controversy in animal behavior. *In*: Greenberg, C & Tobach, E. (*eds*), *Comparative Psychology of Invertebrates: The Field and Laboratory Study of Insect Behavior*. Garland, N.Y. pp. 3-37.

Wenner, A. M. (2007). The honey bee odor-search hypothesis. *Bee Culture* **135**, 25-26.

Wenner, A. M. & Wells, P. H. (1990). *Anatomy of a Controversy: The Question of a "Language" Among Bees*. Columbia University Press, N.Y.

Wenner, A. M. (2002). The elusive honey bee dance "language" hypothesis. *Journal of Insect Behavior* **15**, 859-878.

Wenner, A. M., Meade, D. E. & Friesen, L .J. (1991). Recruitment, search behavior, and flight ranges of honey bees. *American Zoologist* 31, 768-782.

Wenner, A. M., Wells, P. H. & Johnson, D. L. (1969). Honeybee recruitment to food sources: Olfaction or language? *Science* **164**, 84-86.

Werner, S. J., Cummings, J. L., Tupper, S. K., Hurley, J. C., Stahl, R. S. & Primus, T. M. (2007). Caffeine formulation for avian repellency. *Journal of Wildlife Management* **71**, 1676-1681.

Whitehead, A.T. & Larsen, J. (1976). Ultrastructure of the contact chemoreceptors of *Apis mellifera* l. (Hymenoptera, Apidae). *International Journal of Insect Morphology and Embryology* **5**, 301-315.

Willmer, P. G. & Stone, G. N. (2004). Behavioral, ecological, and physiological determinants of the activity patterns of bees. *Advances in the Study of Behavior* **34**, 347-466.

Xu. F. Q., Liu , N., Kida, L., Rothman, D. L., Hyder, F., *et al.* (2003). Odor maps of aldehydes and esters revealed by functional MRI in the glomerular layer of the mouse olfactory bulb. *Proceedings of the National Academy of Sciences* USA **100**, 11029-11034.

Xue, R. D., Barnard, D. R. & Ali, A. (2001). Laboratory and field evaluation of insect repellents as oviposition deterrents against the mosquito *Aedes albopictus*. *Medical and Veterinary Entomology* **15**, 126-131.

Yang, E. C., Lin, H. C. & Hung, Y. S. (2004). Patterns of chromatic information processing in the lobula of the honeybee, *Apis mellifera* L. *Journal of Insect Physiology* **50**, 913-925.

In: Bees
Editor: Richard M. Florio

ISBN: 978-1-61324-825-6

Chapter 4

THREATS TO THE STINGLESS BEES IN THE BRAZILIAN AMAZON: HOW TO DEAL WITH SCARCE BIOLOGICAL DATA AND AN INCREASING RATE OF DESTRUCTION

***Gislene Almeida Carvalho-Zilse*[1,a]**
***and Carlos Gustavo Nunes-Silva*[2,b]**
[a]Grupo de Pesquisas em Abelhas, Instituto Nacional de Pesquisas da Amazônia, Manaus, AM, Brazil
[b]Departamento de Biologia, Universidade Federal do Amazonas, Manaus, AM, Brazil

Abstract

Deforestation and nest destruction by "professional" honey gatherers has drastically reduced the population of stingless bees in the Amazon, especially species of the *Melipona* genera, which are very good honey producers among the native stingless bees group. Here, we present some of the peculiar aspects of *Melipona* nesting biology, its ecology and its puzzling sex and caste determining system and show how these aspects are correlated with the rapidly growing rates of nest extinction by human activity. Conservation efforts are needed to promote more knowledge of the *Melipona* species and its biology as well as to enhance the beekeeping of stingless bees (Meliponiculture) in many communities and villages throughout the Brazilian Amazon in time to slow down nest destruction. We discuss issues and results from cryptic species as well as the logistical challenges of doing research in the Brazilian Amazon. Finally, we present suggestions for the development of appropriate and advanced technical training for beekeepers that can be used to strengthen and expand a rational way of exploiting the honey from stingless bees in the region.

[1] E-mail address: gislene@inpa.gov.br.
[2] E-mail address: cgmanaus@gmail.com.

Keywords: Stingless bee, honey, beekeeping, sex-determination, sustainability, environmental politics.

1. Introduction

Like many remaining areas of rainforest in the world, the Amazon biome and its fauna are in danger, with many species in danger of extinction or regarded as threatened. However, very little information is available about native bees, especially stingless bees, of this region, despite the fact that these species are essential players in the maintenance of genetic diversity of the countless plant species that create the vast mosaic of plants that exist there.

In this chapter, ecological and biological data on stingless bees will be presented along with genetic information that emphasizes sex determination and caste, and the relationship of this information to the scenario of nest extinction due to honey extrativism. As well as beekeeping of native bees (Meliponiculture) in the Amazon will be discussed as a rational and sustainable way to conserve this important fauna. Three lines of action that in our opinion are necessary to quell the threats to the disappearance of these species in the Brazilian Amazon and that may also serve as a suggestion to academics, governments and societies that want and understand the importance of doing so, will be proposed. First, for this study, there is an assumption that little is known about these species and their real significance in the context of various ecosystems of this vast forest area, so there must be emphasis in studies on taxonomy, richness and abundance of species and ecology of stingless bees in the Amazon, making use of the advanced technology available, such as tools for genomic and related sciences. Second, continuing this reasoning, the need to knowing the population genetics and the reproduction mechanism of stingless bees is addressed. To consolidate training and studies in this field and thereby establish partnerships, networking projects are suggested. Finally, the incentive and support of Meliponiculture with sustainable management and production are recommended as a commercial and species conservation alternative, as long as there is use of various technologies to add value and characterize the products, as accompanied by the necessary human training to multiply this knowledge-attitude shift that is proposed.

2. The Bees and their Relationship with the Environment

The bees (Apidae: Hymenoptera: Insecta) are the main pollinators not only of autochthonous flora but also of various agricultural crops that occupy key positions in different ecosystems, and these ecosystems may eventually collapse without the presence of these insects (Kerr *et al.*, 1996; Kerr *et al.*, 2001; Michener, 2007). Bees represent part of what is called bioindicator fauna; as such, they promote gene flow in forest, agro-forestry and even in urban systems (Neff and Simpson, 1993; Matheson *et al.*, 1996; Allen-Wardell *et al.*, 1998; Kearns *et al.*, 1998).

From an economic point of view, the *Apis mellifera* species is the most widely used in pollination services; it is estimated that this species generates an annual impact of 200 billion dollars. This raises global concern for the potential losses that would have if there is a decline of bee colonies both wild and managed (Slaa *et al.*, 2006).

Given this information, the stingless bees, which are excellent pollinators of several native plants, represent an alternative, in the medium to long-term perspective, for commercial pollination (Slaa *et al.*, 2006). There is no doubt that some species grown in the Amazon, such as *Euterpe oleracea* (açaí), *Paullinia cupana* (guaraná), *Spondia* spp. (taperebá), *Bixa orellana* (urucum) and *Tapirira guianensis* (tapiririca), plants from the Myrtaceae family (*Psidium guajava* L., *Psidium littorali*, *Eugenia uniflora)*, among others, benefit from visits of the stingless bees to their flowers and that the stingless bees contribute to an increase in production of fruit (Rosa *et al.*, 1998; Cortopassi-Laurino, 2003; Venturieri *et al.*, 2010; Nunes-Silva *et al.*, 2010) in Amazonian home gardens. Stingless bees also function in the dispersal of seeds of important timber species for the Amazon region, as in the case of *Zygia racemosa* (Duke) Barneby and Grimes (angelim rajado) (Bacelar-Lima *et al.*, 2006; Nunez *et al.*, 2008).

Stingless bees are so named because they lack a functional sting and each species uses a different strategy (other than stinging) to defend its nests (Kerr and Lello, 1962; Nogueira Neto, 1997). They comprise the tribe Meliponini Lepeletier, 1836, which shows tropical and subtropical distribution worldwide (Michener, 2007). Moure *et al.* (2007) consider 54 genera, of which 33 are exclusively neotropical (one extinct), and a total of 397 valid species, with emphasis on the genus *Melipona* (Figure 1).

The genus *Melipona* Illiger, 1806, comprising a total of 69 species (Kerr, 1948; Camargo and Pedro, 2008) shows its greatest diversification in the Amazon basin (Silveira *et al.*, 2002). To better picture that is worth to inform this is a vast region that influences and comprises a large part of the Brazilian territory (the fourth largest country in continuous territorial extension) and being an area with a complexity of ecosystems which build an extensive mosaic.

Figure 1. Stingless bee *Melipona seminigra* visiting flower *Cosmos* sp.

Overall, knowledge of the fauna of Amazonian bees suffers from large gaps due to the availability of only occasional and non-systematic collections, and there is need for taxonomic revision of many groups, as in the case of the bees of the genus *Melipona*. Nevertheless, surveys in the Amazon have revealed a rich diversity of species and some areas of endemism (Roubik, 1989; Oliveira *et al.*, 1995; Camargo and Pedro, 2003; Moure *et al.*, 2007). Investigations into the phylogeny and biogeography of some genera of Meliponinae (Camargo and Moure, 1994, 1996; Michener, 2007; Danforth *et al.*, 2006) indicate the potential for the use of this group in the characterization of local fauna, as well as in broader contexts that involve aspects of the recent evolutionary history of biotas.

Although stingless bees are capable of establishing their nests in the most varied substrates, such as cracks in houses, floors, and the nests of other insects, pre-existing cavities in tree trunks and branches have been identified as the preferred nesting sites used by these organisms (Michener, 1974; Eltz *et al.*, 2002; 2003). Surveys carried out in different regions of Brazil suggest the existence of a certain loyalty by stingless bees in terms of the tree species used for nesting. In these preliminary studies, it was not determined for certain whether the selection of nesting place was due to some characteristic of the tree species or whether the selection occurred because there were no other available places to build the nest (Rego and Brito, 1996; Castro and Silva, 2000; Martins *et al.*, 2004; Antonini and Martins, 2003; Serra *et al.*, 2009). Other studies have considered stingless bees to be opportunists, colonizing any trees that have suitable cavities (Roubik, 1989; Eltz *et al.*, 2003). The diameter of the trees, however, seems to be an important factor in the probability of finding stingless bee nests in a given area, in that there is a positive correlation between the diameter of the trees and the densities of stingless bee nests (Eltz *et al.*, 2002; Samejima *et al.*, 2004). Notwithstanding the details on nesting in hollow trees, it is striking that, in forests with little or no management, the majority of nests and species are found in this type of environment (hollow trees). This fact raises a particular concern: if stingless bees nest in such places, any removal of trees for agricultural or urban expansion or for logging purposes is extremely detrimental to the diversity of populations, and consequently, to the plants that depend on their pollination, as well as the animals that benefit from this flora.

Besides the foregoing, other factors may accentuate this issue, making the destruction of nests even more harmful to the population of stingless bees than is currently believed. These factors are as follows:

1) The diversity of these bees in the Amazon is very large and is distributed without an apparent pattern. For example, Oliveira *et al.* (1995) found 15 nests belonging to nine species of stingless bees in an area of 100 ha located within a solid ground forest (called "terra firme" paths) in the Central Amazon (this corresponds to 6.5 nests/ha of different species). On the other hand, Camargo (1990) found more than 50 nests of stingless bees belonging to two genera (*Frieseomellita* and *Partamona*) in a single hectare of forest in floodplains in the margins of black water rivers and lakes in the Amazon. Despite some areas of endemism, the lack of knowledge of the geographical distribution of species and their patterns complicates the proposal of an effective strategy for the conservation of populations and species and emphasizes the need for further studies. Brown and Albrecht (2001), studying seven species of

Melipona in the state of Rondonia (Southeast of the Amazon basin) revealed that the richness and abundance of Meliponinae are directly related to the area of plant cover and inversely proportional to the size of the deforested area.

2) The intense deforestation of the Amazon that has occurred as a result of logging, creation of pastures and, more recently, the expansion of soybean cultivation in the northern region of Brazil (Fearnside, 2008) has led to a decrease in the number of trees that are sufficiently large and that offer the best conditions for bee nesting. Such "clearings" may represent an extension that is often greater than the flying range of the bees (Kerr *et al.*, 1996), thereby reducing the availability of bee pasture and potential crosses (Carvalho and Kerr, 2004). Considering that forest management for timber production has been indicated as the best alternative in the process of occupation of the Amazon (De Graaf *et al.*, 2003; Schulze *et al.*, 2008), it is necessary to conduct quantitative studies to assess the implications of selective exploitation of plant-origin products on fauna. Overall, it is known that changes in the structure of local vegetation may lead to changes in the diversity of pollinating and/or dispersing agents, which, in turn, can impair the reproduction of some tree species (Samejima *et al.*, 2004).
3) From Mexico to South America, long before the arrival of European colonizers, stingless bees have been managed by Amerindians. Like the Mayas of Central America, the indigenous peoples of the Amazon kept these bees or at least managed their nests and collected their honey. Indigenous peoples of the Brazilian Amazon still consider the honey and pollen of stingless bees to have great medicinal value and to be a delicacy (Kerr *et al.*, 1994). For example, the Enawene Nawê tribe (upper Xingu River, in the Amazon) use honey mixed with water every day because it is their only source of sugar (Kerr, personal communication). According to Kerr *et al.* (1996, 2001), some extractive practices collective (destruction of up to 1200 colonies/month) or individual (200 nests/20 years) show destructive effects on populations and species of stingless bees. With the arrival of non-indigenous settlers to these regions, honey extraction ceased to be an activity with low environmental impact and began to have major effects on the conservation of populations of native bees as well as on the flora and fauna that they influence. Currently, the gathering of honey from stingless bees is commonly accompanied by the destruction of the bee colonies; if it is true that these insects are the most important pollinators of the native flora, this causes a serious environmental impact (Kerr *et al.*, 1999; 2001). Today, collecting honey from these bees is a major profitable business; the marketing of native bee honey has notoriously emerged in recent years, and now there seems to be a race in search of this "liquid gold". The "professional" honey gatherers are responsible for the elimination of hundreds of nests annually; their harvests often involve the destruction of centennial trees in order to obtain small amounts of honey. This activity is prohibited by Brazilian environmental legislation; however, with the countless bee communities in this vast area of thousands of square miles, effective monitoring is an enormous task that is incompatible with the infrastructure of the government.

Figure 2. Overview of a meliponarie in Amazon, Brazil.

The extractive harvesting of honey occurs mainly because of a lack of information about the biology of native stingless bees. For example, many harvesters believe that once the honey is removed, the bees choose another site for nesting. Due to the inability of the queen to fly, this is a false belief. The disturbed colony is sentenced to death at the hands of many predators in the forest and disappears without a trace, leading people to think that the bees fled, seeking refuge in another tree. With dissemination of accurate information on this subject, it is hoped that there will be a reduction in nest destruction and an increase in the use of more ecological techniques of honey collection, as well as an increase in meliponaries (apiaries for stingless bees – Figure 2) in which proper management for the production of honey, pollen and propolis, as well as the multiplication of colonies, is employed. In general, the better the understanding of the biology and ecology of stingless bees, the better are the chances of using these bees and their products in a planned and sustainable manner.

3. Reproduction Biology and Genetics in Bees

Colonies of stingless bees (Figure 3) consist of individuals that perform specific functions in society. The queen is the fertile female responsible for laying eggs; the workers perform all the functions related to the provision and maintenance of the nest, in addition to helping the queen in the generation of new individuals, and the male drones are responsible for fertilizing the queen and in some cases helping in the maintenance of the nest.

As previously mentioned, a peculiar characteristic of *Melipona* is that the physogastric queen is not able to fly, unlike bees of *Apis* species. This makes swarming behavior impossible. The laying of fertilized eggs is carried out exclusively by the queen, who, in most

Figure 3. Internal structure of a colony of stingless bee *Melipona seminigra*.

cases, breeds with only one male (Carvalho, 2001) or up to three males, as shown by Paxton *et al.* (1999) for *Melipona beecheii*. In some species, worker bees are able to collaborate in the production of haploid males (Alves *et al.*, 2010). In the vast majority of species, there is only one laying queen; however, it is common to observe two queens laying eggs during the natural process of queen replacement (Carvalho and Kerr, 2004). It has long been known that *M. bicolor* develops polygynous colonies with simultaneous egg-laying by up to six queens (Velthius *et al.*, 2006). Recently, Alves *et al.* (2010) and Carvalho *et al.* (2011) showed the existence, although uncommon, of polygynous colonies also for *M. quadrifasciata* and *M. scutellaris*.

The workers are totally responsible for the food supply for the colony. Nectar is collected by the foragers, dehydrated (usually inside the colony, although an exception occurs in colonies of *M. rufiventris*, which dehydrate the nectar in the external part of the hive), and then deposited in wax pots. One of the main honey-producing species in the Amazon, *Melipona seminigra*, produces between 1.87 and 5.00 kg of honey per beehive/year (Carvalho-Zilse, 2006). Carvalho *et al.* (2003) estimate production to be between 1.0 and 10.00 liters/beehive/year in some species of *Melipona*; this productivity is variable according to the species of bee and the flowers available, as well as depending on the type of management employed by the beekeeper. The bees also store pollen, water and plant resins in the colony. Honey and pollen provide the food base for these bees; as with nectar, the workers process the pollen, storing it in pots made of wax to which they add glandular secretions that promote the maturation of the honey in a natural fermentation process. Unlike *Apis*, these bees structure their colonies into sections, clearly separating the brood in combs or disks and the food in pots.

The colonies necessarily depend on the maintenance of temperature of the colony; hence, the bees seal the cracks with clay, wax, geopropolis (resin + clay), plant remains (Nogueira Neto, 1997) and even seeds (Bacelar-Lima *et al.*, 2006).

As the colony develops and the availability of food increases, there is a rapid growth in the population of bees, leading to the process of migration by part of the colony to nest in a not-so-distant site from the colony of origin, because, at this moment, there is direct dependence of the newly forming colony on the colony of origin. The workers prepare the new nesting site with material available in the colony of origin; subsequently, a new queen migrates with part of the worker population to this new site (Campos, 1987). Due to the initial dependence on the colony of origin, which limits the distance of migration (about 200 meters) and the limited flight capacity of the queen and male drones for mating (about 1,000 meters), the area of reproduction of the organisms is limited to about a 1.2 km radius (Carvalho and Kerr, 2004).

Along with these biological-reproductive peculiarities, the intriguing genetic system of sex and caste determination that exists in the stingless bees must be emphasized. As in most members of the Hymenoptera order, gender in bees is determined by haplodiploidy, more specifically by arrhenotokous parthenogenesis, a system in which fertilized eggs develop into females and non-fertilized eggs into males (Carvalho, 2001). The discovery of the existence of diploid males in some species showed that sex is also determined by a multiallelic locus, a process known as complementary sex determination or CSD (Whiting, 1943). Individuals heterozygous for this locus are female, whereas hemizygotes or homozygotes are male. However, the diploid males are reproductively unviable and are killed by the workers during development in the brood cells or following hatching, depending on the species (Kerr, 1997; Carvalho, 2001).

The identification of molecular markers flanking the sex-determination locus (Hunt and Page, 1994, 1995; Beye and Moritz, 1994; Beye *et al.*, 1999) and the use of chromosomal walking and fine-scale mapping techniques (Hasselmann *et al.*, 2001) enabled the isolation and characterization of the sex determination gene of *Apis*, which is called *csd* (Beye *et al.*, 2003). The *csd* gene regulates sex determination by allele combination. There are no sex-specific alleles or transcripts. It is estimated that there are 15 allelic forms of this gene in *A. mellifera* (Gempe *et al.*, 2009), supporting the data of Adams *et al.* (1977) based on segregating progeny. For *Melipona*, up to 36 alleles have been reported (Carvalho, 2001), and the estimates are variable in species where the CSD system has been confirmed (Cook, 1993; Kerr, 1997; van Wilgenburg *et al.*, 2006). In *Melipona*, this mechanism directly influences the stability of populations.

Caste determination, an important factor in the maintenance of eusociality in bees, is determined by biochemical pathways that are regulated by gene expression during development. These pathways are capable of generating two alternative phenotypes, the worker (sterile female) and queen (full female) (Kerr, 1950; Evans e Wheller, 2000). Among bees with high levels of eusociality, there are different mechanisms that determine caste (Kerr, 1946, 1948; Michener, 1979); these mechanisms have been classified in a review by Hartfelder *et al.* (2006). However, despite the many studies on the subject, one of the greatest challenges of caste determination is the explanation of its intricate regulatory cascade, especially in stingless bees of the genus *Melipona*.

Caste determination in bees generally occurs via differential feeding of larvae. In *A. mellifera*, the eggs produce queens or workers depending on the quantity and quality of the larval diet. Brood cells of different sizes guide the queen to lay fertilized or non-fertilized eggs and indicate to the workers the type of care that must be given to its occupant. The larger cells, known as queen cupules, receive a larger amount and better quality of food – royal jelly (food with a high nutrient content) and will give rise to the queen (Gullan and Cranston, 2007). A similar mechanism to that of *Apis* was observed in *Trigona* and *Scaptotrigona* (genera belonging to the Trigonini tribe, whose members are known as stingless bees). In these genera, the larger cells in the brood comb (queen cupules) receive a larger amount of food that is qualitatively the same and produce queens (Kerr *et al.*, 1996). Another mechanism was observed in *Frieseomellita* (Trigonini tribe), which builds its brood combs in a cluster shape. In this species, an older larva from a lower cell pierces the cell above and sucks the food therein contained; thus receiving double of amount of food, it develops into a queen (Terada, 1974 cited by Kerr *et al.*, 1996).

In contrast, caste determination in *Melipona*, which occurs in the prepupal phase, takes place via a genotypic mechanism first proposed by Kerr (1950). Called the nutritional genetic system of caste determination, this mechanism suggests that caste determination could result from the combination of two genes (*xa* and *xb*), each possessing two alleles, coupled with proper nutrition of larvae. When a larva that is a double heterozygote for these genes is fed sufficiently, it gives rise to a queen, while when there is homozygosis for one or both genes; the larva develops into a worker.

Based on the above, *Melipona* bees have two key moments of genetic differentiation; these involve sex (male or female) and caste (queen or worker) determination. In the first embryonic hours, sex is determined by the haplodiploid system directed by the multiallelic *csd* locus, while caste is defined later, during the last larval stage, as has been previously described.

The *csd* locus in Meliponini and its system for caste determination have been studied by Kerr and collaborators (Kerr, 1946, 1987, 1997; Bonetti and Kerr, 1987) for decades. The two systems are related in an intricate temporal regulatory cascade that is sensitive to the bees' nutritional and hormonal condition. The possibility of molecular screening of the gene(s) region(s) involved in this mechanism using more recently developed techniques is one of the next challenges for the academic community. Although the genetic mechanisms of sex determination in social Hymenoptera as well as the demographic conditions and population structure of these species remain poorly explored, research in these areas is vitally necessary given the fact that bees, wasps and ants are included in the group of insects that contribute most beneficially to the Earth, providing crucial services to ecosystems. The study of mechanisms involved in sex and caste determination in organisms that reproduce sexually has also been highly significant for the understanding of population and evolution questions (Salin *et al.*, 2004), thus contributing to conservation programs.

One of the great challenges in the conservation of species *in situ* is the maintenance of biodiversity and genetic viability of populations in one location. Based on the model of a multiallelic locus, Kerr and Vencovsk (1982) estimated that, with at least 44 colonies of *Melipona* in a mating area, there is sufficient diversity of sexual alleles (minimum of 6 alleles) for the maintenance of the population.

It is known that fragmentation of the forest contributes to conditions that can isolate bee populations and drive these populations to endogamy and, consequently, to the production of diploid males (DMP). The behavior of workers in exterminating these males and the queens that produced them (Carvalho, 2001) leads to a reduction in population viability in some species (Zayed and Parker, 2005). Carvalho (2001) showed the consequences of this in an isolated population of *Melipona scutellaris* and stressed the need for at least 44 colonies of one species in an area of reproduction as a valid strategy to avoid DMP.

As a consequence of the mechanism of sexual determination in stingless bees, DMP therefore has an important role in their population structure, and the destabilization of alleles may lead to a population decline. Thus, genetic-population estimates of natural populations as well as of populations in captivity may greatly add to the understanding of natural population dynamics and contribute to the development of conservation strategies for species that are excessively managed or in danger of extinction. For example, using microsatellite markers, Costa-Pinto and Carvalho-Zilse (2007) showed that in *M. compressipes,* beehives managed in meliponaries where there was occasional exchange of queens and brood disks between hives of different meliponaries showed greater rates of heterozygosity (Ho) than natural beehives. These results indicate that the occurrence of genotypic recombination and heterozygosity in captive beehives is accelerated by human action in comparison to natural colonies.

A similar situation was observed by Carvalho-Zilse *et al.* (2009) upon studying populations of *M. scutellaris* in meliponaries. These authors observed that variation within populations was greater than variation between populations, indicating a lack of genetic structure in these populations and leading to the speculation that the recurrent multiplication of selected colonies may homogenize the populations of meliponaries. This, in turn, reinforces the suggestion of a need for exchanging genetic material between meliponaries as an alternative to promote genetic variability in captive populations, as proposed by Carvalho (2001).

In any case, there is a vital need for more studies that monitor managed populations and natural populations in order to guide programs for the management and conservation of these native pollinators. Ensuring conditions for the reproductive success of *Melipona* bees when faced with forest fragmentation and reduction of populations is a constant challenge for scientists, governments and societies.

4. Meliponiculture: Reality and Perspectives

"Preservation of native Brazilian bees: a question of historical and ecological conscience" is the title of an interesting article by Kerr *et al.* (1999a). This article presents a detailed history of the use of native Brazilian bees since the discovery of Brazil. Kerr *et al.* (2001) also stress that bees have three core values that Norton (1997) attributed to the species: commodity, convenience and moral value. These observations confirm the close relationship of bees with human life in its broadest historical, cultural, ecological and economic aspects.

As a consequence of this close relationship, it can be said that the raising of stingless bees or Meliponiculture (Nogueira-Neto, 1970) is potentially an alternative for the conservation of stingless bees. Meliponiculture is a low- environmental-impact activity that requires little space, does not need large investments and can be performed in agricultural areas as well as in areas that are not suitable for agriculture with the use of local flora (Venturieri, 2009).

Cortopassi-Laurino *et al.* (2006) consider that the exploration of the pollinating potential of stingless bees widens the economic scope of this activity. There are diverse beekeeping initiatives in the world, but beekeeping on a commercial scale is still a challenge (Cortopassi-Laurino *et al.*, 2006).

In Brazil, the first record in favor of commercial exploration of stingless bees dates from 1910, showing the proposal of "a type of beehive for commercial enjoyment" by Marianno-Filho (1910b). This initiative was intended to maintain *M. scutellaris* bees; the pressure on this main species from the Northeast of Brazil was very intense, leading to the death of dozens of hives annually.

In most states that make up the Brazilian Amazon, Meliponiculture initiatives have been identified (Carvalho-Zilse and Nunes-Silva, 2010). These initiatives have been strengthened such that, since 2004, the Brazilian Confederation of Beekeeping has incorporated the Brazilian Congress on Meliponiculture within the Brazilian Congress on Apiculture.

Historical records of Meliponiculture in the state of Amazonas are rare, although it is almost common to have two or more beehives located close to houses in indigenous or riverine communities, given that folk remedies are traditionally swallowed with a glass of stingless bee honey diluted in water. Recently detected in field studies has been the extractive practice of several local communities in search of honey and wax; these communities often discard the pollen because they consider it bee feces. With the goals of environmental education, technical training and support for the growing demand that exists among the farmers and beekeepers of the state, the Association of Beekeepers of the Amazon state was created in 2000; in the following year, the first Meeting of Amazonian Beekeepers was held in the town of Presidente Figueiredo.

From then on, some Meliponiculture programs in the Amazon have been initiated, but with activities spread throughout the region. In the state of Amazonas, some projects developed by organizations connected to local and state governments (for example, the State Office for the Environment and Sustainable Development of Amazonas state [Secretaria de Estado do Meio Ambiente e Desenvolvimento Sustentável do Amazonas – SDS], the Forestry and Sustainable Business Agency of Amazonas State [Agência de Florestas e Negócios Sustentáveis do Amazonas – AFLORAM], and the Institute of Sustainable Agricultural and Forestry Development of the state of Amazonas [Instituto de desenvolvimento Agropecuário e Florestal Sustentável do Estado do Amazonas – IDAM]), by Stingless Beekeeper Associations (for example, the Association of Beekeepers of the state of Amazonas [ACAM – Associação de Criadores de Abelhas do Amazonas] and the Honey Cooperative of Boa Vista do Ramos [COOPMEL – Cooperativa de Mel de Boa Vista do Ramos]), and also by companies (for example, the Company Specialized in the Amazon Meliponiculture [EMAZON – Empresa Especializada em Meliponicultura na Amazônia]), have been investing in Meliponiculture with the goal of transforming the honey into a commercially attractive product. Nevertheless, given that there are countless transportation and handling problems due to the large distances between the centers of the state, there is a need for priority incentives for the establishment of partnerships aimed at the flow of production.

One of the most successful examples of Meliponiculture is found in the indigenous communities of Saterê Mawé, in Maués, Amazonas state; these communities are working to promote the pollination of guarana flowers (*Paullinia cupana* (Sapindaceae)) in order to increase the production of such pollen and export it to Europe as an energy food supplement. They are also increasing the production of honey by native bees and selling it as "honey from Amazonian guarana flowers". It is worth noting that this honey is produced in small indigenous family farms. The development of the ProVarzea and the Abelha (Bee) Projects in these communities has generated market demand for this honey. Nevertheless, while Meliponiculture faces various problems, the commercialization of products faces an even greater problem, which is the urgency for the development of legislation to transform the honey of stingless bees into a product with established technical standards of composition, harvest and storage and possessing an ensured commercial competitiveness.

Any incentive and work in favor of Meliponiculture with sustainable management and production as a commercial alternative and for species conservation must meet the specific economic calling of each location in which it is carried out. Furthermore, such incentives must ensure the use of various technologies to increase the value and characterization of products and must provide the necessary human training for the multiplication of this entire knowledge-attitude movement.

Currently, as a result of collaborations organized by the Bee Research Group of National Institute for Amazonian Research [Grupo de Pesquisas em Abelhas do INPA], there is an estimate of more than 700 stingless beekeepers (with approximately 8,000 beehives under technical management) in 19 municipalities of Amazonas state (Figure 4).

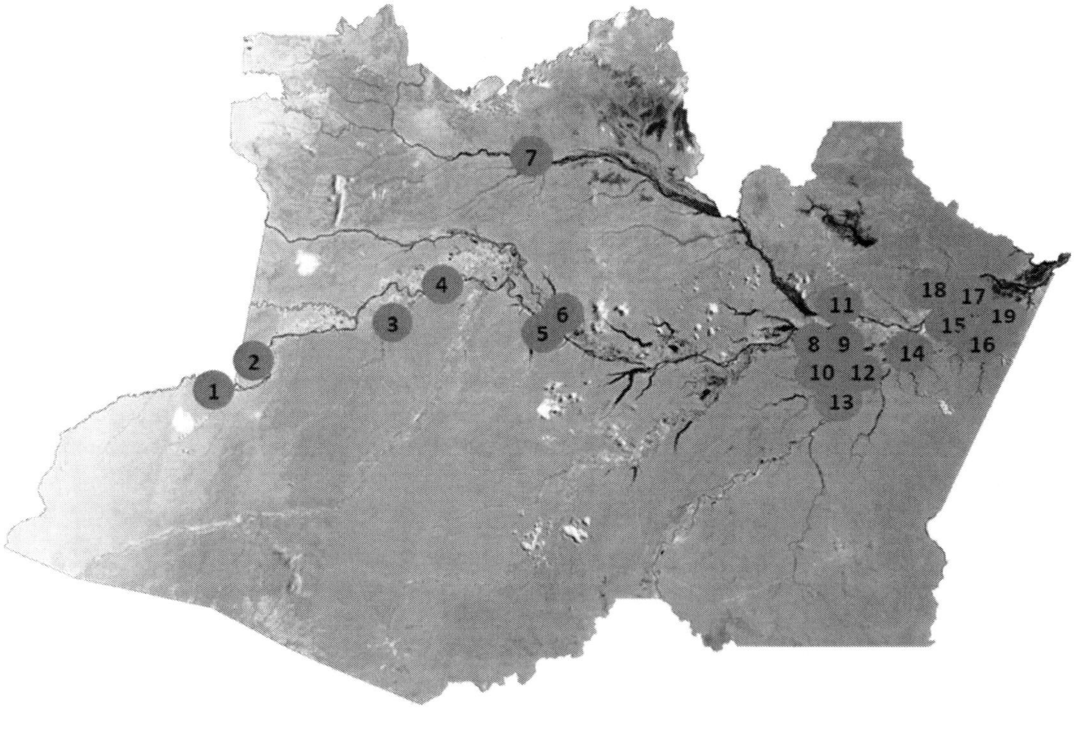

Figure 4. Initiatives of Meliponiculture in Amazon. 1 – Benjamin Constant; 2 – Tabatinga; 3 – Jutaí; 4 – Fonte Boa; 5 – Tefé; 6 – Alvarães; 7 – São Gabriel da Cachoeira; 8 – Careiro Castanho; 9 – Cacau Pirera; 10 – Iranduba; 11 – Manaus; 12 – Manacapuru; 13 – Novo Airão; 14 – Autazes; 15 – Barreirinha; 16 – Maués; 17 – Parintins; 18 – Urucará. 19 – Boa Vista do Ramos.

The capacitation process is essential to the success of Meliponiculture activity; this process should result in a virtually complete elimination of the exploitation of forest nests and in the establishment of an alternative income supplement for the family and of a socio-educational activity for the maintenance of biodiversity, in addition to action in favor of an increase in the number of pollinators and in fruit and food production.

However, there are still challenges for national Meliponiculture, especially for the Amazon region. The characterization and standardization of honey according to bee species, vegetation, factors related to soil and climate of the regions where it is produced, and, in particular, the use of appropriate hygiene and sanitation techniques are goals that need to be achieved for the improvement of the quality of honey produced and for product quality assurance for the consumer.

It must be remembered that current Brazilian legislation regulating the standardization of honey for commercial purposes only applies to the characteristics of honey from *Apis mellifera* and does not take into account honey from native bees. Such legislation is based on international standards; this makes it difficult, in many cases, to introduce the honey of native bees into the market. In addition, processes for the conservation of honey (dehumidification, pasteurization or refrigeration) are difficult to access by the members of the rural communities where the bees are kept and produce honey. Subsidies are needed to structure associations and cooperatives that can set up depots or communal houses in which to process honey or to provide access to centers for carrying out such procedures in strategic regions. It is also worth noting that it is vital to create a mechanism for fast flow of production, given that the honey of stingless bees does not last as long as that of *Apis* bees because it has a high humidity content, which significantly favors its fermentation and deterioration.

It is envisioned that Meliponiculture can rescue the "family business" through income generation that involves the entire family, keeping the man at home and operating in harmony with the ecosystem while promoting profitable involvement in financial and institutional value scales without damage to nature and men.

5. Lines of Action: Perspectives

In general terms, any loss of biodiversity is a question of public concern. However, the loss of pollinating insects may become a particular problem because insects are directly involved in plant reproduction, and consequently, in the feeding of humans and animals. This perspective requires immediate action; the better the understanding of the biology and ecology of stingless bees, the better will be the chances of using these bees and their products in a planned and sustainable fashion. It is believed that the prioritization of three specific lines of action is essential for the effective mitigation of threats described here and for truly achieving conservation and sustainable use of the main pollinators of the Amazonian flora.

Line of Action 1: Research Focused on the Taxonomy, Ecology, Biology and Genetics of Amazonian Stingless Bees and Formation of Human Resources

Bees have been considered one of the key groups for the study of the biological diversity of the Amazon (Overal, 2001). The prioritization of taxonomic studies is of extreme significance in studies on any organism and its biology. The current state of knowledge on our biodiversity has been recorded in the wise words of priest Jesus Santiago Moure, one of the best Brazilian zoologists, "One of the most serious problems to study our fauna and flora is in the lack of proper bibliography" (Silveira *et al.*, 2002).

From the point of view of the collection of materials for identification, reports on the abundance and richness of species, biogeographical studies, and bioprospecting of new bioproducts all point to the Amazon as the target of countless collections of material on the region by foreign institutions. The resulting knowledge represents a piece of literature that is inaccessible to Amazonian society. Amazonians are nevertheless very much interested in participating in the sharing of this benefit of intangible value because knowing the forest and its biodiversity is a *sine qua non* condition for its conservation and sustainable and innovative use and can be expected to yield environmental, social and economical dividends.

Activities proposed in studies that inventory the biodiversity of stingless bees must seek the knowledge of species and its distribution. These data would provide a zoning of species to be used in management and conservation programs. They would also provide support to Brazilian governmental bodies, state organizations for the environment and other related organizations in developing legislation that is pertinent to this wildlife, which is so very different from the macrofauna (reptiles, mammals, birds, etc.) and which is devoid of standardization as to its conservation, domestication, management and exploitation. In this line of action, clear attention must be given not only to identification by species but also to the careful characterization and correlation of habitat, ecological niches, climate, trees for nesting, associated flora, among others.

In recent years, numerous tools for taxonomical analyses, georeferencing models, monitoring, and genetic studies have been made available to researchers, generating opportunities to explore the complex biological system of insects (Heckel, 2003). Furthermore, studies related to the flora visited by Amazonian bees can promote the better usage of resources of the region, aid in the formation and enrichment of agroforests in the areas around the meliponaries, and also provide food to the bees throughout the year. This knowledge can also be used in studies focused on greenhouse pollination, a practice that is employed in several countries with bees from the *Apis* genus and which contributes to the production of fruit and seeds in greater quantity and quality (Malagodi-Braga *et al*, 2000; Absy *et al.*, 2003; Imperatriz-Fonseca *et al.*, 2006). Thus, biological aspects are extremely important for any conservation or commercial exploitation of these insects.

Strategically, the formation of human resources must also be prioritized from the technical to the postgraduate levels; thus, an initial training program that involves the topics outlined in this article is recommended. Courses must emphasize the biology and ecology of bees, as well as training for the management and sustainable use of these insects. These efforts will achieve a high impact both in academia and in the productive sector.

Line of Action 2: Consolidation of Partnerships via Networking Projects

It is believed that the building of skills, which is currently fragmented among different Amazonian institutions with limited resources and personnel, is key to improving the knowledge production chain involving Amazonian biodiversity, and in particular, stingless bees. Despite these limitations, there is knowledge production in accordance with specialties that have been established in different research groups working with bees in the Amazon. These groups include researchers, students and technicians who have been generating knowledge on the biology, genetics, management and conservation of the region, creating a large knowledge mosaic that must be organized, updated and integrated. For this, the establishment of a network that excels at multidisciplinary study and has a participatory character is essential.

Any type of activity that involves several approaches must involve a transverse, interdisciplinary and inter-institutional approach in order to leverage the knowledge chain. The main purpose of such action is to increase the mobility and collaboration between regional centers and to stimulate leadership ideas and integration of Amazonian bee research groups, instead of continuing to promote the mere provision of raw materials for studies in other non-Amazonian institutions. Thus, the current challenge is to break the virtual cycle among the geographically distant groups and ensure functionality for the consolidation of a network of research that generates a legacy for the region. It is important to remember that the shortest distance between state capitals in the Brazilian Amazon is at least 800 kilometers and without land access.

Ultimately, the application of the results of studies carried out in programs for the conservation and management of these pollinators are recommended, such as the endeavors of the Brazilian Beekeepers Initiative in line with the International Initiative for the same topic. This knowledge will also potentially provide scientific bases to government programs for the use of the forest, such as the "Green Free Zone (Zona Franca Verde)" and "Forest Allowance (Bolsa Floresta)", currently active in the Amazon. These political actions must have technical support in order to provide concrete alternatives to keep the forest "standing" without demagoguery.

Within research lines about these insects, the challenge is to implement a network of studies that consider the guidelines for any program acting on biodiversity. These guidelines include: a) the knowledge of biodiversity and its interrelations; b) biodiversity conservation; c) sustainable use of biodiversity components; d) monitoring, assessment and mitigation of impacts on biodiversity; and e) development of human resources and institutional consolidation of the players of the project.

Line of Action 3: Commodity Chain of Meliponiculture

Tropical forest areas are increasingly being destroyed by factors related to social disorganization and to environmental imbalance; these factors have been generated both by human intervention and by natural phenomena. There are mechanisms that can attenuate, if not reverse, the rapid extermination of these insects and that can contribute to the maintenance of genetic diversity. Meliponiculture is an intelligent option for the sustainable management of these natural resources because it promotes the increase of bee populations and consequently promotes pollination, resulting in the perpetuation of plant species and providing food and shelter for animal species.

As previously mentioned, Meliponiculture is an activity that is compatible with degraded areas because it can be implemented in areas that are not suitable for agriculture. Furthermore, the existence of a range of products for commercialization, including honey, pollen, propolis and wax, have been gradually attracting various national and international investors, thus arousing biotechnological interest in these products (Vit *et al*., 2006). In addition, Meliponiculture is an activity that fits perfectly within the guidelines for conservation and sustainable use of biodiversity that guide the development of the Amazon Region. These guidelines are based on the rational use of forest resources and the balancing of environmental interests with social interests in order to improve the quality of life of the populations that live in the region (Cavalcante *et al*., 2009). The cascade of Amazon biodiversity conservation that would occur due to the primary environmental services provided by these insects are also emphasized.

To professionalize the activity of this commodity chain, it is necessary to involve all the interested parties (beekeepers, farmers, entrepreneurs), the supporters (technical support, legalization and inspection agencies) and the developers (development agencies, banks and governments). Efforts must be concentrated on technical training programs in the practice of Meliponiculture, utilizing the skills of keepers and farmers and the available natural resources. Meliponiculture training programs provide opportunities to teach simple techniques that maximize the production of honey and minimize environmental impacts.

The first beneficiaries and users of the results of studies related to Meliponiculture are expected to be the farmers and the local Amazon communities, who will benefit from the implementation of actions by development and environmental agencies (in all spheres). Furthermore, the accumulation of data, products and processes resulting from this initiative could change the status of the activity in the region, making it more attractive, profitable and sustainable. As a result, institutes, companies and farmers could benefit from the adequacy of their activity and obtain certification, thus adding value to an original non-timber product of the Amazon.

Conclusion

This book comes at an opportune time when the players involved have been called to work together in order to mitigate the threats to bees, to ensure suitable conditions for the conservation of native Amazon species, and to provide relevant environmental services. To this end, efforts must focus on the expansion of our knowledge of stingless bees in the Amazon, the estimation of anthropogenic effects on the genetic variability of stingless bee populations and species, the inclusion of a component of wildlife (bees) among the criteria for the assessment of sustainability of forest management plans for timber, provision to the appropriate agencies of subsidies that allow the elaboration of specific legislation for industrial and sanitary inspection of honey and other stingless bee products that are appropriate to the reality of the local communities, and on recommendations that can be used in the development of public conservation and management policies for stingless bees.

Without a shadow of a doubt, there exists the potential for an integrated development program that is based on the idea that the development and conservation of the environment must constitute an indivisible binomial, and as such, must be incorporated in all our public policies and social practices. Such a program would serve two of our greatest aspirations, the right to development and the right to an environmentally health life for this and subsequent generations.

REFERENCES

Absy, M. L., Ferreira, M. N., Kaminski, A. C., Silva, S. J. R. (2003). A importância da flora apícola na biodiversidade amazônica.. In: Jardim, M.A.G., Bastos, M. N. C., Santos, J. U. M. (eds.). *Desafios da Botânica no Novo Milênio: Inventário, Sistematização e Conservação da Diversidade Vegetal*. Belém, MPEG, UFRA; EMBRAPA, pp. 268-270.

Adams J.; Rothman E. D.; Kerr. W. E.; Paulino, Z. L. (1977). Estimation of the number of sex alleles and queen matings from diploid male frequencies in a population of *Apis mellifera*. *Genetics* 86(3):583-596.

Allen-Wardell, G.; Bernhardt, P.; Bitner, R.; Burquez, A.; Buchmann, S.; Cane, J.; Cox, P.; Dalton, V.; Feinsinger, P.; Ingarm, M.; Inouye, D.; Jones, C.; Kennedy, K.; Kevan, P.; Koopowitz, H.; Medellin, R.; Medellin-Morales, S.; Nabhan, G. (1998). The potencial consequences of pollinator decline on the conservation of biodiversity and stability of food crop yields. *Conservation Biology* 12(1):8-17.

Alves, D. A.; Menezes, C.; Imperatriz-Fonseca, V. L.; Wenseleers, T. (2010). First discovery of a rare polygyne colony in the stingless bee *Melipona quadrifasciata* (Apidae, Meliponini). *Apidologie* 1-3. DOI: 10.1051/apido/2010053.

Antonini, Y.; Martins, R. P. (2003). The value of a tree species (*Caryocar brasiliense*) for stingless bee *Melipona quadrifascita quadrifasciata*. *Journal of Insect Conservation* 43(3/4):191-223.

Bacelar-Lima, C. G.; Freire, D. C. B.; Coletto-Silva, A.; Costa, K. B.; Laray, J. P. B.; Vilas-Boas, H. C.; Carvalho-Zilse, G. A. (2006). Melitocoria de *Zygia racemosa* (Ducke) Barneby & Grimes por *Melipona seminigra merrillae* Cockerell, 1919 y *Melipona compressipes manaosensis* Schwarz, (Hymenoptera, Meliponina) en la Amazonia Central, Brasil. *Acta Amazonica* 36(3):343-348.

Beye, M.; Moritz, R. F. A. (1994). Sex linkage in the honeybee *Apis mellifera* detected by multilocus DNA fingerprinting. *Naturwissenschaften*, 81:460-462.

Beye, M.; Hunt, G. J.; Page, R. E.; Fondrk, M. K.; Grohmann, L.; Moritz, R. F. A. (1999). Unusually high recombination rate detected in the sex locus region of the honey bee (*Apis mellifera*). *Genetics,* 153:1701-1708.

Beye, M.; Hasselmann, M.; Fondrk, M. K.; Page Jr., R. E.; Omholt, S. W. (2003). The gene *csd* is the primary signal for sexual development in the honeybee and encodes an SR-type protein. *Cell*, 114:419-429.

Bonetti, A. M.; Kerr, W. E. (1987). Sex determination in bees. XXIII. Broad of a laying queen with Juvenile hormone III. *Revista Brasileira de Genética* 3:593-597.

Brown, J. C.; Albrecht, C. (2001). The effect of tropical deforestation on stingless bees of the genus *Melipona* (Insecta: Hymenoptera: Apidae: Meliponini) in central Rondonia, Brazil. *Journal of Biogeography*, 28:623-634.

Camargo, J. M. F. (1990). Stingless bees of the Amazon. *In:* Proceedings of the 11th International Congress of the IUSSI (eds G. K. Veeresh, B. Mallik and C. A. Viraktamath), pp. 736±738. IUSSI, Bangalore, India.

Camargo, J. M. F.; Moure, J. S. (1994). Meliponinae Neotropicais: Os gêneros *Paratrigona* Schwarz, 1938 e *Aparatrigona* Moure, 1951 (Hymenoptera, Apidae). *Arq. Zool. (São Paulo)* 32(2):33-109.

Camargo, J. M. F.; Moure, J. S. (1996). Meliponini Neotropicais: o gênero *Geotrigona* Moure, 1943 (Apinae, Apidae, Hymenoptera), com especial referência à filogenia e biogeografia. *Arq. Zool. (São Paulo)* 33(3): 95-161.

Camargo, J. M. F.; Pedro, S. R. M. (2003). Meliponini neotropicais: o gênero *Partamona* Schwarz, 1939 (Hymenoptera, Apidae, Apinae) - bionomia e biogeografia. *Rev. Bras. Entomol.* 47(3):311-372.

Camargo, J. M. F.; Pedro, S. R. M. (2008). Meliponini Lepeletier, 1836. *In*: Moure, J. S.; Urban, D. e Melo, G.A.R. (Orgs). *Catalogue of Bees (Hymenoptera, Apoidea) in the Neotropical Region*. Online version, available at http://www.moure.cria.org.br/catalogue. Accessed Fev/22/2010.

Campos, L. A. O. (1987). Abelhas indígenas sem ferrão: o que são? *Informe Agropecuário*: 3-6.

Carvalho, G. A.; Kerr, W. E. (2004). Substituição natural de rainhas fisogástricas e distância de vôo dos machos em Tiuba (*Melipona compressipes fasciculata* Smith, 1854) e Uruçu (*Melipona scutellaris* Latreille, 1811) (Apidae, Meliponini). *Acta Amazonica* 34(4):649-652.

Carvalho, C. A. L.; Alves, R. M. O.; Souza, B. A. (2003). *Criação de abelhas sem ferrão: aspectos práticos*. Serie Meliponicultura N°1, UFB/SEAGRI-BA, 42 p.

Carvalho, C. A. L.; Souza, B. A.; Dias, C. S.; Alves, R. M. O.; Melo, A. F. L.; Soares, A. C. F.; Carvalho-Zilse. G. A. (2011). Five egg-laying queens in a single colony of brazilian stingless bees (*Melipona scutellaris* Latreille). *Acta Amazonica* 41(1):123-126.

Carvalho, G. A. (2001). The number of sex alleles (CSD) in a bee population and its practical important (Hymenoptera: Apidae). *Journal of Hymenoptera Research* 10(1):10-15.

Carvalho-Zilse, G. A.; Nunes-Silva, C. G. (2010). Biodiversidade de abelhas amazônicas: a atuação dos Grupos de Pesquisa da região Norte do Brasil. *In*: Anais do IX Encontro sobre Abelhas (Ribeirão Preto – SP), pp. 317-323. CD ROM.

Carvalho-Zilse, G. A. (2006). Meliponicultura na Amazônia. *In*: Anais do VII Encontro sobre Abelhas (Ribeirão Preto – SP). CD ROM.

Carvalho-Zilse, G. A.; Costa-Pinto, M. F. F.; Nunes-Silva, C. G.; Kerr, W. E. (2009). Does beekeeping reduce genetic variability in *Melipona scutellaris* (Apidae, Meliponini)? *Genetics and Molecular Research* 8(2):758-765.

Castro, M. S; Silva, L. G. S. (2000). Árvores da caatinga utilizadas para nidificação de abelhas eussociais. *In*: Anais do IV Encontro sobre Abelhas (Ribeirão Preto-SP): 290.

Cavalcante, D. G. Matos, I. I.; Pinheiro, E. S. (2009). Meliponicultura como fator de permanência do homem nas zonas de várzea do munícipio de Boa Vista do Ramos/AM. *In*: Anais do XIX Encontro Nacional de Geografia Agrária, São Paulo: 1-12.

Cook, J. M. (1993). Sex determination in the Hymenoptera: a review of models and evidence. *Heredity* 71:421-435.

Cortopassi-Laurino, M. (2003). Potencial Apícola de Plantas Amazônicas. *In*: Anais do 54º Congresso Nacional de Botânica, Belém-PA.

Cortopassi-Laurino, M.; Imperatriz-Fonseca, V. L.; Roubik, D. W.; Dollin, A.; Heard, T.; Aguilar, I.; Venturieri, G. C.; Eardley, C.; Nogueira-Neto, P. (2006). Global meliponiculture: challenges and opportunities. *Apidologie* 37(2): 275-292.

Costa-Pinto, M. F. F.; Carvalho-Zilse, G. A. (2007). Comparação da variabilidade genética entre colméias naturais e manejadas da abelha sem ferrão *Melipona compressipes* no Estado do Amazonas. *In:* Resumos do 53º Congresso Brasileiro de Genética (Águas de Lindóia-SP): 33.

Danforth, B. N.; Sipes, S.; Fang, J.; Brady, S. G. (2006). The history of early bee diversification based on five genes plus morphology. *PNAS, October* 10, 103(41):15118-15123.

De Graaf, N. R.; Filius, A. M.; Huesca Santos, A. R. (2003). Financial analysis of sustained forest management for timber Perspectives for application of the CELOS management system in Brazilian Amazonia. *Forest Ecology and Management* 177:287-299.

Eltz, T.; Brühl, C. A.; Kaars, S.; Linsenmair, E. R. (2002). Determinants of stingless bees nest density in lowland dipterocarp forests of Sabah, Malaysia, *Oecologia* 131:27-34.

Eltz, T.; Brühl, C. A.; Imiyabir, Z.; Linsenmair, K. E. (2003). Nesting and nest trees of stingless bees (Apidae: Meliponini) in lowland dipterocarp forests in Sabah, Malaysia, with implications for forest management. *For. Ecol. Manage.* 172:301-313.

Evans, J. D.; Wheeler, D. E. (2000). Expression profiles during honeybee caste determination. *Genome Biol.,* 2(1):research0001.1-0001.6.

Fearnside, P. M. (2008). The roles and movements of actors in the deforestation of Brazilian Amazonia. *Ecology and Society* 13(1):23. [online] URL: http://www.ecologyandsociety.org/vol13/iss1/art23/.

Gempe, T.; Hasselmann, M.; Schiott, M.; Hause, G., Otte, M.; Beye, M. (2009). Sex determination in honeybees: two separate mechanisms induce and maintain the female pathway. *PLoS Biol* 7:e1000222.

Gullan, P. J.; Craston, P. S. (2007). *Os insetos: um resumo de entomologia.* 3.ed. São Paulo: Roca, 440 p.

Hartfelder, K.; Makert, G. R.; Judice, C. C.; Pereira, A. G.; Santana, W. C.; Dallacqua, R. P.; Bitondi, M. M. G. (2006). Physiological and genetic mechanisms underlying caste development, reproduction and division of labor in stingless bees. *Apidologie* 37:144-163.

Hasselmann, M.; Fondrk, M. K.; Page, R. E.; Beye, M. (2001). Fine scale mapping in the sex locus region of the honey bee (*Apis mellifera*). *Insect Molecular Biology* 10(6):605-608.

Heckel, D. G. (2003). Genomics in pure and applied entomology. *Annu. Rev. Entomol.* 48:235-260.

Hunt, G. J., Page, R. E. (1994). Linkage analysis of sex determination in the honey bee (*Apis mellifera*). *Molecular Genetics* 244:512-518.

Hunt, G. J.; Page, R. E. (1995). Linkage Map of the Honeybee, *Apis mellifera*, Based on RAPD Markers. *Genetics* 139:1371-1382.

Imperatriz-Fonseca, V. L.; Saraiva, A. M.; De Jong, D. (2006). *Bees as pollinators in Brazil: assessing the status and suggesting Best practices*. Ed. Holos. Ribeirão Preto-SP. 96 p.

Kearns, C. A.; Inouye, D. W.; Waser N. M. (1998). Endangered mutualisms: The conservation biology of plant-pollinator interactions. *Annu. Rev. Ecol. Syst.* 29:83-112.

Kerr, W. E. (1946). Formação de castas no gênero *Melipona*. Nota Prévia. *Anais da Escola Superior de Agricultura Luiz de Queiroz* 2:299-313.

Kerr, W. E. (1948). Estudos Sobre o Gênero *Melipona*. *Anais da Escola Superior de Agricultura "Luiz de Queiroz"* 5:182-276.

Kerr, W. E. (1950). Genetic determination of castes in the genus *Melipona*. *Genetics* 35:143-152.

Kerr, W. E. (1987). Sex determination in bees. XXI. Number of XO-heteroalleles in a natural population of *Melipona compressipes fasciculata*. *Insects Sociaux* 34:274-279.

Kerr, W. E. (1997). Sex determination in honey bees (Apinae and Meliponinae) and its consequences. *Brazilian Journal of Genetics* 4:601-611.

Kerr, W. E.; Lello, E. (1962). Sting glands in stingless bees a vestigial character (Hymenoptera: Apoidea). *Journal of the New York Entomological Society* 70:190-214.

Kerr, W. E.; Vencovsky R. (1982). Melhoramento genético em abelhas. I. Efeito do número de colônias sobre o melhoramento. *Brazilian Journal of Genetics* 5:279-285.

Kerr, W. E.; Nascimento, V. A.; Carvalho, G., A. (1994). Há salvação para os meliponínios? *In*: Anais do 1° Encontro sobre abelhas de Ribeirão Preto (Ribeirão Preto-SP):60-65.

Kerr, W. E.; Carvalho, G. A.; Nascimento, V. A. (1996). *Abelha Uruçu: biologia, manejo e conservação*. Ed. Fundação Acangaú, Paracatu, MG, 144 p.

Kerr, W. E. ; Carvalho, G. A. e Nascimento, V. A. (1999). The probable consequences of the destruction of Brazilian stingless bees. Section 6. pp. 393-403. *In*: Várzea: diversity, development and conservation of Amazonia's whitewater flood plains. Edited by Christine Padoch; José Márcio Ayres; Miguel Pinedo-Vasquez; Andrew Henderson.

Kerr, W. E.; Nascimento, V. A.; Carvalho, G. A. (1999a). Preservation of native Brazilian bees: a question of historical and ecological conscience. *Ciência e Cultura (Journal of Brazilian Association for the Advancement of Science)* 51(5/6):390-393.

Kerr, W. E.; Carvalho, G. A; Silva, A. C.; Assis, M. G. P. (2001). Aspectos pouco mencionados da biodiversidade amazônica. *Parcerias Estratégicas* 12:20-41.

Malagodi-Braga, K. S.; Kleinert, A. M. P.; Imperatriz-Fonseca V. L. (2000). Stingless bees: greenhouse pollination and meliponiculture. *In*: Anais do IV Encontro sobre Abelhas (Ribeirão Preto-SP): 145-150.

Marianno-Filho, J. (1910). O Cultivo das abelhas indígenas e um tipo de colméia para o seu desfrutamento industrial. *O Entomologista Brasileiro* 3(1):14-18.

Martins, C. F.; Cortopassi-Laurino, M.; Koedam, D.; Imperatriz-Fonseca, V. L. (2004). Espécies arbóreas utilizadas para nidificação por abelhas sem ferrão na caatinga (Seridó, PB; João Câmara, RN). *Biota Neotropica* 4:1-8.

Matheson, A., Buchmann, S.L., O'Toole, C., Westrich, P., Williams, I. H. (1996). *The Conservation of Bees*. Academic Press for the International Bee Research Association and the Linnaean Society; London, UK, 254 p.

Michener, C. D. (1974). *The Social Behavior of the Bees*. Harvard University Press. 404 p.

Michener, C. D. (1979). Biogeography of the bees. *Annals of the Missouri Botanical Gardens* 66:277–347.

Michener, C. D. (2007). *The Bees of the World*, 2nd Edition. The Johns Hopkins University Press, Baltimore, 953 p.

Moure, J. S.; Urban, D.; Melo, G. A. R. (org.). (2007). *Catalogue of Bees (Hymenoptera, Apoidea) in the Neotropical Region*. Ed. Sociedade Brasileira de Entomologia, Curitiba, XIV, 1058 p.

Neff, J. L.; Simpson, B. B. (1993). Bees, pollination system and plant diversity. *In*: La Salle J., Gauld I.D. (eds) Hymenoptera and Biodiversity. C.A.B. International, Wallingford, UK, pp. 143-167.

Nogueira-Neto, P. (1997). *Vida e Criação de Abelhas Indígenas sem Ferrão*. Ed. Nogueirapis. São Paulo, SP, 446 p.

Norton, B. (1997). Mercadoria, comodidade e moralidade: os limites da quantificação na avaliação da biodiversidade. *In*: 22° capítulo de *Biodiversidade*. Ed. por E. O. Wilson e Frances M. Peter. Editora Nova Fronteira, Rio de Janeiro. pp. 260-263.

Nunes-Silva, P.; Hrncir, M.; Venturieri, G. C.; Imperatriz-Fonseca, V. L. (2010). O potencial das abelhas *Melipona* na polinização por vibração, considerando as características físicas da vibração. *In*: Anais do IX Encontro sobre Abelhas (Ribeirão Preto-SP): 76-84.

Nunez, C. V.; Oliveira, M. L.; Lima, R. D.; Diaz, I. E. C.; Sargentini, E. Jr.; Pereira, O. L. Jr.; Araújo, L. M.. (2008). Chemical analyses confirm a rare case of seed dispersal by bees. *Apidologie* 39:618-626.

Oliveira, M. L.; Morato, E. F.; Garcia, M. V. B. (1995). Diversidade de espécies e densidade de ninhos de abelhas sociais sem ferrão (Hymenoptera, Apidae, Meliponinae) em floresta de terra firme na Amazônia central. *Revista Bras. Zool.* 12:13-24.

Overal, W. L. (2001). O peso dos invertebrados na balança de conservação biológica da Amazônia. In: Veríssimo, A. et al. (Eds.). *Biodiversidade na Amazônia Brasileira*. São Paulo: Estação Liberdade e Instituto Socioambiental, pp. 50-59.

Rêgo, M. M. C.; Brito, C. (1996). Abelhas sociais (Apidae: Meliponini) em um ecossistema de cerrado s.l. (Chapadinha - MA, BR): distribuição dos ninhos. *In*: Anais do III Encontro Sobre Abelhas (Ribeirão Preto-SP): 238-247.

Rosa, L.; Castellani, T. T.; Reis, A. (1998). Biologia reprodutiva de *Butia capitata* (Martius) Beccari var. *odorata* (Palmae) na região de restinga do município de Laguna, SC. *Revista Brasileira de Botânica* 21:131-142.

Roubik, D. W. (1989). *Ecology and natural history of tropical bees*. Cambridge, University Press, New York, 514 p.

Salin, C.; Deprez, B.; Van Bockstaele, D. R.; Mahillon, j; Hance, T. (2004). Sex determination Mechanism in the Hymenopteran *Aphilus rhopalosiphi* De Stefani-Pers (Braconidae: Aphidiinae). *Belg. J. Zool.*, 134 (supplement 1):155-21.

Samejima, H.; Marzuki, M.; Nagamitsu, T.; Nakasizuka, T. (2004). The effects of human disturbance on a stingless bee community in a tropical rainforest. *Biol. Conserv.* 120:577-587.

Serra, B. D. V.; Drummond, M. S.; Lacerda, L. M.; Akatsu, I. P. (2009). Abundância, distribuição espacial de ninhos de abelhas Meliponina (Hymenoptera, Apidae, Apini) e espécies vegetais utilizadas para nidificação em áreas de cerrado do Maranhão. *Iheringia, Sér. Zool.* 99(1):12-17. DOI: 10.1590/S0073-47212009000100002.

Silveira, F. A.; Melo, G. A. R.; Almeida, E. A. B. (2002). *Abelhas brasileiras: sistemática e identificação.* Ed. MMA e Fundação Araucária, Curitiba-PR, 1ª ed, 253 p.

Slaa E. J., Sánchez Chaves L. A., Malagodi-Braga K., Hofstede F. E. (2006). Stingless bees in applied pollination: practice and perspectives. *Apidologie* 37:293-315.

Velthuis, H. H. W.; Vries, H. D.; Imperatriz-Fonseca, V. L. (2006). The polygyny of *Melipona bicolor*: scramble competition among queens. *Apidologie* 37:222-239.

Venturieri, G. C. (2009). The impact of forest exploitation on Amazonian stingless bees (Apidae, Meliponini). *Genet. Mol. Res.* 8(2):684-689.

Venturieri, G. C.; Venturieri, G. R.; Pires, N. V. C. R.; Contrera, F. A. L. (2010). Uso de *Melipona* (Apidae, Meliponini) na polinização de solanáceas em casa de vegetação. *In*: Anais do IX Encontro sobre Abelhas (Ribeirão Preto-SP): 220-224.

Schulze, M.; Grogan, J.; Vidal, R. (2008). Forest certification in Amazonia: standards matter. *Fauna & Flora International, Oryx* 42(2):229-239 DOI:10.1017/S0030605308000689.

Vit, P.; Enriquez, E.; Barth, O. M.; Matsuda, A. H.; Almeida-Muradian, L. B. (2006). Necesidad del control de calidad de la miel de abejas sin aguijon. MedULA. *Revista de la Facultad de Medicina, Universidad de los Andes* 15(2):89-95.

Whiting, P. W. (1943). Multiple alleles in complementary sex determination of *Habrobracon. Genetics* 28:365-382.

van Wilgenburg E., Driessen G., Beukeboom L. (2006). Single locus complementary sex determination in Hymenoptera: an "unintelligent" design? *Front. Zool.* 3:1-15.

Zayed, A.; Packer, L. (2005). Complementary sex determination substantially increases extinction proneness of haplodiploid populations. *Proceedings of the National Academy of Sciences USA* 102:10742-10746.

In: Bees
Editor: Richard M. Florio

ISBN: 978-1-61324-825-6

Chapter 5

THE IMPORTANT HONEY BEE VIRUSES: A SHORT DESCRIPTIVE REVIEW ENHANCED WITH RECENT DATA

Nicolaos Bacandritsos[1], Eriffili Roinioti and Iosif Papanastasiou
National Agricultural Research Foundation, Institute of Veterinary Research of Athens, Neapoleos, Agia Paraskevi, Greece

Abstract

The Chronic bee paralysis virus (CBPV), Acute bee paralysis virus (ABPV), Israeli acute paralysis virus (IAPV), Deformed wing virus (DWV), Kashmir bee virus (KBV), Black queen cell virus (BQCV) and Sacbrood virus (SBV) are considered to be able to cause severe disease in honeybees and hence they have been objects of active research currently.

In this chapter, information about the geographical spread of the virus, its genome (briefly) and close relatives, the transmission pathways in nature and especially the significant relationships of the virus with other bee enemies are presented. Tissue tropism along with the symptoms are also noted. Old and modern methods of bee viruses diagnosis are quoted. The recent case of colony depopulation in Greece and the first detection of viruses in Greek adult bee populations with the use of molecular techniques is presented separately.

INTRODUCTION

The honey bee *Apis mellifera L.* has been reported to harbor at least 18 viruses (Bailey & Ball, 1991; Allen & Ball 1996; Chen & Siede, 2007). Usually they persist as inapparent infections in colonies although in certain cases they may cause serious or lethal disease in individual bees or the collapse of entire colonies (Berényi *et al.*, 2006). Many insect species

[1] E-mail address: bac.ivra@nagref.gr.

have been found to be inapparently infected with viruses (Scotti *et al.,* 1981). These infections differ from acute infections in that they may persist in their hosts for many generations causing little or no harm or recognizable symptoms, yet in certain circumstances they may be stimulated or activated to yield greater numbers of particles and initiate acute infections

Infestation with the ectoparasitic mite *Varroa destructor* is the major predisposing factor (Shimanuki *et al.,* 1994; Nordström *et al.*, 1999; Yang & Cox-Foster, 2005) however, a variety of other weakening circumstances may play a role in clinical manifestation of bee virus infections (e.g. *Nosema apis* and *Nosema ceranae* infestations, intoxications, environmental pollution, cold weather).

Seven viruses are considered to be able to cause severe disease in honeybees and hence they have been objects of active research currently. These are the chronic bee paralysis virus (CBPV) , the acute bee paralysis virus (ABPV) , the deformed wing virus (DWV), the black queen cell virus (BQCV), the sacbrood virus (SBV) and the Kashmir bee virus (KBV). The Israeli acute bee paralysis virus (IABPV) was characterised only recently (Maori *et al.,* 2007). It is a close relative to KBV and ABPV, but sufficiently different to be discerned by PCR and serology.

These viruses are 30 nm isometric particles containing a single positive sense genomic RNA. Sacbrood virus and deformed wing virus are assigned to the genus *Iflavirus* while Kashmir bee virus, acute bee paralysis virus and black queen cell virus are classified as members of the genus *Cripavirus* (family *Dicistroviridae*) (Mayo, 2002) while the Israeli acute paralysis virus is the new member of the family. The chronic bee paralysis virus remains unclassified.

Studies show that viruses are widely distributed in honey bees throughout the world (Nordström *et al.,* 1999; Allen & Ball, 1996) most likely resulting from intensive exchanges of honey bee stocks (Tentcheva *et al.,* 2004).

THE CASE OF GREECE

Five bee viruses (CBPV, ABPV, DWV, BQCV, SBV) have been recently detected and identified in Greek colonies (Bacandritsos *et al.*, 2010). There are no data concerning Kashmir bee virus and Israeli acute bee paralysis virus because they have not been included in the study.

In five apiaries in the Peloponnesus region (Mt. Mainalo) of Greece and during summer period between 05/06/09 and 08/07/09 the following clinical signs were recorded:

- large numbers of dead bees in front of the hive entrance
- large numbers of trembling bees clustered outside the hive entrance
- apiary losses between 50% and 70%
- queenless colonies , colonies with non-laying queens

According to the history, symptoms were manifested in colonies with low level mite infestation. In total, five samples – each from every apiary- of symptomatic adult bees and ten samples of asymptomatic young bees from the inside of the hives were assessed for viruses' presence using the RT-PCR method. All fifteen bee samples were tested positive for multiple viruses. BQCV was found to be the most prevalent, present in all samples. SBV and DWV exhibited the same prevalence of 87%. CBPV was detected in 73% of samples while ABPV in 67%.

All the samples collected from outside the hive, tested positive for CBPV and DWV and most tested positive for ABPV and SBV. In samples collected from the inside of the hive, SBV was detected in 90%, while DWV, CBPV and ABPV exhibited a prevalence of 80%, 60% and 60% respectively (Bacandritsos *et al.*, 2010).

In total, five different viruses were detected in the fifteen bee samples. Although several bee viruses had been detected in extracts of infected brood by electron microscopy in Greece (Allen & Ball 1996), Bacandritsos *et al.* (2010) made the first detection of viruses in Greek adult bee populations with the use of molecular techniques. Whilst it is impossible to interpret these results in the absence of a large-scale virus survey including healthy and diseased apiaries across Greece, the high proportion (73%) of CBPV observed in this study appears unusual when compared to reports from other countries where 5-28% is more typical (Tentcheva *et al.*, 2004; Nielsen *et al.*, 2008).

Chronic Bee Paralysis Virus (CBPV)

Chronic bee paralysis is an infectious and contagious disease affecting adult honey bees (Ball & Bailey, 1997). Although the symptoms of paralysis were probably recognized more than two thousand years ago by Aristotle, who described hairless black bees that he called "thieves", the causative agent was not confirmed until 1963 when Bailey and colleagues isolated and characterized Chronic bee paralysis virus (CBPV) (Bailey *et al.*,1963). The distribution of the virus is worldwide (Allen & Ball, 1996).

The genome is composed of two major RNA fragments, RNA 1 (3674 bases) and RNA 2 (2305 bases). Three minor RNA fragments (1100 bases each) have also been described. Western blot analysis revealed four CBPV-associated polypeptides of about 75, 50, 30 and 20 kDa. Purified preparations contain anisometric particles of about 30-65 nm long and 20 nm wide (Olivier *et al.,* 2008).

The virus can persist as an inapparent infection, however it can multiply to high levels in honey bees causing significant losses in colonies (Allen & Ball, 1996). According to Tentcheva *et al.* (2004) the virus is being detected in healthy colonies during the year and outbreaks of severe disease are erratic and exhibit no seasonal pattern. No links were established between CBPV dissemination or paralysis outbreaks and the important mite parasite *Varroa destructor,* which is associated with numerous other honey bee viruses (Tentcheva *et al.*, 2004; Ribière *et al.*, 2010).

Virus particles, round to oval in shape, have been observed in the nerve ganglia of the adult honey bee since 1965, while no particles were observed in fat or muscle tissue of the afflicted bees (Lee & Furgala, 1965). The virus seems to exhibit neurotropism combined with symptoms reflective of damage to the nervous system in diseased bees. Furthermore, of the many millions of virus particles which can be extracted from one diseased bee, half of these are concentrated in the head, which accounts for about only one tenth of the total body weight (Ribière *et al.,* 2010).

Nervous symptoms such as ataxia, trembling and incapacity to fly, observed in both artificial and natural CBPV infections, may be explained by the presence and multiplication of the virus in the central nervous system, especially in the neuronal cells of the higher order integration centers of the insect's brain and in regions involved in sensory information (optic and antennal lobes) (Olivier *et al.*, 2008). Some individuals become almost hairless and dark in appearance, suffering nibbling attacks from healthy bees in the colony. The bloated abdomen is caused by distension of the honey sac with fluid, leading to the so-called "dysentery" symptom (Ribière *et al.*, 2007). Sick individuals die within a few days of the onset of the symptoms (Bailey & Ball, 1991; Ball & Bailey, 1997).

Bailey *et al.* (1983) demonstrated that the virus can be easily transmitted through direct contact of healthy bees crowded with paralysed individuals. The presence of infectious CBPV particles in the feces of bees with active disease has been demonstrated (Ribière *et al.*, 2007). Furthermore the researchers demonstrated that these virus particles are sufficiently infectious to contaminate and induce overt disease when injecting into naïve bees and also when naïve bees are kept in soiled cages. Better knowledge of virus half-lives outside the host is needed to see if decontamination could provide a resource-efficient means of disease control (Ribière *et al.*, 2007).

Further study is needed to see if there is a relationship between CBPV excretion and nosema infections, paying special attention to distinguish between *N. apis* and *N.ceranae* (Ribière *et al.*, 2007), which are both present in European colonies but may induce different disorders in colony health (Higes *et al.*, 2006). The CBPV has also been detected in adult queen bees. Although no obvious disease was observed in the queens, this detection raises the possibility of a vertical transmission pathway wherein infected queens can pass virus through their eggs to their offspring (Chen *et al.*, 2005).

ACUTE BEE PARALYSIS VIRUS (ABPV)

The ABPV is very common and spread in Europe (Allen & Ball, 1996). In a research carried out in Austria the ABPV was found to be the second most prevalent identified virus in adult bees after DWV, present in 68% of samples (Berényi *et al.*, 2006). Nielsen *et al.* (2008) reported a low frequency (14%) of the virus in Denmark while in France the ABPV was present in adult bee samples in 21 of the 36 apiaries examined (Tentcheva *et al.,* 2004). In Greece the ABPV was detected in 67% of the adult bee samples examined (Bacandritsos *et al.,* 2010).

The complete genome sequence was determined (Govan *et al.*, 2000). The 9470 nucleotide, polyadenylated RNA genome encoded two open reading frames (ORF 1 and ORF 2), which were separated by 184 nucleotides.

Although ABPV causes the same symptoms of trembling and the inability to fly in infected bees that CBPV does, the two viruses are different in several ways: ABPV is more virulent of the two viruses, as CBPV takes several days to kill the diseased bees while ABPV takes only one day; the shapes of the two viruses are different – ABPV particles are isometric and CBPV are asymmetric; there are many more virus particles of CBPV than of ABPV in naturally paralyzed bees (Chen & Siede, 2007).

ABPV can be detected in both brood and adult stages of bee development. Spread in the colonies is probably via salivary gland secretion of infected adult bees when glandular secretions are fed to young larvae or mixed in the pollen. Infected larvae either die before they are sealed in brood cell if large amounts of virus particles were ingested, or survive to emerge as inapparently infected adult bees (Bailey & Ball, 1991).

Detection of ABPV in varroa mites further supports the possible role of varroa mites in the virus transmission (Bakonyi *et al.,* 2002; Tentcheva *et al.,* 2004). However, the presence of the virus in bees from apiaries where no ABPV- positive varroa mites were detected, suggests that the transmission may also occur by contact between individual bees (Tentcheva *et al.*, 2004).

Faucon *et al.* (1992) isolated the ABPV from brood heavily infested with varroa mites presenting symptoms of foulbroods although the microbiological exams did not reveal the presence of microbes. Inoculation of the virus into healthy bees resulted in their death within 3 days. The symptoms of the ABPV in brood and adult bees resemble to those observed in American foulbrood, European foulbrood, sacbrood and to a lesser degree to the symptoms of *Acarapis woodi* (Faucon *et al.*, 1992). The virus seems to have a seasonal incidence. According to Tentcheva *et al.* (2004), ABPV infections were prominently found during the summer in adult bees.

Israeli Acute Bee Paralysis Virus (IAPV)

Close relative to ABPV and KBV has been recently characterized as a new member of the *Dicistroviridae* family. So far, the only known host of IAPV is *A. mellifera* (de Miranda *et al.*, 2010).

The RNA genome was cloned and sequenced and the capsid proteins were analysed. The viral RNA is 9487 nt long (excluding the poly-A tail) and carries two ORFs both coding for polyproteins. The two ORFs are separated by a 184-nucleotide-long intergenic region. The frequent reciprocal exchange of genetic material between IAPV and the honey bee genome, and in particular the correlation of such genome- integrated IAPV with resistance to IAPV infection was reported (Maori *et al.*, 2007).

The virus has been isolated from bees in Israel (Maori *et al.*, 2007), different states of USA such as Florida, California, Maryland and Pennsylvania where it has been strongly correlated with the Colony Collapse Disorder (Chen & Evans 2007; Cox-Foster *et al.*, 2007), France (Blanchard *et al.*, 2008), several provinces within Canada (Currie *et al.*, 2010). The virus has also been found in package bees imported from Australia and isolates of royal jelly imported from China (Chen & Evans, 2007). IAPV is likely present in Russia (Palacios *et al.*, 2008).

The virus is considered to be a newly emerging pathogen. Little is known specifically about the transmission routes of IAPV. Nearly 80% of adult bees infected orally with IAPV die within a week, with little difference across a 1000-fold range of inoculum concentrations (Maori *et al.*, 2009)

The virus has been associated with symptoms reminiscent of those inflicted by ABPV, severe bee mortality and heavy losses on Israeli apiculture (Maori *et al.,* 2007). In France IAPV was detected in a significant number of the surveyed apiaries (14%). Although severe honey bee colony losses and mortalities had occurred during winter 2008, it was not possible to establish a causal relationship between them and IAPV (Blanchard *et al.*, 2008). The prevalence of IAPV in Spain was found to be 18%, however, no relationship between the virus and depopulated colonies in professional Spanish apiares was established (Garrido-Bailón *et al.*, 2010).

DEFORMED WING VIRUS (DWV)

Commonly detected in honeybees, the virus has a global spread. DWV was by far the most frequently detected in seemingly healthy French colonies (adults, brood, varroa mites) (Tentcheva *et al.*, 2004). Some seasonal variation in virus incidence was observed and the frequency of DWV infection in both adult bees and pupae increased considerably from summer to autumn during the year (Tentcheva *et al.*, 2004). It was also found to be the most prevalent virus, present in 91% of samples from sick Austrian honeybee colonies (Berényi *et al.*, 2006). Chen & Siede (2007) reported that DWV infection occurred in 100% of the apiaries investigated in Beltsville, Maryland. In Greece, in a survey which was carried out during summer 2009, the virus was detected in 100% of samples of symptomatic bees and in 80% of samples from asymptomatic ones (Bacandritsos *et al.*, 2010).

The complete genome sequence has been determined, having a 97% sequence homology with Kakugo virus. The genomic RNA is 10,140 nucleotides in length and contains a single large open reading frame encoding a 328-kDa polyprotein. The three major structural proteins VP1 (44kDa), VP2 (32kDa) and VP3 (28kDa) were identified and their genes were mapped to the N-terminal section of the polyprotein (Lanzi *et al.*, 2006).

DWV normally persists at low levels within the bee colony with no detrimental effect (Yue & Genersch, 2005; Chen *et al.*, 2005; Berényi *et al.*, 2006). The virus has been implicated in overwintering honeybee colony losses (Highfield *et al.*, 2009).

It has been detected in worker bees, pupae, larvae, drones and queens (Chen *et al.*, 2006b) as well as within the varroa mites (Tentcheva *et al.*, 2004; Yue & Genersh, 2005) and more recently within the mite *Tropilaelaps mercedesae.* Dainat *et al.* (2009) showed for the first time that the DWV can infect *Tropilaelaps mercedesae* (Acari, Laelapidae) and replicate in this host. According to Eyer *et al.* (2009a) the small hive beetle (SHB) *Aethina tumida,* can become infected with DWV through oral contamination pathways. The detection of the virus in SHB suggests that these beetles are potential hosts for DWV and may also be involved in virus transmission among honeybees The fact that SHBs are active flyers, could considerably favor the spread of the virus (Eyer *et al.,* 2009a).

Bowen-Walker *et al.* (1999) demonstrated that *V. destructor* is a highly effective vector of DWV. Nevertheless, the transmission of the virus within the colony can also be independent of *V. destructor* (Bowen-Walker *et al.*, 1999; Highfield *et al.*, 2009). This would explain the occasional outbreaks of DWV in the UK before the arrival of *V. destructor* in 1992 (Bowen-Walker *et al.*, 1999). DWV is thought to have an intricate relationship with varroa mites such that immunosuppression of the honeybee pupae by the mites results in increased DWV replication when the honeybees are exposed to other bacterial factors (Yang & Cox-Foster, 2005).

Several bee tissues can be infected by the DWV, particular in the digestive and the reproductive organs. In drones, the seminal vesicles, the mucus glands and testis epithelia were shown to be infected. The presence of DWV in these tissues explains the detection of viral RNA in the sperm, through which drones could contaminate queens and the next generation's worker brood following fertilization (Fievet *et al.*, 2006). The experiments of Yue *et al.* (2007) demonstrated the presence of viral RNA in the reproductive organs of the queens and furthermore the possibility of vertical transmission of DWV from queens and drones to drone and worker offspring through unfertilized and fertilized eggs (Yue *et al.*, 2007).

Shah *et al.* (2009) demonstrated that in bees infected with DWV the virus is replicating in critical regions of the brain, including the neuropils responsible for vision and olfaction. Therefore DWV infection of the brain could adversely affect critical sensory functions and alter normal bee behavior such as flight behavior and homing performance (Shah *et al.*, 2009).

Typical symptoms of DWV infection are vestigial and crumpled wings, bloated abdomens, paralysis, learning deficits_and a drastically shortened life span (Chen & Siede, 2007; Shah *et al.*, 2009). There is a direct correlation between the viral titer and the symptoms displayed, with viral loads significantly higher in symptomatic than in asymptomatic bees (Highfield *et al.*, 2009).

Symptoms of the viral infection are most prevalent during stressful environmental conditions, leading to reduced performance in infected colonies. Therefore as the number of infected worker bees increases, the colony is less likely to survive (Shah *et al.*, 2009). According to Yue *et al.* (2007), in the absence of *V. destructor,* DWV causes covert infections with the virus being present in the absence of disease symptoms and being transmitted vertically from parents to offspring. Such a colony will develop normally and eventually swarm, transmitting the virus vertically to the next colony generation, allowing long-term persistence of DWV in the honeybee population.

Kashmir Bee Virus (KBV)

KBV was first identified in 1975 in adults of the Eastern hive bee, *Apis cerana,* from the northen and western regions of India, but several serologically related strains were later isolated from the European bee, *Apis mellifera,* from widely separated areas of Australia (Ball & Bailey, 1991). Later, strains of KBV have been found in *Apis mellifera* from Canada and New Zealand, Fiji, Spain and the United States (Chen & Siede, 2007). During 2002 the virus was detected for the first time in France, both in adult bees and the pupae. It was also present in varroa samples (Tentcheva *et al.,* 2004). In 2003 the virus was detected in Hessian (Germany) bee colonies. This was the first documented detection of KBV in Middle Europe (Siede & Büchler, 2004). Interestingly, in a research which was carried out the same year, the virus was not identified either in Austrian apiaries or in a small number of bee samples from Poland, Hungary and Slovenia (Berényi *et al.,* 2006). The first record of KBV in Denmark was presented by Nielsen *et al.* (2008).

Kashmir bee virus, Acute bee paralysis virus and Israeli acute paralysis virus share a close genetic relationship. The basic genome organization of them is typical of the Dicistroviridae: single positive strand RNA containing two open reading frames (ORF), separated by an intergenic region (IGR) and flanked by non-translated regions. The larger ORF is located in the 5′half of the genome and encodes the non-structural proteins involved in virus replication and processing. The shorter ORF is located towards towards the 3′end of the genome and encodes the structural capsid proteins. Although these viruses are closely related, they are not identical. They can be distinguished by serology, capsid protein profiles and by RT-PCR (de Miranda *et al.,* 2010).

KBV attacks all stages of the bee life cycle and commonly persists within brood and adult bees as an inapparent infection (Dall, 1987). Occasionally it is associated with severe mortality of bees and brood in the field (Ball & Bailey, 1991). Allen & Ball (1995) suggested that KBV under laboratory conditions, is the most virulent of all known honeybee viruses even though infected bees have no clearly defined disease symptoms (Shen *et al.*, 2005a; Chen & Siede, 2007). Dall (1987) investigated the virulence of KBV in pupae. Observation by electron microscopy showed that the virus multiplies in fore- and hindgut epithelial tissue, alimentary canal musculature, epidermis and tracheal epithelium and in hemocytes, oenocytes and tracheal end cells. Infection was associated with major cytopathological damage to the host cell. No evidence of KBV multiplication was found in tissues of the nervous system, even though degeneraration of glial cells was observed (Dall, 1987).

Viral transmission between bees is poorly understood. The virus was detected in feces of worker and queen honeybees (*A. mellifera*) (Hung, 2000). Shen *et al.,* (2005a) demonstrated that KBV could potentially be transmitted either vertically from the queen to the eggs via transovarial transmission, or horizontally from workers to larvae or other bees via food resources containing glandular secretions. In their study, 33.3-80% of the food sources tested were positive for KBV. Further studies of colony food resources as possible resources in the transmission of bee viruses, showed that six viruses including KBV were found in pollen samples (Chen *et al.,* 2006a).

Hung *et al.,* (1996) proved that transmission of KBV does not require *Varroa*. Their study supported the finding that *Varroa* infestation activates and replicates to detectable concentrations the inapparent infection. (Hung *et al.*, 1996). The direct evidence for the ability of the mites to vector the KBV was the detection of the KBV capsid proteins in varroa-mite saliva. Therefore, varroa mites may transmit the virus to bees during feeding (Shen *et al.,*2005b). Studies carried out subsequently by Chen *et al.*, (2006a) confirmed previous findings about the role of *Varroa* mite in transmitting KBV.

Black Queen Cell Virus (BQCV)

Black queen-cell virus (BQCV) was first isolated from queen larvae and pupae of honey bees found dead in their cells (Bailey & Woods, 1977). The name of the virus was derived from darkened areas on the walls of the cells containing infected pupae.

The presence of the virus has been ascertained in all continents (Allen & Ball, 1996). According to Anderson (1993) BQCV has been shown to be the most common cause of death of queen larvae in Australia. The virus has also been detected in Denmark in honey bee colonies not showing obvious problems (Nordström *et al.*, 1999; Nielsen *et al.*, 2008).

In France BQCV was detected in 86% of the investigated apiaries and found to be the second-most prevalent bee virus (Tentcheva *et al.*, 2004). It has also been detected in Austrian apiaries (Berenyi *et al.*, 2006) and in Spain (Higes *et al.*, 2007). In Greece, BQCV was detected with the use of molecular techniques, in all worker bee samples tested. (Bacandritsos *et al.*, 2010).

The complete genome sequence of the BQCV has been analysed. According to Leat *et al.*, (2000), nucleotide sequence analysis revealed an 8550 nt polyadenylated genome containing two large ORFs. The 5'-proximal ORF, encoding a putative replicase protein, represented 4968 nt while the 3'-proximal ORF, encoding a capsid polyprotein, represented 2562 nt. The ORFs were separated by a 208 nt intergenic region and were flanked by a 657 nt 5'-untraslated region and a 155 nt 3'-untraslated region

BQCV transmission seems to be largely independent of *V. destructor.* However, an intimate association between the virus and *Nosema apis* infection seems to exist both in nature and in the laboratory (Ribière *et al.*, 2008). A survey in Britain from 1977 to 1979 showed that both virus and parasite followed the same annual cycle of incidence with an increase in infection in late winter reaching a peak in May or June and declining rapidly in August (Bailey *et al.*, 1981).

Evidence of food-borne transmission of the virus in honey bees has been provided by its detection in pollen samples and the honey (Chen *et al.*, 2006a), in queen feces and the tissues of the gut (Chen *et al.*, 2006b). The virus could be ingested by queens from contaminated foods and passed into the digestive tract. The tissue of the gut may be a major reservoir for replication of BQCV while the detection of the virus in hemolymph, ovaries and spermatheca with relatively lower virus titers, suggests the possibility that the virus could penetrate the wall of the gut and move into the insect hemocoel to spread infection to other tissues (Chen *et al.*, 2006b). The detection of BQCV in the spermatheca of queens, implies that venereal transmission may play an important role.

The main symptoms consist of blackened cell walls of sealed queen cells, containing dead propupae (Leat *et al.*, 2000). Diseased larvae have a pale yellow appearance and tough sac-like skin, much like sacbrood. The virus is present in adult bees but without obvious symptoms. According to Allen & Ball (1996) BQCV shortens the life of adult bees that are also infected with *N. apis.* BQCV could also disturb the queen's performance which is the overall effect of queen behavior on colony productivity. Experimental work indicated that, in certain circumstances, BQCV may be a major cause of early supersedure of young queens (Anderson, 1993).

Sacbrood Virus (SBV)

SBV appears to be the most widely distributed of all the honeybee viruses, occurring in colonies of *Apis mellifera* on every continent (Allen & Ball, 1996). The Chinese sacbrood virus (CSBV) with poses a serious threat to honey bee *Apis cerana*, was first observed in

China 1972 and spread to the countries of the Southeast Asia. CSBV is similar to SBV in physiological and biochemical characters but different in the antigen; these two viruses had no cross infection to the Chinese honeybees. Analyses of the sequenced CSBV RNA fragment revealed 87.6% homology of the nucleotide sequence and 94.6% homology of the deduced amino acid sequence with that of the SBV (Liu *et al.*, 2010). The presence of the SBV has recently been reported in the North Western Himalayan region of India (Bachitter & Rakesh, 2008) while a SBV Korean strain with high percentages of homology with the published virus sequence has been described in Korea (Kim *et al.*, 2008). SBV was detected with the use of molecular techniques in adult bee samples in Greece (Bacandritsos *et al.*, 2010).

The genomic RNA is 8832 nucleotides long and contains a single, large open reading frame encoding a polyprotein of 2858 amino-acids. Amongst the insect picornavirus-like agents, SBV resembles Infectious Flacherie virus (IFV) of the silk worm, in both genome length and gene order and shows an overall 23.2% identity and 45.4% similarity in amino acid sequence to this vitus (Ghosh *et al.*, 1999).

The phylogenetic tree shows that at least three distinct groups of SBV exist: a continental European genotype, including Spanish, French and Austrian isolates; a group that is formed by the Uruguayan isolate and the reference SBVgenome in the UK; and a third group, comprising the Chinese isolate. The European strains formed an homogeneous cluster, wherein nucleotide sequence identities ranged between 95% and 97%. The identity between the British and French isolates was 90%. The British reference strain and the Chinese strain were 90% identical (Kukielka & Sánchez-Vizcaíno, 2009).

SBV attacks both developing brood and adult stages of bees (*Apis mellifera, Apis cerana*) but larvae about 2-day old are the most susceptible. The virus affects adult bees without causing obvious signs of disease, although they may have a decreased metabolic rate and a shortened life span (Ball, 1999). Shen *et al.* (2005a) tested several food sources from two bee colonies, for the presence of SBV. The virus was detected in all bee products (honey, pollen, royal jelly, brood food).

The initial spread of SBV within a colony occurs when nurse bees become infected while removing larvae killed by SBV. Virus particles accumulate in the hypopharyngeal glands of the nurse bees. Then the infected bees can spread the virus throughout the colony by feeding larvae with their glandular secretion and exchanging food with other adult bees including foraging bees. Infected foraging bees spread the virus by passing it from their glandular secretions to the pollen loads as they collect pollen. Young larvae become infected with the virus by ingesting virus-contaminated food. The SVB starts to replicate in the larva and the infected larva turns pale yellow after the brood cell is caped. As the disease progresses, the skin of the larva becomes leathery and the larva fails to pupate because it cannot digest the old cuticle. A large amount of fluid containing millions of SBV particles accumulates between the body of a diseased larva and its saclike skin. Affected larvae appear to be a water-filled sac when removed from the cell (Chen & Siede, 2007). SBV was also detected in both queens and eggs by RT-PCR (Shen *et al.*, 2005a). Transovarial transmission from the queen to the eggs, could potentially be a route of SBV transmission in the hive.

SBV infection has been associated with varroa mite infestation. Although varroa mite as a vector in transmitting SBV has not yet been experimentally demonstrated, the detection of SBV in varroa mites indicates the potential role of the mites in transmitting SBV (Tentcheva *et al.,* 2004). According to Shen *et al,* (2005a) varroa mites can be vectors for SBV. The detection of viral RNA in mites indicated that the mites may also be infected with the virus.

Recently, the results of the laboratory experiments of Eyer *et al.* (2009b) showed that the small hive beetle (SHB) *Aethina tumida* (Coleoptera: Nitidulidae) can become naturally infected by SBV via feeding on bee brood. Virus replication in SHB is also very possible (Eyer *et al.,* 2009b).

Prevalence of SBV in honey bees has been found to be prominently seasonal. Frequencies of SBV infection in spring and summer in both adults and pupae were significantly higher than in autumn (Tentcheva *et al.,* 2004).

DIAGNOSTIC TECHNIQUES

Various tools and technologies for the detection of honey bee viruses have been referred papers throughout the last three decades. Diagnostics for honeybee viruses have been developed very rapidly over the last years. This starts from the traditional diagnostic methods (symptomatology, microscopy, and bioassays) and continues to the most recent molecular detection methods.

Although symptomatology (record of the symptoms in the apiary) is a basic method, it has some noteworthy disadvantages. The recorded symptoms are not clear and relative for any pathogen, something that makes the method not enough accurate and sensitive. Moreover, .many viruses' diseases are symptomless, or a colony can suffer from more than one disease at the same time. This situation could easily confuse any upshot. Another shortcoming of the symptomatology method is that when the symptoms appear in a colony the pathogen has already spread in the colony and often to the other colonies of the apiary. All these led the researchers to develop more specific diagnostic methods.

From the 80s, honey bee viruses were detected by immunodiffussion tests either indirectly or directly (Baily *et al.,* 1981) The indirect way involved live adults bees and pupae that were injected with extract of samples of live bees and after a multiplication period the extract tested against suitable dilutions of antisera. The direct way was the detection of the viruses in extracts of samples of dead bees by electron microscopy and tested also against antisera. The use of microscopy in the honey bee virus detection was decisive. Through the microscopy procedure or in combination with stains (histology), serology (immunohistochemistry, immuno-electron microscopy) or nucleic acid technologies (in situ hybridisation, in situ RT-PCR), researchers managed to categorize novel viruses according to particle morphology, tissue distribution, inclusion bodies or crystallization properties (Thomson and Smirk, 1966; Lee and Fungala, 1965; Bailey, 1968; Bailey and Milne, 1969) or to distinguish viruses that have similar shape (Fievet *et al.*, 2006).

Protein and serology techniques are the first molecular tools developed for the characterization of new viruses. These techniques gave to the research community a linkage between the symptomatic and physical characteristics of a virus and its genetic identity (de Miranda *et al.*, 2004; de Miranda, 2008). It is known that the number and the size of the nucleoproteins are characteristic for each virus. Thus, it is a very useful tool in order to make the identification, classification and differentiation between virus species (Madriz *et al.*, 2000). Polyacrylamide gel electrophoresis (PAGE and SDS-PAGE) is a technique where the proteins are denatured and put into an electric field in an environment of a polymer of acrylamide monomers. The proteins are visualized by staining (Coomasie blue) and compared with molecular weight standards of known sizes running in the same gel. For the extra characterisation the N-terminal and C-terminal sequencing have been used. Protein sequencing is a technique to determine the amino acid sequence of a protein, as well as which conformation the protein adopts and the extent to which it is complexed with any non-peptide molecules. The two major direct methods of protein sequencing are mass spectrometry and the Edman degradation reaction.

Enzyme linked immunosorbet assay (ELISA) has also been used as a diagnostic tool. In simple terms, in ELISA, an unknown amount of antigen is affixed to a surface, and then a specific antibody is applied over the surface so that it can bind to the antigen. This antibody is linked to an enzyme, and in the final step a substance is added that the enzyme can convert to some detectable signal, most commonly a colour change in a chemical substrate. There are two types of this method, the direct and the indirect ELIZA. Until the late 1990s ELISA was the main tool for diagnosing honeybee viruses (Anderson and Gibbs, 1988; 1989; Allen and Ball, 1995; Nordström *et al.,* 1999)

Many tags, markers, dyes and probes were used by molecular detection methods. These could be radioactive isotopes of common elements (phospate, sulfur, nitrogen), gold particles, enzymes (alkaline phosphatase, horse radish peroxidase, luciferase), fluorophores (green flurescent protein) and colourful dyes (ethidium bromide, propidium iodine, acridine). The labeling is done either thought chemical modification or through incorporation of modified nucleotides by a polymerase.

The most popular and powerful method that implicates the nucleic acids in the overall procedure is the polymerase chain reaction widely known as PCR. With this method we have the ability to detect the nucleic acid. This detection is either qualitative or quantitative. The original molecule of DNA is replicated by the DNA polymerase enzyme and doubling the number of DNA molecules. Afterwards, each of these molecules is replicated in a second "cycle" of replication, resulting in four times the number of the original molecules. Again, each of these molecules is replicated in a third cycle of replication. The method relies on thermal cycling, consisting of cycles of repeated heating and cooling of the reaction for DNA melting and enzymatic replication of the DNA. Primers (short DNA fragments) containing sequences complementary to the target region along with a DNA polymerase are key components to enable selective and repeated amplification. This repetitive process is the chain reaction in which the original DNA template is exponentially amplified. With the method of PCR it is possible to amplify a single piece of DNA, or a very small number of pieces of DNA, over many cycles, generating millions of copies of the original DNA molecule.. PCR has been extensively modified to perform a wide array of genetic manipulations, diagnostic tests, and for many other uses.

The PCR technology is characterized by three major advantages. Firstly, is the the speed and ease of use. PCR procedure can be performed in a few hours, using relatively unsophisticated equipment. Typically, a PCR reaction consists of 30 cycles containing a denaturation, synthesis and re-annealing step, with an individual cycle typically taking 3-5 minutes in an automated thermal cycler. Some time is required for designing and synthesizing oligonucleotide primers, but this has been simplified by the availability of computer software for primer design and rapid commercial synthesis of custom oligonucleotides. Once the conditions for a reaction have been tested, the reaction can be repeated simply. Secondly the sensitivity of the method. PCR is capable of amplifying sequences from minute amounts of target DNA, even the DNA from a single cell. Such exquisite sensitivity has afforded new methods of studying molecular pathogenesis and has found numerous applications in diagnosis and other scientific sections where samples may contain extremely low numbers of cells. However, the extreme sensitivity of the method means that great care has to be taken to avoid contamination of the sample under investigation by external DNA, such as from minute amounts of cells from the operator. Thirdly, the reliableness. PCR can permit amplification of specific sequences from material in which the nucleic acid is badly degraded or embedded in a medium from which conventional DNA isolation is problematic. Taking into account these advantages along with the ever reducing cost of the procedure, PCR is a very popular method for routine diagnostics.

Reverse transcription polymerase chain reaction (RT-PCR) is a variant of polymerase chain reaction (PCR). This laboratory technique is commonly used for the detection of the main honeybee viruses (BQCV, SBV, CBPV, KBV, ABPV, DWV). Many RT-RCR protocols for this purpose have been published during the last ten years (Evans & Hung 2000; Benjeddou *et al.*, 2001; Evans & Wheeler, 2001; Bakonyi *et al.*, 2002; Ribiere *et al.*, 2002; Ongus *et al.*, 2004; Genersch, 2005; Yue & Genersch, 2005; Topley *et al.*, 2005; Antunez *et al.*,2006; Berenyi *et al.*, 2006; Siede & Büchler, 2006; Tentcheva *et al.*, 2006; de Miranda *et al.*, 2007; de Miranda *et al.*, 2010). The RT_PCR procedure includes tree major steps. The first step is reverse transcription (RT), in which RNA is reverse transcribed to cDNA using reverse transcriptase. This step is very important in order to perform PCR since DNA polymerase can act only on DNA templates. The RT step can be performed either in the same tube with PCR (one-step PCR) or in a separate one (two-step PCR) using a temperature between 40°C and 50°C, depending on the properties of the reverse transcriptase used. The next step involves the denaturation of the dsDNA at 95°C, so that the two strands separate and the primers can bind again at lower temperatures and begin a new chain reaction. Then, the temperature is decreased until it reaches the annealing temperature which can vary. The main consideration, of course, when choosing the optimal annealing temperature is the melting temperature (Tm) of the primers and probes. The final step of PCR amplification is DNA extension from the primers. This is done with thermostable Taq DNA polymerase, usually between 60 and 72°C, the temperature at which the enzyme works optimally. The length of the incubation at each temperature, the temperature alterations, and the number of cycles are controlled by a programmable thermal cycler. The analysis of the PCR products depends on the type of PCR applied. If a conventional PCR is used, the PCR product is detected using agarose gel electrophoresis and ethidium bromide. In order to quantify RCR reaction Real Time RT-PCR or Quantitative RT-PCR is used. In this version of PCR the amplicons can be visualized as the amplification progresses using a fluorescent reporter molecule. This function is operated through a light-detection system which is integrated into the thermocycler machine.

In conclusion, this section represents a short description of the major diagnostic techniques concerning the honeybee viruses' detection. All techniques that are referred to the scientific articles have their advantages and disadvantages. But in general terms, the ideal protocol of a method has to be sensitive, accurate, reliable, simple, fast and cheap. Based on these components, researchers are trying to find the best combination of the above characteristics.

REFERENCES

Allen M., Ball B. V. (1995). Characterisation and serological relationships of strains of Kashmir bee virus. *Annals of applied biology* 126:471-484.

Allen M., Ball B. V. (1996). The incidence and world distribution of honey bee viruses. *Bee World* 77:141-162.

Anderson D. (1993). Pathogens and Queen Bees. The Australian Beekeeper, pp. 292-296.

Anderson D. L., Gibbs A. J. (1988). Inapparent virus infections and their interactions in pupae of the honey bee (*Apis mellifera* Limnaeus) in Australia. *Journal of general virology* 69:1617-1625.

Anderson D. L., Gibbs A. J. (1989). Tanspuparial transmission of Kashmir bee virus and sacbrood virus in the honey bee (*Apis mellifera*). *Annals of applied biology* 114(1):1-7.

Antúnez K., D'Alessandro B., Corbella E., Ramallo G., Zunino P. (2006). Honeybee viruses in Uruguay. *Journal of Invertebrate Pathology* 91:67-70.

Bacandritsos N., Granato A., Budge G., Papanastasiou I., Roinioti E., Caldon M., Falcaro C., Gallina A., Mutinelli F. (2010). Sudden deaths and colony depopulation decline in Greek honey bee colonies. *Journal of Invartabrate Pathology* 105:335-340.

Bachitter S. R., Rakesh R. (2008). Detection of sacbrood virus and the incidence of sacbrood disease in *Apis mellifera* colonies in the North-Western Himalayas. *Journal of Apicultural Research* 47(1):58-82.

Bakonyi T., Farkas R., Szendroi A., Dobos-Kovacs M., Rusvai M. (2002). Detection of acute bee paralysis virus by RT-PCR in honey bee and *Varroa destructor* field samples: rapid screening of representative Hungarian apiaries. *Apidologie* 33:63-74.

Bailey, L. (1968). The purification and properties of chronic bee-paralysis virus. *Journal of General Virology* 2:251-260.

Bailey L., Ball B. V. (1991). Honeybee pathology. Academic Press, 1991, 193 p.

Bailey L., Ball B. V. and Perry J. N. (1983). Association of viruses with two protozoal pathogens of the honey bee. *Annals of Applied Biology* 103:13-20.

Bailey L., Brenda B. V. and Perry J. N. (1981). The prevalence of viruses of honey bees in Britain. *Annals of Applied Biology* 97:109-118.

Bailey L., Gibbs A. J. and Woods R. D. (1963). Two viruses from adult honey bees (*Apis mellifera* Linnaeus). *Virology* 21:390-395.

Bailey L., Milne R. G. (1969). The multiplication regions and interactions of acute and chronic bee paralysis viruses in adult honey bees. Journal of General Virology 4:9-14.

Bailey L., Woods R. D. (1977). Two more small RNA viruses from honey bees and further observations on sacbrood and acute bee-paralysis viruses. *Journal of general virology* 37:175-182.

Ball B. V. (1999). Sacbrood. In Colin M. E. (ed.), Ball B. V. (ed.), Kilani M. (ed.). *Bee disease diagnosis*. Zaragoza: CIHEAM-IAMZ, 1999, pp. 91-97: 1 ill. 22 ref. (Options Méditerranéennes: Série B. Etudes et Recherches; n. 25). Course on Bee Disease Diagnosis, 1997/05/19-30, Tunis (Tunisia).

Ball B. V., Bailey L. (1991). Viruses of honey bees. *In*: Atlas of Invertebrate viruses, Florida Boca Traton USA CRC Press, Adams J. R., Bonami J. R. (Eds.), pp. 525-551.

Ball B. V., Bailey L. (1997). Viruses. In: R. A. Morse and K. Flottum, Editors, Honey Bee Pests, Predators, & Diseases (third ed.), A. L. Root Company, Medina, 1997, pp. 11-32.

Benjeddou M., Leat N., Allsopp M., Davison S. (2001). Detection of acute bee paralysis virus and black queen cell virus from honeybees by reverse transcriptase PCR. *Applied and Environmental Microbiology* 67:2384-2387.

Berényi O., Bakonyi T., Derakhshifar I., Koglberger H. and Nowotny N. (2006). Occurence of six honeybee viruses in diseased Austrian apiaries. *Applied and environmental microbiology* 72(4):2414-2420.

Blanchart Ph., Schurr F., Celle O., Cougoule N., Drajnudel P., Thiéry R., Faucon J. P., Ribière M. (2008). First detection of Israeli acute paralysis virus (IAPV) in France, a dicistrovirus affecting honeybees (*Apis mellifera*). *Journal of Invertebrate Pathology* 99(3):348-350.

Bowen-Walker P. L., Martin S. J. and Gunn A. (1999). The transmission of deformed wing virus between honeybees (*Apis mellifera L.)* by the ectoparasitic mite *Varroa jacobsoni* Oud. *Journal of Invertebrate Pathology* 73:101-106.

Cox-Foster D. L., Conlan S., Holmes E. C, Palacios G., Evans J. D, Moran N. A, Quan P. L., Briese T., Hornig M., Geiser D. M., Martinson V., vanEngelsdorp D., Kalkstein A. L., Drysdale A., Hui J., Zhai J., Cui L., Hutchison S. K., Simons J. F., Egholm M., Pettis J. S., Lipkin W. I. (2007). A Metagenomic Survey of Microbes in Honey Bee Colony Collapse Disorder. *Science* 318(5848):283-287.

Chen Y., Evans J. D. (2007). Historical presence of Israeli acute paralysis virus in the United States. *American Bee Journal* 147:1027-1028.

Chen Y., Evans J., Feldlaufer M. (2006a). Horizontal and vertical transmission of viruses in the honey bee, *Apis mellifera*. *Journal of Invertebrate Pathology* 92:152-159.

Chen Y. P., Higgings J. A., Feldlaufer M. F. (2005). Quantitative real-time reverse transcription-PCR analysis of deformed wing virus infection in the honeybee (*Apis mellifera L.*). *Applied and Environmental Microbiology* 71:436-441.

Chen Y. P., Pettis J., Collins A., Feldlaufer M. (2006b). Prevalence and transmission of honeybee viruses. *Applied and Environmental Microbiology* 72(1):606-611.

Chen Y. P., Siede R. (2007). Honey bee viruses. *Advances in Virus Research* 70:33-80.

Currie R.W., Pernal S.F., Guzmán-Novoa E. (2010). Honey bee colony losses in Canada. *Journal of Apicultural Research* 49:104-106.

Dainat B., Ken T., Berthoud H., Neumann P. (2009). The ectoparasitic mite *Tropilaelaps mercedesae* (Acari, Laelapidae) as a vector of honeybee viruses. *Insect Soc* 56:40-43.

Dall D. J. (1987). Multiplication of Kashmir Bee Virus in pupae of the honeybee, *Apis mellifera. Journal of Invertabrate Pathology* 49:279-290.

de Miranda J. R. (2008). Diagnostic techniques for virus detection in honey bees. In: Virology and the Honey Bee, pp. 125-232. Edited by M. Aubert, B. Ball, I. Fries, R.Moritz, N.Milani & I. Bernardinelli. Luxembourg: European Communities Press.

de Miranda J. R., Ball B. V., Todd J. H. (2007). Incidence and molecular characterization of viruses found in dying New Zealand honey bee (*Apis mellifera*) colonies infested with *Varroa destructor*. *Apidologie* 38(4):354-367.

de Miranda J. R., Cordoni G., Budge G. (2010). The acute bee paralysis virus-Kashmir bee virus-Israeli acute paralysis virus complex. *Journal of Invertebrate Pathology* 103:S30-S47.

de Miranda J. R., Drebot M., Tyler S., Cameron C. E., Stoltz D. B. and Camazine S. M. (2004). Complete nucleotide sequence of Kashmir bee virus and comparison with acute bee paralysis virus. *Journal of General Virology* 85:2263-2270.

Eyer M., Chen Y. P., Schafer M. O., Pettis J., Neumann P. (2009a). Small hive beetle, *Aethina tumida*, as a potential biological vector of honey viruses. *Apidologie* 40:419-428.

Eyer M., Chen Y. P., Schafer M. O., Pettis J., Neumann P. (2009b). Honey bee sacbrood virus infects adults small hive beetles, *Aethina tumida* (Coleoptera: Nitidulidae). *Journal of Apicultural Research and Bee World* 48(4):296-297.

Evans J. D., Hung A. C. F. (2000). Molecular phylogenetics and classification of honey bee viruses. *Archives of Virology* 145:2015-2026.

Evans J. D., Wheeler D. E. (2001). Gene expression and the evolution of insect polyphenisms. *Bioessays* 23:62-68.

Faucon J. P., Vitu C., Russo P., Vignoni M. (1992). Diagnostic de la paralysie aigue: application a l'epidemiologie des maladies virales de l'abeille en France en 1990. *Apidologie* 23:139-146.

Fievet J., Tentcheva D., Gauthier L., de Miranda J. R., Cousserans F., Colin M. E. Bergoin M. (2006). Localization of deformed wing virus infection in queen and drone *Apis mellifera* L. *Journal of Virology* 3:16.

Garrido-Bailón E., Martín-Hernández R., Bernal J., Bernal J. L., Martínez-Salvador A., Barrios L., Meana A., Higes M. (2010). Short communication. The detection of Israeli Acute Paralysis virus (IAPV), fipronil and imidacloprid in professional apiaries are not related with massive honey bee colony loss in Spain. *Spanish Journal of Agricultural Research* 8(3):658-661.

Genersch E. (2005). Development of a rapid and sensitive RT-PCR method for the detection of deformed wing virus, a pathogen of the honeybee (*Apis mellifera*). *The Veterinary Journal* 169:121-123.

Ghosh R. C., Ball B. V., Willcocks M. M., Carter M. L. (1999). The nucleotide sequence of sacbrood virus of the honey bee: an insect picorna-like virus. *Journal of General Virology* 80:1541-1549.

Govan V. A., Leat N., Allsopp M., Davison S. (2000). Analysis of the complete genome sequence of Acute Bee Paralysis Virus shows that it belongs to the novel group of insect-infecting RNA viruses. *Virology* 277:457-463.

Highfield A. C., Nagar A. El., Mackinder L. C. M. Noel L. M. L. J., Hall M. J., Martin S. J., Schoeder D. C. (2009). Deformed wing virus implicated in overwintering honeybee colony losses. *Applied and Environmental Microbiology* 75(22):7212-7220.

Higes M., Esperon F., Sanchez-Vizcaino J. M. (2007). Short communication. First report of black queen-cell virus detection in honey bee (*Apis mellifera*) in Spain. *Spanish Journal of Agricultural Research* 5(3):322-325.

Higes M., Martin R., Meana A. (2006). *Nosema ceranae*, a new microsporidian parasite in honeybees in Europe. *Journal of invertebrate pathology* 92:93-95.

Hung A. C. F. (2000). PCR detection of Kashmir bee virus in honey bee excreta. *Journal of Apicultural Research* 39:103-106.

Hung, A. C. F., Shimanuki, H., and Knox D. A. (1996). Inapparent infection of acute paralysis virus and Kashmir bee virus in the U.S. honey bees. *American Bee Journal* 136:874-876.

Kim H. K., Choi Y. S., Lee M. L., Lee M. Y., Lee K. G., Ahn N. H. (2008). Detection of sacbrood virus (SBV) from the Honeybee in Korea. *Korean Journal of Apiculture* 23(2):103-109.

Kukiela D., Sánchez-Vizcaino J. M. (2009). One-step real-time quantitative PCR assays for the detection and field study of the Sacbrood honeybee and Acute bee paralysis viruses. *Journal of Virological Methods* 161:240-246.

Lanzi G., de Miranda J. R., Boniotti M. B., Cameron C. E., Lavazza A., Cappuci L., Camazine S. M., Rossi C. (2006). Molecular and biological characterization of deformed wing virus of honeybees (*Apis mellifera L.). Journal of virology* 80(10):4998-5009.

Leat N., Ball B., Govan V., Davison S. (2000). Analysis of the complete genome sequence of black queen-cel virus, a picorna-like virus of honey bees. *Journal of General Virology* 81:2111-2119.

Lee P. E., Furgala B. (1965). Chronic bee paralysis virus in the nerve ganglia of the adult honey bee. *Journal of Invertebrate Pathology* 7: 170-174.

Liu X., Zhanh Y., Yan X., Han R. (2010). Prevention of Chinese Sacbrood virus infection in *Apis cerana* using RNA interference. *Current Microbiology* 61:422-428.

Madriz J., Bonilla J. A., de Miranda J. R. Espinoza A. M. (2000). Serological relationships between rice hoja blanca virus and Echinochloa hoja blanca virus. *Uniciencia* 17:21-26.

Maori E., Lavi S., Mozes-Koch R., Gantman Y., Reretz Y., Edelbaum O., Tanne E., Sela I. (2007). Isolation and characterization of Israeli acute paralysis virus, adicistrovirus affecting honeybees in Israel: evidence for diversity due to intra- and inter- species recombination. *Journal of General Virology* 88:3428-3438.

Maori E., Paldi N., Shafir S., Kalev H., Tsur E., Glick E., Sela I. (2009). IAPV, a bee-affecting virus associated with colony colapse disorder can be silenced by dsRNA ingestion. *Insect Molecular Biology Journal* 18:55-60.

Mayo M. A. (2002). A summary of taxonomic changes recently approved by ICTV. *Archives of Virology* 147(8):1655-1656.

Nielsen S. L., Nicolaisen M., Kryger P. (2008). Incidence of acute bee paralysis virus, black queen cell virus, chronic bee paralysis virus, deformed wing virus, Kashmir bee virus and sacbrood virus in honey bees (*Apis mellifera*) in Denmark. *Apidologie* 39:310-314.

Nordström S., Fries I., Aarhus A., Hansen H., Korpela S. (1999). Virus infections in Nordic honey bee colonies with no, low or severe *Varroa jacobsoni* infestations. *Apidologie* 30:475-484.

Olivier V., Massou I., Celle O., Blanchard P., Schurr F., Ribiere M., Gauthier M. (2008). *In situ* hybridization assays for localization of the chronic bee paralysis virus in the honey bee (*Apis mellifera*) brain. *Journal of Virological Methods* 153:232-237.

Ongus J. R., Peters D., Bonmatin J. M., Bengsch E., Vlak J. M., van Oers M. M. (2004). Complete sequence of a picorna-like virus of the genus Iflavirus replicating in the mite *Varroa destructor*. *Journal of General Virology* 85:3747-3755.

Palacios G., Hui J., Quan P.., Kalkstein A., Honkavuori K. S., Busseti A. V., Conlan S., Evans J., Chen Y. P., vanEngelsdorp D., Efrat H., Pettis J., Cox-Foster D., Holmes E. C., Briese T., Lipkin W. I. (2008). Genetic analysis of Israel acute pararysis virus: distinct clusters are circulating in the United States. *Journal of Virology* 82(13):6209-6217.

Ribière M., Triboulot C., Mathieu L., Aurières C., Faucon J. P., Pépin M. (2002). Molecular diagnosis of chronic bee paralysis virus infection. *Apidologie* 33:339-351.

Ribiere M., Lallemand P., Iscache A.-L., Schurr F., Celle O., Blanchard P., Olivier V., Faucon J.-P. (2007). Spread of infectious chronic bee paralysis virus by honeybee (*Apis mellifera L.*) feces. *Applied and environmental microbiology* 73(23):7711-7716.

Rebière M., Ball B., Aubert M. F. A. (2008). Natural history and geographical distribution of honey bee viruses. In: Virology and the Honey Bee, pp. 125-232. Edited by M. Aubert, B. Ball, I. Fries, R.Moritz, N.Milani & I. Bernardinelli. Luxembourg: European Communities Press.

Ribière M., Olivier V., Blanchard Ph. (2010). Chronic bee paralysis: a disease and a virus like no other? *Journal of Invertebrate Pathology* 103:120-131.

Scotti P. D., Longworth J. F., Plus N., Croizier G., Reinganum C. (1981). The biology and ecology of strains of an insect small RNA virus complex. *Advances in Virus Research* 26:117-127.

Shah K. S., Evans E. C., Pizzorno M. C. (2009). Localization of deformed wing virus (DWV) in the brains of the honeybee, *Apis mellifera* Linnaeous. *Virology Journal* 6:182-188.

Shen M., Cui L., Ostiguy N., Cox-Foster D. (2005a). Intricate transmission routes and interactions between picorna-like viruses (Kashmir bee virus and sacbrood virus) with the honeybee host and the parasitic varroa mite. *Journal of general virology* 86:2281-2289.

Shen M., Yang X., Cox-Foster D., Cui L. (2005b). The role of varroa mites in infections of Kashmir bee virus (KBV) and deformed wing virus (DWV) in honeybees. *Virology* 342(1):141-149.

Shimanuki H., Calderone N. W., Knox D. A. (1994). Parasitic mite syndrome: the symptoms. *American Bee Journal* 134:827-828.

Siede R., Buchler R. (2004). First detection of Kashmir bee virus in Hesse (German). *Berliner und Munchener Tierarztliche Wochenschrift* 117:12-15.

Siede R., Büchler R. (2006). Spatial distribution patterns of acute bee paralysis virus, black queen cell virus and sacbrood virus in Hesse, Germany Wien. Tierätrzl. Mschr. 93:90-93.

Tentcheva D., Gauthier L., Bagny L., Fievet J., Dainat B., Cousserans F., Colin M. E., Bergoin M. (2006). Comparative analysis of deformed wing virus (DWV) RNA in *Apis mellifera* L. and *Varroa Destructor. Apidologie* 37:1-7.

Tentcheva D., Gauthier L., Zappulla N., Dainat B., Cousserans F., Colin M. E., Bergoin M. (2004). Prevalence and seasonal variations of six bee viruses in *Apis mellifera L.* and *Varroa destructor* mite populations in France. *Applied and Environmental Microbiology* 70(12):7185-7191.

Thomson A. D., Smirk B. A. (1966). Rod-shaped particles associated with a disease of alkali bees. *Journal of Virology* 28:348-350.

Topley E. Davison S., Leat N., Benjeddou M. (2005). Detection of three honeybee viruses simultaneously by a single multiplex reverse transcriptase PCR. *African Journal of Biotechnology* 4:763-767.

Yue C., Genersch E. (2005). RT-PCR analysis of deformed wing virus (DWV) in honey bees (*Apis mellifera*) and mites (*Varroa destructor*). *Journal of General Virology* 86:3419-3424.

Yang X., Cox-Foster D. L. (2005). Impact of an ectoparasite on the immunity and pathology of an invertebrate: Evidence for host immunosuppression and viral amplification. *PNAS* 102(21):7470-7475.

Yue C., Schröder M., Gisder S., Genersch E. (2007). Vertical transmission routes for deformed wing virus of honeybees (*Apis mellifera*). *Journal of general virology* 88:2329-2336.

In: Bees
Editor: Richard M. Florio

ISBN: 978-1-61324-825-6

Chapter 6

Why Massive Honeybee Colony Losses Do Not Occur in Uruguay?

Karina Antúnez[1,a], Ciro Invernizzi[b] and Pablo Zunino[a]
[a]Departamento de Microbiología,
Instituto de Investigaciones Biológicas Clemente Estable,
Montevideo, Uruguay
[b]Sección Etología, Facultad de Ciencias,
Universidad de la República, Montevideo, Uruguay

Abstract

In recent years massive honeybee colony losses have been reported mainly in the north hemisphere, including North America and several European countries. Scientists and beekeepers from all over the world are working to elucidate the causes of this phenomenon. Among different factors potential involved, the presence of different pathogens such as the mite *Varroa destructor*, the microsporidium *Nosema ceranae* and the presence of different RNA viruses may have a role in honeybee losses. It has been proposed that these events cannot be due to a single factor and a combination of causes should be considered.

In Uruguay (South America) there are about 4,000 beekeepers and more than 400,000 beehives. The most important pathogens of honeybees are present and widely distributed in the country.

V. destructor is found along the country, but their virulence varies from region to region. Almost every beekeeper applies acaricides on the end of the summer or the beginning of the fall, owing to avoid colony loses. The acaricides fluvalinate, flumetrin, amitraz and coumaphos are very efficient, except for some regions where resistant mites to some of these molecules have been detected.

N. ceranae is present in Uruguay before 1990 and nowadays is the main species detected of *Nosema* spp.. However, there have not been reports of colony loses whose cause was without doubts nosemosis. This is confirmed by thousands of colonies that are

[1] E-mail address: karina@iibce.edu.uy.

moved to *Eucalyptus grandis* plantations in the north of the country at the end of the summer, where every colony suffers severe infections with *Nosema* spp. and even in that way, reach the following spring without difficulties.

On the other side, the presence of acute bee paralysis virus (ABPV), chronic bee paralysis virus (CBPV), black queen cell virus (BQCV), sacbrood virus (SBV) and deformed wing virus (DWV) have been detected in Uruguay since 2005. Their detection in different geographic regions, the simultaneous co-infection of colonies by several viruses, the high prevalence found in the country and a high incidence in different seasons indicates that they are widely spread. Interestingly, the seasonal incidence of DWV around the year was similar to the incidence of *V. destructor*, according with previous reports. Kashmir virus and Israeli Acute paralysis virus, other honeybee viruses potentially related to honeybee looses have not been detected so far.

The presence and co-existence of these pathogens are maybe involved in the reduction of colonies production detected during the last years. However, massive honeybee looses have not been reported. In the present chapter we present in detail the sanitary situation of Uruguay focussed on the presence of *V. destructor*, *N. ceranae* and RNA viruses and their prevalence and coexistence along the year, and discuss possible reasons why honeybee looses episodes have not been detected.

HONEYBEE COLONY LOSSES

During the last years, surprising high losses of honeybees have devastated a large number of colonies worldwide, mainly in the north hemisphere including North America and several European countries. Reports about colony losses have been carried out in the USA and different European countries [Stokstad, 2007; van Engelsdorp *et al.*, 2008; Potts *et al.*, 2010; Neumann and Carreck, 2010]. These colony losses seriously affect beekeepers production, but more important, also affect all agricultural activities that rely on the honeybee pollination [Stokstad, 2007].

Scientists and beekeepers from all over the world are working to elucidate the causes of this phenomenon. However, to date there is no consensus about the origin and symptoms of such losses. In the USA, a percentage of colony losses are explained by the "Colony collapse disorder" (CCD). This syndrome is characterized by the rapid loss in a colony of its adult bee population, occurring mostly because bees are failing to return to the hive, leaving behind their brood, their queen and a small cluster of adults. No dead adult bees are found inside or in close proximity to the colony. Colony losses have been rapid, and are occurring in large numbers [van Engelsdorp *et al.*, 2008; van Engelsdorp *et al.*, 2010; Stokstad, 2007]. However, since honeybee host and pathogens are genetically diverse, the symptoms and causes of colony losses in different countries may be different [Neumann and Carreck, 2010].

Among the causes that are being studied are the presence of pathogens in the bees and brood, emergence of new pathogens or more virulent ones, chemical contamination with pesticides or agrotoxics, or a combination of these factors that weaken the colonies and allowing infection by other pathogens.

BEEKEEPING IN URUGUAY

Uruguay is a South American country of 176,214 km2 that lies between latitudes 30° and 35°S, and longitudes 53° and 59°W. It has a smooth landscape and a relative temperate weather and in general it offers very appropriate conditions for apiculture.

According to data from the Ministry of Agriculture, Livestock and Fisheries (MGAP) in 2010 there are 2869 registered beekeepers that owed 457625 hives (Dirección General de la Granja, DIGEGRA-MGAP, 2010).

Almost all the honey produced in Uruguay is exported mainly to Germany and other European countries like Spain or the United Kingdom, and USA.

A signal of alarm for beekeepers has risen since the volume of honey produced in the country is decreasing in the last years. Although in 2009 the international price of honey was higher than in previous years, total honey production (6,032 tons) was about 30% lower compared to traditional rates obtained at the middle of the decade.

It is interesting to note that this drop in the volume of honey production was not associated to a decrease in the number of hives that in spite of minor variations remained relatively constant through all these years (Table 1) [Errea and Licandro, 2009].

In consequence this important reduction can be explained by the drop in honey production by hive, as it can be seen in Table 2.

Considering the whole figure, it can be concluded that in recent years massive honey bee losses like those reported in the USA or Western Europe [Neumann and Carreck, 2010; Stokstad, 2007] did not occur in Uruguay. A similar conclusion was obtained by Vandame and Palacio [2010] when analyzed the situation of different Latin American countries. However, in our country, a strong drop in honey yields was noticed. This fact reflects that there are several problems affecting beekeeping in Uruguay. Among these we can mention climatic factors like drought or frosts, fast expansion of land area dedicated to crops like soy,

Table 1. Evolution of the number of hives and beekeepers in Uruguay in recent years

	Number of hives	Number of beekeepers
2007	517,245	4,039
2008	491,430	3,314
2009	486,492	3,114
2010	457,625	2,869

Table 2. Produced honey by beehive in Uruguay from 2006 to 2009

Year	Honey production per hive
2006	38,66 kg
2007	35,54 kg
2008	26,64 kg
2009	16,74 kg

barley or wheat in the western littoral where about the half of all beehives in the country are located or the increasing use of agrochemicals related to commercial cultures [Errea and Licandro, 2009; Vandame and Palacio, 2010]. Another problem that must be taken into account is the influence of bee pathogens, issue that will be treated in this chapter, focussed on the more important ones, *Varroa destructor*, *Nosema ceranae* and RNA viruses.

V. Destructor, the Main h\Honeybee Health Threat

The ectoparasitic mite *V. destructor* is the major pest of honey bees worldwide, especially in temperate regions, causing great economic losses to beekeeping [Rosenkranz *et al.*, 2010]. Adult female mites have a phoretic phase on the body of workers and drones and reproduce inside the brood cells feeding on the hemolymph of pupae. Infected adult bees have lower longevity, reduced learning ability and show difficulties when returning to the hive [Rosenkranz *et al.*, 2010].

V. destructor dispersal through different European, Asian and South American countries occurred during the 1970 and 1980 [De Jong, 1997; Rosenkranz *et al.*, 2010]. In Uruguay, the mite was first reported in 1978 in Montevideo province and quickly spread throughout the whole territory [Invernizzi *et al.*, 2011]. Since then, two stages can be distinguished in relation to the damage that the mite has caused to the colonies. From its arrival until the late '90s, colonies were not significantly affected and many beekeepers had no need to apply acaricides. In contrast, over the last decade the situation changed dramatically and the vast majority of beekeepers had to routinely use synthetic acaricides to prevent massive colony losses. However, surprisingly in some particular areas of the country there are beekeepers that still do not use acaricides without having significant colony losses.

The good tolerance that bees had to *V. destructor* for nearly 20 years motivated different studies to determine its causes. Thus, Ruttner *et al.* [1984] suggested that in Uruguay *V. destructor* presented a reproductive rate and a juvenile hormone level different from those found in other countries where the mite caused important damage. Kirsch and Rosenkranz [1998] found that the damage caused by the mite varied along different country's regions also noting a reduced descent of male and female mites.

Another factor that may have influenced the reduced damage caused by *V. destructor*, is the level of africanization (hibridization with *Apis mellifera scutellata*) that European bees (mostly *A. mellifera mellifera*) have in Uruguay. The greater tolerance to the mite of africanized bees compared to European bees was verified in several studies. It could be explained by many factors that vary between africanized and European bees including behavioral mechanisms of resistance (hygienic behavior, grooming), larvae attractiveness, duration of the brood capping period, cells size, tendency to swarm, mite's fertility and fecundity, among others factors [Rosenkranz *et al.*, 2010; De Jong, 1997; Büchler, 1994].

In the region, Issa *et al.* [2000] found a variation in *V. destructor* infestation rates in the transition area of africanized bees and European bees (30° to 35° south), being lower toward areas with greater africanization. In Uruguay there are two studies about bees' africanization. Burgett *et al.* [1995] using morphometric and molecular (mtDNA) analysis found that 30% of the colonies belonged to africanized bees and 53% to africanized hybrids. Some years later, Diniz *et al.* [2003] based on analysis of allozyme loci and mtDNA, confirmed that in Uruguay bees have a high degree of africanization. They found almost 100% of africanized haplotypes in the region bordering Brazil and a decreasing gradient from north to south.

The increase in the level of damage caused by *V. destructor* since the late '90s could be due to the enhanced virulence of the mite populations. Virulence of a pathogen is promoted when the probability of horizontal transmission rises [Fries and Camazine, 2001]. In the case of *V. destructor*, colonies proximity in apiaries in addition to apiaries closeness in some regions, could favour mite's horizontal transmission over vertical transmission through swarming that would occur in natural conditions. Coincidentally, the greatest colonies losses caused by *V. destructor* occurred in the west provinces where historically a high density of colonies can be seen. To study this aspect, *V. destructor* infestation was compared between an apiary in Colonia, a western province and an apiary in Treinta y Tres, a south eastern province where colonies density has been historically much lower. Moreover, colonies of this latter (apiary) had never received acaricides treatment. In June Colonia colonies had an average of 15.9% phoretic mites and bee's adult population covered an average of only 2.8 frames, while in Treinta y Tres colonies the values were 6.9 and 6.4 respectively. In Colonia's apiary it was necessary to apply acaricides to prevent the loss of the colonies while in Treinta y Tres there were no significant losses during the winter. This situation is best explained by a difference in *V. destructor* virulence than by a greater resistance of southeastern bees because when they are transferred to the west they do not survive winter unless they receive an acaricide. Seeley [2007] obtained a similar result when he compared resistance to *V. destructor* of wild bee colonies from a reserve without beekeeping, with years of co living with the mite, with colonies used in apiculture (New World Carniola). He found that the population of *V. destructor* increased similarly in both groups showing that bees did not show differential resistance to the mite. The author suggests that the stable relationship between bees and *V. destructor* at the reserve is a product of adaptive avirulence of the parasite and not of bees' resistance.

Currently, Uruguayan beekeepers rely on four active ingredients to control the mite: the pyrethroids fluvalinate and flumetrine, the formamidine amitraz and the organophosphate coumaphos. The fluvalinate was practically the only product used from 1989 until around 2000 when *V. destructor* resistance started to become evident in colonies of western areas. At present it offers acceptable efficiency across the country except in some western and south central areas where it has been used for/over several years [Campa *et al.*, 2007]. Coumaphos, valued by beekeepers for its acaricide effectiveness, has recently presented problems in the western Colonia province where in some apiaries *V. destructor* showed a significant resistance verified in the field and by laboratory tests [Maggi *et al.*, 2010]. However, most beekeepers get to control *V. destructor* rotating between years the active ingredients available. Beekeepers of western and south central areas of the country must make tighter controls and eventually strengthen the fall treatment using an organic acaricide in the spring, being oxalic acid the most commonly used.

Increased damage caused by *V. destructor* coincided with the entry into the country in 1998 of the bacterium *Paenibacillus larvae*, etiological agent of American foulbrood, probably from Argentina [Harriet *et al.*, 2000; Piccini and Zunino, 2001; Piccini *et al.*, 2002; Antunez *et al.*, 2004]. This determined that around 2000, Uruguayan apiculture went through a difficult period under a health point of view. Colony losses caused by both pathogens, in addition to colonies with American foulbrood eliminated by beekeepers following health officials recommendations, probably acted as a strong selection pressure to increase bees' hygienic behavior in Uruguay. This is clear according to assessments of the hygienic behavior made over 15 years [Invernizzi *et al.*, 2011] (Figure 1). Several studies have shown that this

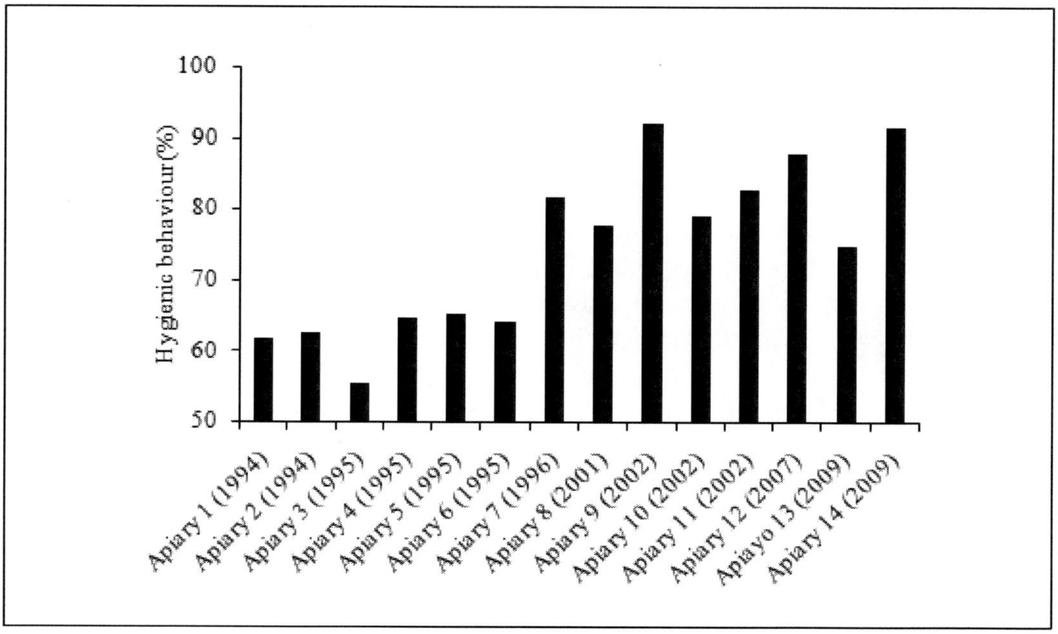

Figure 1. Average of hygienic behavior of *A. mellifera* colonies measured along 15 years in apiaries without selection [Invernizzi *et al.* 2011].

social behavior is an important general mechanism of resistance against pathogens such as *V. destructor* [Boecking and Drescher, 1992; Spivak and Reuter, 2001b; Spivak, 1996] and *P. larvae* [Rothenbuhler, 1964; Spivak and Reuter, 2001a]. The increase in hygienic behavior expressed by bees in recent years may explain the very low prevalence of brood diseases such as American foulbrood, European foulbrood, Chalkbrood disease and Sacbrood [Invernizzi *et al.*, 2011]. It is possible that more hygienic colonies are more able to limit the growth of *V. destructor* population facilitating the control of the parasite.

As a conclusion it can be established that *V. destructor* is the main health problem of bees in Uruguay. The different impact on different areas of the country is probably due to mite populations that vary in virulence and resistance to some acaricides. *V. destructor* associations with different viruses and the likelihood of colonies reinfection due to the high density in some regions may also explain the differential level of damage caused by this parasite. However, the application of conventional synthetic acaricides in late summer or early fall, and the possibility to complement it with an organic acaricide before spring, is sufficient to control *V. destructor* avoiding colony losses.

Nosema ceranae, a Problem Associated to Plantations of *Eucalyptus Grandis*

The microsporidium *N. ceranae*, whose original host is the Asian honeybee *Apis cerana*, was found in European honeybees in Spain in 2005 [Higes *et al.*, 2006]. Until then it had been believed that *Nosema apis* was the only species causing nosemosis in European honeybees [Bailey and Ball, 1991; Fries, 1997; Hornitzky, 2008]. The discovery of *N. ceranae* had a huge impact on beekeeping and on the scientific community worldwide since it could be a cause of the massive colony losses. Studies conducted in recent years found that this emergent pathogen is distributed worldwide and that its "jump" from *A. cerana* to *A.*

mellifera occurred more than 20 years ago [Invernizzi *et al.*, 2009; Higes *et al.*, 2006; Klee *et al.*, 2007; Huang *et al.*, 2007; Chen *et al.*, 2008; Higes *et al.*, 2009; Giersch *et al.*, 2009; Fries, 2010]. *Nosema* spp. that affect honeybees reproduce within the epithelial ventricular cells of adult bees affecting digestive functions, leading to malnutrition, physiological aging, reduction of hypopharyngeal glands and premature death [Bailey and Ball, 1991; Fries, 1997; Hornitzky, 2008].

Comparative studies found that *N. ceranae* is more virulent and tolerates higher temperatures than *N. apis*, which would explain expansion of the latter species through different regions worldwide [Higes *et al.*, 2007; Martin-Hernandez *et al.*, 2007; Paxton *et al.*, 2007; Higes *et al.*, 2008; Higes *et al.*, 2010]. Recently Antúnez *et al.* [2009] found that *N. ceranae* suppresses the immune response of bees while this does not happen with *N. apis*. These results may explain the different virulence of both species. However, the role of *N. ceranae* as a cause of massive colony losses in recent years in the Northern Hemisphere is questioned [Cox-Foster *et al.*, 2007; Chen *et al.*, 2008; Gómez Pajuelo *et al.*, 2008; Invernizzi *et al.*, 2009; Mayack and Naug, 2009; Forsgren and Fries, 2010].

In Uruguay, nosemosis is recognized since 1940 and, as in other countries, it was believed that *N. apis* was the parasite that caused the disease. However, Invernizzi *et al.* [2009] recently found that *N. ceranae* is spread through the country, while *N. apis* was not detected until date. This study also reported the presence of *N. ceranae* in bees obtained before 1990, being the earliest record of this species [Paxton, 2010].

The DILAVE (MGAP) has analyzed 61916 bee samples submitted by beekeepers from 1964 to 2007 finding that the proportion of infected samples was variable along the analyzed period. According to these records the highest incidence of the disease occurred between the years 1964-1967 with more than 40% of positive samples. In the following years, values decreased until they reached an average of less than 10% in the period 1978-1983. The second period of increased incidence of nosemosis went through the years 1998-2002 with more than 35% of positive samples. Thereafter, values remained below 30%. It is noted that from 1990 onwards [with the certain presence of *N. ceranae*) nosemosis has not increased steadily [Invernizzi *et al.*, 2009]. This behaviour of the disease is very different from the epidemiological pattern found in Spain in the 1999-2005 period characterized by a significant increase in positive samples [Martin-Hernandez *et al.*, 2007].

In Uruguay nosemosis occurs inevitably in colonies that are moved to commercial plantations of *Eucalyptus grandis* in late summer to exploit the enormous potential for honey production of these trees. For example, colonies virtually free of *Nosema* spp. were transferred to an *E. grandis* plantation (in late summer), and three days later showed a 27.6 % average of infected foragers. In may (autumn), at the end of the flowering period, colonies exhibited an average of 90.8 % infected foragers [Invernizzi, unpublished data].

Despite colonies in *E. grandis* plantations acquire high levels of infection beekeepers do not lose colonies if these are removed at the end of the flowering period (autumn) and these colonies frequently survived in good conditions until the following spring.

Mendoza (unpublished data) studied this aspect in two apiaries that were removed from *E. grandis* plantations in autumn and transferred to two different zones, one in the north and one south of the country. In each apiary colonies were divided into four groups with similar levels of *Nosema* spp. infection and treatments were performed as follows: 1) application of 100 mg of fulmagilin/colony, 2) supply two 150 g protein supplement cakes, 3) application of 100 mg of fulmagilin/colony and supply of two 150 g protein supplement cakes and 4)

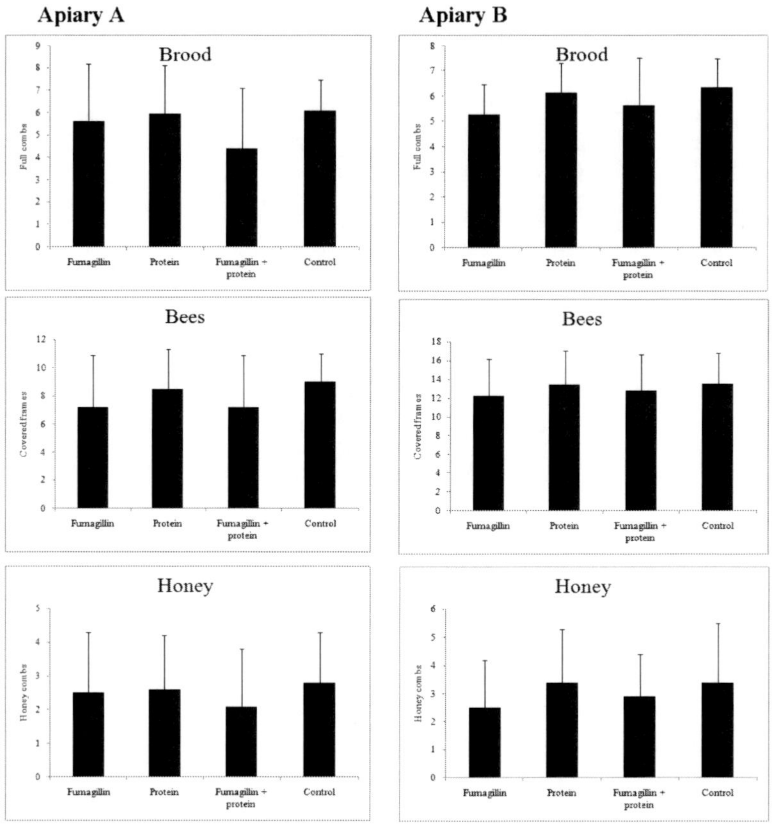

Figura 2. Adult honeybee population, breeding area and honey stores from two apiaries removed from *E. grandis* plantations in autumn and subjected to different treatments: 1) application of fumagillin (100 mg), 2) protein cakes (2 x 150 g), 3) application of fumagillin (100 mg) and protein cakes (2 x 150 g), and 4) control.

control. In the assessment of the colonies at the beginning of the spring he found no significant differences in adult honeybee population, breeding area and honey stores between the groups indicating that *Nosema* spp. does not affect wintering conditions (Figure 2).

The situation changes when colonies remain in the *E. grandis* plantations after the flowering period (autumn). Santos *et al.* [2005] found that in an apiary of 27 colonies that remained more than two months after that, 20 colonies were heavily depopulated with less than four frames covered with bees and 14 days later, when colonies were removed, 8 of them were already dead. According to the authors both *N. ceranae* and *V. destructor* could be responsible for the weakening of the colonies. This aspect was also confirmed by Mendoza *et al.* [2010] in an apiary located in an *E. grandis* plantation that remained there until spring. In autumn, colonies were classified into two groups according to whether they had more or less than a million spores on average per bee. At spring adult population was recorded and colonies were separated into three groups: dead, weak (up to 5 frames covered with bees) and strong (more than 5 frames of bees). It was found that the level of *N. ceranae* infection in autumn was significantly associated to colony condition in spring (Figure 3). Thus, *E. grandis* plantations are a very interesting scenario to investigate how *N. ceranae* affects colonies.

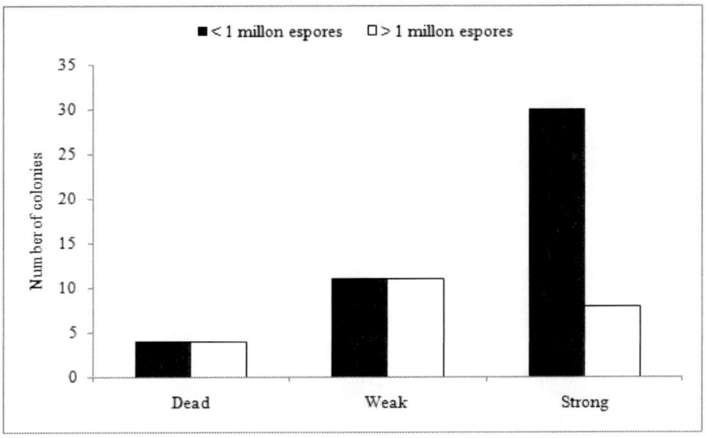

Figure 3. Relation between the number of *Nosema* spp. spores in autumn (more and less than a million spores on average per bee; white and black respectively) and the status of the colonies (dead, weak or strong) in spring (Figure 3), in colonies located at *E. grandis* plantations.

According to the information available we can state that colonies that are removed of the plantations at the end of the flowering of the trees survive the winter without the need to apply fumagillin. However, if the colonies remain within plantations they face the risk of depopulation and eventually death. One factor that may explain the evolution of colonies infected with *N. ceranae* at the end of the *E. grandis* flowering period is the availability of pollen from different botanical origins.

Santos *et al.* [2005] moved to a plantation colonies of equal size separated into three groups according to the availability of pollen: colonies without pollen stores, colonies with abundant pollen stores from different botanical origin and colonies in the same condition as the previous group but regularly supplemented with pollen from different botanical origin. In the middle of *E. grandis* flowering period colonies that lacked pollen stores had significantly more infected foragers than those colonies with varied pollen available (Figure 4). These differences disappeared at the end of the flowering period when the colonies had consumed the stores and *E. grandis* was the sole source of pollen.

Recently Alaux *et al.* [2010b] found that pollen botanic diversity induced higher glucose oxidase activity compared with monofloral diets, including protein-richer diets. Nevertheless, it did not improve the expression of other components of humoral and cellular responses possibly involved in the *Nosema* spp. resistance [Antunez *et al.*, 2009; Alaux *et al.*, 2010b]. Thus, it is possible that colonies that are removed from the plantations have access to new sources of pollen which allow bees to control *N. ceranae*.

Based on the studies conducted so far we can say that in Uruguay there is no evidence that colonies infected with *N. ceranae* go through the four stages described by Higes *et al.* [2008] (Asyntomatic, Replacement, False Recovery, Depopulation), along approximately 18 months culminating inevitably with the death of the colony. However, the difference could be explained by other factors such as the presence of other pathogens. In this sense, Bromenshenk *et al.* [2010] recently found that coinfection of honeybees with *N. ceranae* and the newly discovered invertebrate iridiscent virus (IIV) could explain the CCD in the USA. Conducting studies to determine if the IIV is present in other countries, including Uruguay, may be useful to know the role of *N. ceranae* in the depopulation of colonies.

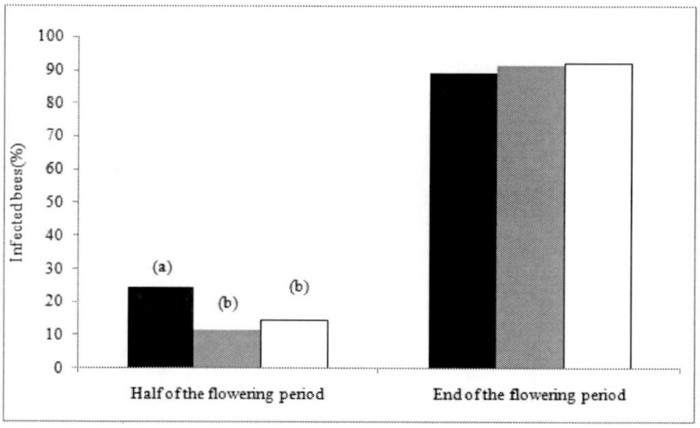

Figure 4. *N. ceranae* infection in colonies with different pollen reserves. Black: colonies without pollen; Gray: colonies with pollen reserves; White: colonies with pollen reserves and pollen supplement. Different letters indicate significant differences at the level $\alpha = 0.05$ for the Mann Whitney test. The lack of letters indicate no significant differences in any comparison.

Occurrence and Distribution of Honeybee Viruses In Uruguay

More than eighteen RNA viruses that affect honeybees have been described. Between the most important are acute bee paralysis virus (ABPV), black queen cell virus (BQCV), chronic bee paralysis virus (CBPV), deformed wing virus (DWV), Israeli acute paralysis virus (IAPV), Kashmir virus (KV) and sacbrood virus (SBV) [Chen and Siede, 2007].

CBPV was one of the first isolated viruses that affect adult honeybees, causing a disease characterized by bee paralysis, trembling, flightlessness and, sometimes, black individuals crawling at the hive entrance [Bailey *et al.*, 1963; Bailey, 1975]. ABPV also affects adult honeybees and causes similar symptoms, but it is more virulent than CBPV [Bailey *et al.*, 1963]. BQCV was first detected in queen larvae and prepupae that became brown to black [Bailey and Woods, 1974]. However, it also affects larvae and pupae from worker bees without developing typical symptoms. SBV affects larvae of honeybees that acquire a pale yellow colour; their skin becomes leathery and the ecsydial fluid accumulates between the body and the skin. It also affects adult bees but without developing typical symptoms [Ball and Bailey, 1997]. DWV affects adult bees causing a well-defined disease, characterized by deformed, crumpled or shrunken wings and decrease of body size. However, it has also been detected in eggs, larvae and pupae [Ball and Bailey, 1997]. KV affects all stages of the bee life cycle, causing mortality, but it does not cause specific symptoms.

Although viruses may be associated with different symptoms, all of them can persist in a latent state in apparently healthy colonies [Allen and Ball, 1996; Bailey, 1967].

The importance of viruses in honeybee health has gained attention in recent years, especially due to the serious problem of honeybee mortality occurring worldwide.

The first approach about the study of the occurrence of honeybee viruses in Uruguay was carried out during 2004- 2005. Samples from different provinces (Colonia and Soriano [west], Canelones and San José [south], Maldonado, Treinta y Tres and Lavalleja [east] and Rivera [north]) were collected and analyzed by RT-PCR [Antunez *et al.*, 2005; Antunez *et al.*, 2006].

During the first year of survey (2004) 78 % of the samples were infected; 61 % presented CBPV, 50 % presented ABPV and 43 % were co-infected by both viruses. The presence of mortality symptoms was not associated to one particular virus, or even with co-infection. However, mortality symptoms were also present in samples where these viruses were not detected [Antunez *et al.*, 2005].

In the second year of study (2005), we included the detection of a broad range of viruses, in an attempt to explain the proportion of non infected samples with symptoms associated to mortality previously observed. Most of the samples were infected with at least one virus (96 %). The first highlight of the results compared to the previous year, is the lower proportion of CBPV and ABPV detection (46 and 11 % respectively) and the absence of co-infection between both viruses. We also found a high prevalence of BQCV, DWV and SBV (92, 100 and 100 %, respectively). Mortality-associated symptoms were detected in 38 % of the samples and most of these samples (98 %) were infected by one, two, three or even four different viruses [Antunez *et al.*, 2006].

The detection of several viruses in honeybees from different and distant locations, and the fact that most of the samples were infected at least with one virus or more than one, suggests that they are widely spread in the country.

Seasonal Dynamics of Honeybee Pathogens

In spite of the presence and wide distribution of the most important bee pathogens in the country, as stated before massive colony losses have not been reported so far. However, a couple of apiaries have been selected in which colony losses were higher than expected. In an attempt to understand the causes of these elevated losses, and to prevent the dissemination of

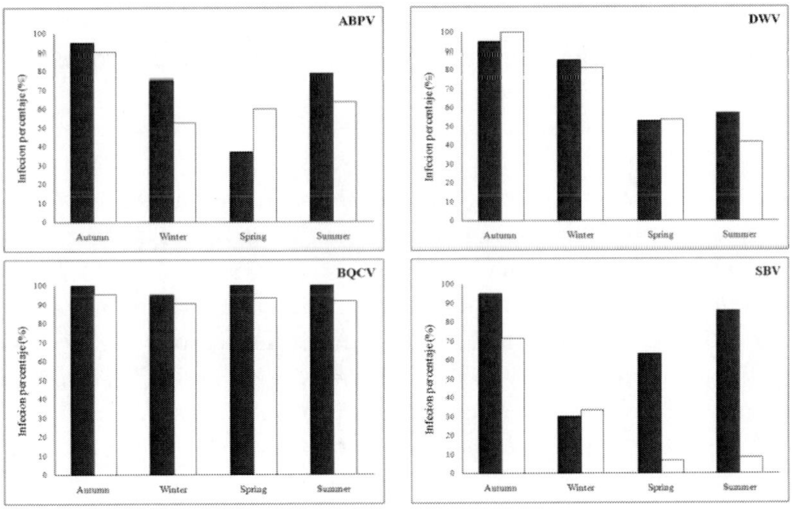

Figure 5. Stational variation in the incidence (percentage of infected colonies per season) of different RNA viruses, ABPV, BQCV, DWV and SBV. Black: apiary located in Colonia (west); White: apiary located in Canelones (south).

this punctual problem to other apiaries around the country, different potential causes are being studied. One of the most important potential causes is the presence of pathogens, so the occurrence and seasonal dynamics of different honeybee pathogens have been under study. Two apiaries located at different geographic locations of Uruguay, Colonia (west) and Canelones (south) were selected, based on the presence of colony losses higher than expected in the region during previous years, although they were subjected to good beekeeping practices and the correct application of treatment against *V. destructor*. Colonies were sampled and analyzed during autumn, winter, spring and summer of 2009. In this case, detection and quantification of *V. destructor* and *N. ceranae* was performed according to OIE standard methods and viruses were detected and quantified using RT and qRT-PCR methods [Anido *et al.*, 2011]. Results are shown in Figure 5.

The incidence (percentage of infected colonies in one season) of ABPV resulted high in autumn (92 %) and it decreased along the year until spring (48 %), when it started to rise again towards the summer. No significant variations were detected between infection levels (viral transcript per colony) around the year.

The incidence of BQCV resulted high during all the year round in both apiaries, being higher than 90 % in both apiaries.

DWV incidence also resulted high in autumn (about 97 % in both apiaries), and also diminished during the year to spring, nearly to 50 %.

Lastly, SBV also presented a high infection rate in autumn, being 83 % and decrease to winter.

In most cases the incidence per season presented a similar behavior than the infection level per colony (viral transcript per colony) at this season. Higher BQCV, DWV infection level was detected in autumn and it diminished along the year.

It is important to notice that IAPV and KBV were not detected in any of the analyzed samples, suggesting that these viruses are not present in our country, or that their prevalence is too low to be detected.

When seasonal variation of *V. destructor* was evaluated, the expected results were found. *V. destructor* was present in 97 % of the colonies in autumn but this percentage decreased to 2 % in winter, after the application of treatment and infection percentage start to rise again to summer close to 40 % (Figura 6).

In the case of *Nosema* spp. results revealed that autumn has the lowest proportion of colonies infected and lower average number of spores per bee. The latter value increases steadily from autumn to summer (Figure 7).

This infestation cycle differs from the commonly described where the peaks of infection occur in autumn and especially in spring, being summer the season with the lower incidence of the disease [Furgala and Mussen, 1990; Hornitzky, 2008; Fries, 2010]. However, these seasonal variations were investigated when the causative agent was *N. apis*. The emergence in recent years of *N. ceranae* could have changed the traditional seasonal pattern of the disease to a strong presence in the summer. In this sense, Higes *et al.* [2010] found that *N. Ceranae* can reproduce at higher temperatures than *N. apis*, so that temperature would not be a limiting factor for the growth of this emerging pathogen. In Spain Martín-Hernández *et al.* [2007] found an increasing number of infected colonies in summer since 2003 and already in 2005 no seasonal change in prevalence was observed. This change was probably associated with the dominant presence of *N. ceranae*.

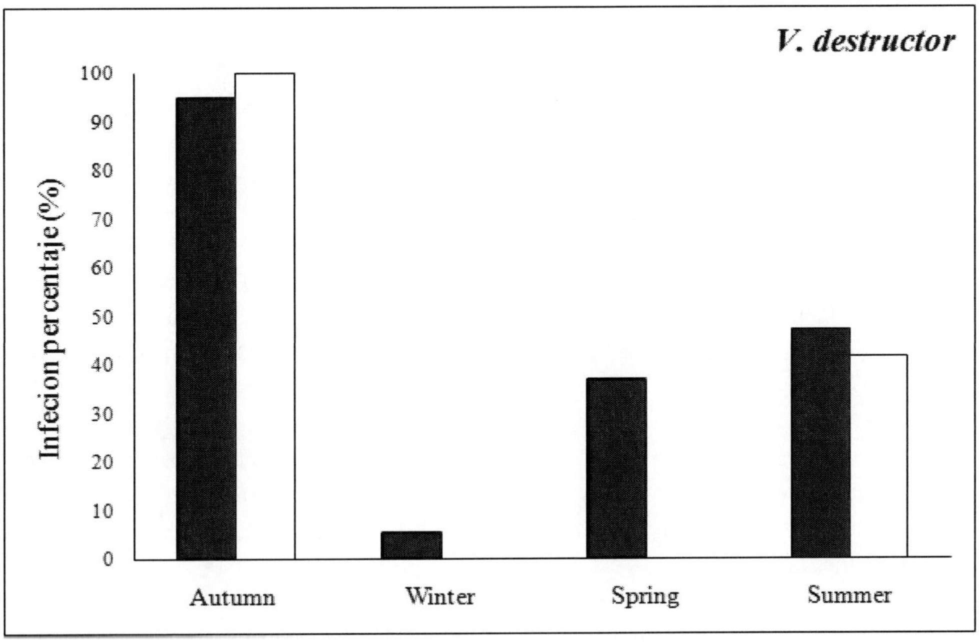

Figure 6. Stational variation of *V. destructor* incidence. Black: apiary located in Colonia (west); White: apiary located in Canelones (south).

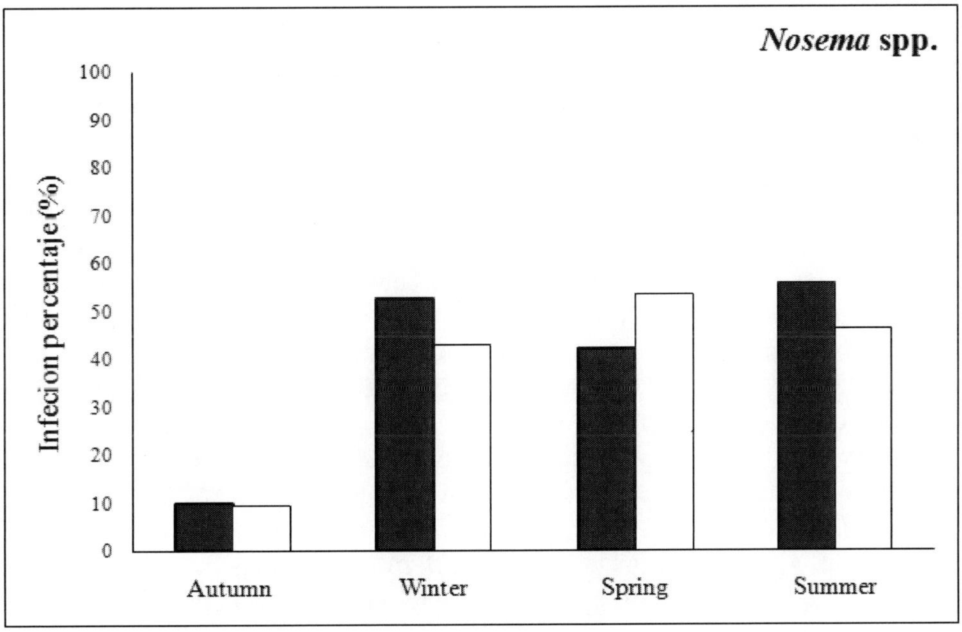

Figure 7. Stational variation of *Nosema* spp incidence. Black: apiary located in Colonia (west); White: apiary located in Canelones (south).

Discussion

Due to the importance of honeybees for the agricultural production and beekeeping, colony losses are a big problem that scientists and beekeepers has to face. Although studies about the potential causes are being carried out all around the world, there is no consensus about them.

It has been accepted that the damage caused by *V. destructor* is a crucial factor in colony losses in Europe and the United States [Rosenkranz *et al.*, 2010]. However, the co-infection of the mite with RNA viruses also have been related to colony losses, since the mite allows the horizontal transmission of different viruses [Shen *et al.*, 2005b; Shen *et al.*, 2005a; de Miranda *et al.*, 2010].

Acute bee paralysis virus, Kashmir bee virus and Israeli acute paralysis virus have a widespread prevalence in honeybee colonies and although can be in a sub-clinical stage, they can be extremely virulent when presented at elevated titres [de Miranda *et al.*, 2010]. DWV also is highly virulent and can potentially act independently of *V. destructor*, causing colony losses [Highfield *et al.*, 2009].

On the other side, it has been proposed that *N. ceranae* can cause the sudden collapse of bee colonies under field conditions in Spain [Higes *et al.*, 2007; Higes *et al.*, 2006; Higes *et al.*, 2008]. In Greece, it has been suggested that the co-infection between viruses and *N. ceranae* can lead to colony losses [Bacandritsos *et al.*, 2010].

In agreement with these reports, it was found in the Unites States, that bees with CCD had higher pathogen loads and were co-infected with a greater number of pathogens than non affected colonies, suggesting either an increased exposure to pathogens or a reduced resistance of bees toward pathogens [Vanengelsdorp *et al.*, 2009].

Lastly, it is important to remind that pesticides and agrotoxics have also been related to colony losses. In Spain, although pesticide residues (fluvalinate and chlorfenvinphos) were detected in stored pollen, no relationship was observed between pesticide residues and colony losses [Bernal *et al.*, 2010]. A similar result was found in France, where residues of imidacloprid and 6- chloronicotinic were detected in pollen loads, honey, and honey bee matrices, but no statistical relationship was found between colony mortality and pesticide residues [Chauzat *et al.*, 2009]. However, although the presence of pesticides apparently is not directly involved in colony losses, the exposure of honeybees to sublethal concentrations of imidacloprid, fenoxycarb and indoxacarb causes a decays of overall colony vitality and disturbing the foraging ability [Belien *et al.*, 2009]. Also, it has been demonstrated that the interaction between *Nosema* spp. and imidacloprid significantly weakened honeybees, causing individual mortality and energetic stress [Alaux *et al.*, 2010a].

In Uruguay, *V. destructor*, *Nosema* spp. and different RNA viruses are present and widely distributed. Although treatments against *V. destructor* diminish the damage associated to this pathogen and conditions of *Nosema* spp. infected colonies improve when moved to environments that provide varied sources of pollen, in apiaries in which punctual high colony mortality is evidenced, high percentage of infection with these pathogens are found. Additionally, most affected colonies presented high loads of several RNA viruses. We propose that although these pathogens can be individually controlled, the co-infection of colonies with several pathogens weakens the bees, increasing the possibility of colony death. Fortunately, this problem is not generally extended in Uruguay. Why massive honeybee colony losses do not occur in Uruguay? Maybe it is related to the exit of treatment against *V. destructor*, the absence of KBV, IAPV or IIV or the higher diversity of pollen resources than in industrialized countries.

REFERENCES

Alaux, C., Brunet, J. L., Dussaubat, C., Mondet, F., Tchamitchan, S., Cousin, M., Brillard, J., Baldy, A., Belzunces, L. P. & Le Conte, Y. (2010a). Interactions between Nosema microspores and a neonicotinoid weaken honeybees (Apis mellifera). *Environ Microbiol*, 12, 774-782.

Alaux, C., Ducloz, F., Crauser, D. & Le Conte, Y. (2010b). Diet effects on honeybee immunocompetence. *Biol. Lett.,* 6, 562-565.

Allen, M. & Ball, B. V. (1996). The incidence and world distribution of honey bee viruses. *Bee World,* 77, 141-162.

Anido, M., Antúnez, K., Brancchicela, B. & Zunino, P. (2011). *Seasonal variation of honeybee viruses in two apiaries with punctual episodes of colony losses in Uruguay.* 42nd Apimondia Congress. Buenos Aires, Argentina.

Antunez, K., B, D. A., Corbella, E. & Zunino, P. (2005). Detection of chronic bee paralysis virus and acute bee paralysis virus in Uruguayan honeybees. *J. Invertebr. Pathol.,* 90, 69-72.

Antunez, K., D'Alessandro, B., Corbella, E., Ramallo, G. & Zunino, P. (2006). Honeybee viruses in Uruguay. *J. Invertebr. Pathol.*, 93, 67-70.

Antunez, K., D'Alessandro, B., Piccini, C., Corbella, E. & Zunino, P. (2004). Paenibacillus larvae larvae spores in honey samples from Uruguay: a nationwide survey. *J. Invertebr. Pathol.,* 86, 56-58.

Antunez, K., Martin-Hernandez, R., Prieto, L., Meana, A., Zunino, P. & Higes, M. (2009). Immune suppression in the honey bee (Apis mellifera) following infection by Nosema ceranae (Microsporidia). *Environ. Microbiol.*, 11, 2284-2290.

Bacandritsos, N., Granato, A., Budge, G., Papanastasiou, I., Roinioti, E., Caldon, M., Falcaro, C., Gallina, A. & Mutinelli, F. (2010). Sudden deaths and colony population decline in Greek honey bee colonies. *J. Invertebr. Pathol.,* 105, 335-340.

Bailey, L. (1967). The incidence of virus diseases in the honey bee. *Ann. Appl. Biol.,* 60, 43-48.

Bailey, L. (1975). Recent research on honey bee viruses. *Bee World*, 56, 55-64.

Bailey, L. & Ball, B. V. (1991). Honey bee pathology. London, Ed. Academic Press. 193pp.

Bailey, L., Gibbs, A. & Wood, R. D. (1963). Two viruses from adult honey bees (Apis mellifera Linnaeus). *Virology*, 21, 390-395.

Bailey, L. & Woods, R. D. (1974). Three previously undescribed viruses from the honey bee. *J. Gen. Virol.*, 25, 175-186.

Ball, B. V. & Bailey, L. (1997). Viruses. In R.A. Morse and K. Flottum (Eds.), *Honey bee pest, predators and diseases* (pp. 11-31). Medina, OH: A.I. Root Co

Belien, T., Kellers, J., Heylen, K., Keulemans, W., Billen, J., Arckens, L., Huybrechts, R. & Gobin, B. (2009). Effects of sublethal doses of crop protection agents on honey bee (Apis mellifera) global colony vitality and its potential link with aberrant foraging activity. *Commun. Agric. Appl. Biol. Sci.*, 74, 245-253.

Bernal, J., Garrido-Bailon, E., Del Nozal, M. J., Gonzalez-Porto, A. V., Martin-Hernandez, R., Diego, J. C., Jimenez, J. J., Bernal, J. L. & Higes, M. (2010). Overview of pesticide residues in stored pollen and their potential effect on bee colony (Apis mellifera) losses in *Spain. J. Econ. Entomol.*, 103, 1964-1971.

Boecking, O. & Drescher, W. (1992). The removal response of Apis mellifera L. colonies to brood in wax and plastic cells after artificial and natural infestation with Varroa jacobsoni Oud. and to freeze killed brood. *Exp. App. Acarol.*, 16, 321-329.

Bromenshenk, J. J., Henderson, C. B., Wick, C. H., Stanford, M. F., Zulich, A. W., Jabbour, R. E., Deshpande, S. V., McCubbin, P. E., Seccomb, R. A., Welch, P. M., Williams, T., Firth, D. R., Skowronski, E., Lehmann, M. M., Bilimoria, S. L., Gress, J., Wanner, K. W. & Cramer, R. A., Jr. (2010). Iridovirus and microsporidian linked to honey bee colony decline. *PLoS One* 5, e13181.

Büchler, R. (1994). Varroa tolerance in honey bees - ocurrence, characters and breeding. *Bee World*, 75, 54-70.

Burgett, M., Shorney, S., Cordara, J., Gardiol, G. & Sheppard, W. S. (1995). The present status of Africanized honey bees in Uruguay. *Am. Bee J.*, 135, 328-330.

Campa, J., Harriet, J., Coll, F., Roth, R., Bounous, C. & Roth, F. (2007). Control de varroa. Evaluación de productos orgánicos y de síntesis. *Serie Actividades de Difusión de INIA*, 500, 13-20.

Cox-Foster, D. L., Conlan, S., Holmes, E. C., Palacios, G., Evans, J. D., Moran, N. A., Quan, P. L., Briese, T., Hornig, M., Geiser, D. M., Martinson, V., van Engelsdorp, D., Kalkstein, A. L., Drysdale, A., Hui, J., Zhai, J., Cui, L., Hutchison, S. K., Simons, J. F., Egholm, M., Pettis, J. S. & Lipkin, W. I. (2007). A metagenomic survey of microbes in honey bee colony collapse disorder. *Science*, 318, 283-287.

Chauzat, M. P., Carpentier, P., Martel, A. C., Bougeard, S., Cougoule, N., Porta, P., Lachaize, J., Madec, F., Aubert, M. & Faucon, J. P. (2009). Influence of pesticide residues on honey bee (Hymenoptera: Apidae) colony health in France. *Environ. Entomol.*, 38, 514-523.

Chen, Y., Evans, J. D., Smith, I. B. & Pettis, J. S. (2008). Nosema ceranae is a long-present and wide-spread microsporidian infection of the European honey bee (Apis mellifera) in the United States. *J. Invertebr. Pathol.*, 97, 186-188.

Chen, Y. & Siede, R. (2007). Honey Bee Viruses. *Adv. Virus. Res.*, 70, 33-80.

De Jong, D. (1997). Mites: varroa and other parasites of brood. In: R.A. Morse and K. Flottum (Eds.), *Honey bee pest, predators and diseases* (pp. 279-327). Medina, OH., A.I. Root Co.

de Miranda, J. R., Cordoni, G. & Budge, G. (2010). The Acute bee paralysis virus-Kashmir bee virus-Israeli acute paralysis virus complex. *J. Invertebr. Pathol.*, 103, 30-47.

Diniz, N. M., Soares, A. E. G., Sheppard, W. S. & Del Lama, M. A. (2003). Genetic structure of honeybee populations from southern Brazil and Uruguay. *Genet. Mol. Biol.*, 26, 47-52.

Errea, E. & Licandro, H. (2009). Apicultura: situación actual y perspectivas. *Anuario 2009 OPYPA, MGAP* (pp. 113-118). Montevideo, Uruguay.

Forsgren, E. & Fries, I. (2010). Comparative virulence of Nosema ceranae and Nosema apis in individual European honey bees. *Vet. Parasitol.*, 170, 212-217.

Fries, I. (1997). Protozoa. In: R.A. Morse and K. Flottum (Eds.), *Honey bee pest, predators and diseases* (pp. 57-76) Medina, OH., A.I. Root Co.

Fries, I. (2010). Nosema ceranae in European honey bees (Apis mellifera). *J. Invertebr. Pathol.*, 103, 73-79.

Fries, I. & Camazine, S. (2001). Implications of horizontal and vertical pathogen transmission for honey bee epidemiology. *Apidologie*, 32, 199-214.

Furgala, B. & Mussen, E. C. (1990). Protozoa. In: *Honey bee pests, predatores, and diseases*. In R. A. Morse and R. Nowogrodzki (Eds.). Cornell University Press, Ithaca.

Giersch, T., Berg, T., Galea, F. & Hornitzky, M. (2009). Nosema ceranae infects honey bees (Apis mellifera) and contaminates honey in Australia. *Apidologie*, 40, 117-123.

Gómez Pajuelo, A., Torres, C. & Orantes Bermejo, F. J. (2008). Colony losses: a double blind trial on the influence of supplementary protein nutrition and preventative treatment with fumagillin agains Nosema ceranae. *J. Apicult. Res.*, 47, 84-86.

Harriet, J., Toscano, H. & Campá, J. (2000). Loque americana. Publicación DILAVE (MGAP).

Higes, M., García-Palencia, P., Botias, C., Meana, A. & Martín-Hernández, R. (2010). The differential development of microsporidia infecting worker honey bee (Apis mellifera) at increasing incubation temperature. *Environ Microbiol*, doi: 10.1111/j.1758-2229.2010.00170.

Higes, M., Garcia-Palencia, P., Martin-Hernandez, R. & Meana, A. (2007). Experimental infection of Apis mellifera honeybees with Nosema ceranae (Microsporidia). *J. Invertebr. Pathol.*, 94, 211-217.

Higes, M., Martin-Hernandez, R., Botias, C., Bailon, E. G., Gonzalez-Porto, A. V., Barrios, L., Del Nozal, M. J., Bernal, J. L., Jimenez, J. J., Palencia, P. G. & Meana, A. (2008). How natural infection by Nosema ceranae causes honeybee colony collapse. *Environ. Microbiol.*, 10, 2659-2669.

Higes, M., Martín-Hernández, R., Botias, C. & Meana, A. (2009). The precence of Nosema ceranae (Microsporidia) in North African honey bees (Apis mellifera intermissa). *J. Apicult. Res.*, 217-219.

Higes, M., Martin, R. & Meana, A. (2006). Nosema ceranae, a new microsporidian parasite in honeybees in Europe. *J. Invertebr. Pathol.*, 92, 93-95.

Highfield, A. C., El Nagar, A., Mackinder, L. C., Noel, L. M., Hall, M. J., Martin, S. J. & Schroeder, D. C. (2009). Deformed wing virus implicated in overwintering honeybee colony losses. *Appl. Environ. Microbiol.*, 75, 7212-7220.

Hornitzky, M. (2008). Nosema disease. Literature review and three year survey of beekeepers. Part 2. Rural Industries Research and Development Corporation Report. *Publication* N° 08/006: 28.

Huang, W. F., Jiang, J. H., Chen, Y. W. & Wang, C. H. (2007). A Nosema ceranae isolate from the honeybee Apis mellifera. *Apidologie*, 38, 30-37.

Invernizzi, C., Abud, C., Tomasco, I. H., Harriet, J., Ramallo, G., Campa, J., Katz, H., Gardiol, G. & Mendoza, Y. (2009). Presence of Nosema ceranae in honeybees (Apis mellifera) in Uruguay. *J. Invertebr. Pathol.*, 101, 150-153.

Invernizzi, C., Antúnez, K., Campa, J. P., Harriet, J., Mendoza, Y., Santos, E. & Zunino, P. (2011). Sanitary situation of the uruguayan honeybees. Veterinaria. In press.

Issa, M. R. C., De Jong, D. & Simões, Z. L. P. (2000). Infestaçao por varroa na zona de transição de abelhas européias e africanizadas (paralelos 30-35). XIII *Congresso Brasileiro de Apicultura, Florianópolis, Brasil.*

Kirsch, R. & Rosenkranz, P. (1998). Population dynamics of honey bees, honey bee brood and Varroa jacobsoni mites in untreated colonies in Uruguay. *Apidologie*, 29, 438-439.

Klee, J., Besana, A. M., Genersch, E., Gisder, S., Nanetti, A., Tam, D. Q., Chinh, T. X., Puerta, F., Ruz, J. M., Kryger, P., Message, D., Hatjina, F., Korpela, S., Fries, I. & Paxton, R. J. (2007). Widespread dispersal of the microsporidian Nosema ceranae, an emergent pathogen of the western honey bee, Apis mellifera. *J. Invertebr. Pathol.*, 96, 1-10.

Maggi, M. D., Ruffinengo, S. R., Mendoza, Y., Ojeda, P., Ramallo, G., Floris, I. & Eguaras, M. J. (2010). Susceptibility of Varroa destructor (Acari: Varroidae) to synthetic acaricides in Uruguay: Varroa mites' potential to develop acaricide resistance. Parasitol Res. DOI: 10.1007/s00436-010-2122-5.

Martin-Hernandez, R., Meana, A., Prieto, L., Salvador, A. M., Garrido-Bailon, E. & Higes, M. (2007). Outcome of colonization of Apis mellifera by Nosema ceranae. *Appl. Environ. Microbiol.*, 73, 6331-6338.

Mayack, C. & Naug, D. (2009). Energetic stress in the honeybee Apis mellifera from Nosema ceranae infection. *J. Invertebr. Pathol.*, 100, 185-188.

Mendoza, Y., Diaz-Cetti, S., Ramallo, G., Katz, H., Campa, J. & Harriet, J. (2010). Daño y control de la nosemosis en colonias de abejas (Apis mellifera). *Agrociencia*, 14, 170.

Neumann, P. & Carreck, N. (2010). Honey bee colony losses. *J. Apicult. Res.*, 49, 1-6.

Paxton, R. J. (2010). Does infections by Nosema ceranae cause "Colony Collapse Disorder" in honey bees (Apis mellifera)? *J. Apicult. Res.*, 49, 80-84.

Paxton, R. J., Klee, J., Korpela, S. & Fries, I. (2007). Nosema ceranae has infected Apis mellifera in Europe since at least 1998 and may be more virulent than Nosema apis. *Apidologie*, 38, 558-565.

Piccini, C., D'Alessandro, B., Antúnez, K. & Zunino, P. (2002). Detection of Paenibacillus larvae subspecies larvae spores in naturally infected bee larvae and artificially contaminated honey by PCR. *W. J. Microbiol. & Biotecnol.*, 18, 761-765.

Piccini, C. & Zunino, P. (2001). American foulbrood in Uruguay: isolation of Paenibacillus larvae larvae from larvae with clinical symptoms and adult honeybees and susceptibility to oxytetracycline. *J. Invertebr. Pathol.*, 78, 176-177.

Potts, S. G., Roberts, S. P., Dean, R., Marris, G., Brown, M. A., Jones, R., Neumann, P. & Settele, J. (2010). Declines of managed honey bees and beekeepers in Europe. *J. Apicult. Res.*, 49, 15-22.

Rosenkranz, P., Aumeier, P. & Ziegelmann, B. (2010). Biology and control of Varroa destructor. *J. Invertebr. Pathol.*, 103, 96-119.

Rothenbuhler, W. C. (1964). Behavior genetics of nest cleaning in honey bees. IV. Responses of F1 and backcross generations to disease killed-brood. *Am. Zool.*, 4, 111-123.

Ruttner, F., Marx, H. & Marx, G. (1984). Beobachtungen ubre eine mögliche Anpassung von Varroa jacobsoni an Apis mellifera L. in Uruguay. *Apidologie*, 15, 43-62.

Santos, E., García, E., Di Landro, R., Daners, G., Saadoun, A., Cabrera, C. & Invernizzi, C. (2005). Variación de la proteína corporal y presencia del protozoario Nosema apis en colonias de abejas melíferas emplazadas en forestaciones de Eucalyptus grandis. VIII *Jornadas de Zoología del Uruguay, Montevideo, Uruguay*, p. 105.

Seeley, T. D. (2007). Honey bees of the Arnot Forest: a population of feral colonies persisting with Varroa destructor in the northeastern United States. *Apidologie*, 38, 19-29.

Shen, M., Cui, L., Ostiguy, N. & Cox-Foster, D. (2005a). Intricate transmission routes and interactions between picorna-like viruses (Kashmir bee virus and sacbrood virus) with the honeybee host and the parasitic varroa mite. *J. Gen. Virol.*, 86, 2281-2289.

Shen, M., Yang, X., Cox-Foster, D. & Cui, L. (2005b). The role of varroa mites in infections of Kashmir bee virus (KBV) and deformed wing virus (DWV) in honey bees. *Virology*, 342, 141-149.

Spivak, M. (1996). Honey bee hygienic behavior and defense against Varroa jacobsoni. *Apidologie*, 27, 245-260.

Spivak, M. & Reuter, G. S. (2001a). Resistance to American foulbrood disease by honey bee colonies Apis mellifera bred for hygienic behavior. *Apidologie*, 32, 555-565.

Spivak, M. & Reuter, G. S. (2001b). Varroa destructor infestation in untreated honey bee (Hymenoptera: Apidae) colonies selected for hygienic behavior. *J. Econ. Entomol.*, 94, 326-331.

Stokstad, E. (2007). The case of the empty hives. *Science*, 316, 970-972.

Van Engelsdorp, D., Evans, J. D., Saegerman, C., Mullin, C., Haubruge, E., Nguyen, B. K., Frazier, M., Frazier, J., Cox-Foster, D., Chen, Y., Underwood, R., Tarpy, D. R. & Pettis, J. S. (2009). Colony collapse disorder: a descriptive study. *PLoS One*, 4, e6481.

Van Engelsdorp, D., Hayes, J., Jr., Underwood, R. M. & Pettis, J. (2008). A survey of honey bee colony losses in the U.S., fall 2007 to spring 2008. *PLoS One*, 3, e4071.

Van Engelsdorp, D., Hayes, J., Underwoos, R. M. & Pettis, J. S. (2010). A survey of honey bee colony losses in United States, fall 2008 to spring 2009. *J. Apicult. Res.*, 49, 7-14

Vandame, R. & Palacio, M. A. (2010). Honey bee preserved health in Latin America: a fragile equilibrium. *Apidologie*, 41, 243-255.

In: Bees
Editor: Richard M. Florio

ISBN: 978-1-61324-825-6

Chapter 7

DISRUPTION IN BIOGENIC AMINES-MEDIATED SIGNALING: A POTENTIAL MOLECULAR MECHANISM ASSOCIATED WITH COLONY COLLAPSE DISORDER

Tahira Farooqui[1]
Department of Entomology/Center for Molecular Neurobiology,
The Ohio State University, Columbus, OH, US

Abstract

The biogenic amines include the three catecholamines (dopamine, norepinephrine, and epinephrine), and the tyrosine metabolites (octopamine and tyramine). An ester of acetic acid and choline (acetylcholine), indoleamine (serotonin), and imidazoleamine (histamine) are also considered as biogenic amines that play important roles in brain functions. Biogenic amines regulate many functions in the brain, including endocrine secretion, cognitive function, aggression, emotional states, motivation, reward circuitry, and learning and memory. Any disruption in the biogenic amines-mediated signaling alters the levels of second messengers, therefore impairs various normal physiological functions. This chapter focuses on biogenic amines-based pesticides (such as acetylcholine mimics and octopamine mimics) that disrupt biogenic amines-mediated signaling in honeybees, impairing their olfactory learning and memory. European honeybees *Apis mellifera* are responsible for the pollination of fruits, vegetables and nuts that accounts ~1/3 of the United States crop. In addition to pollination, honeybee products (honey, propolis, royal jelly and bee venom) are known to have many therapeutic uses. A continuous decline in western honeybee colonies "Colony Collapse Disorder" (CCD) in Europe and in North America is lately drawing a lot of attention due to pollination crisis. If the problem of CCD is not resolved soon enough then this could have a major impact on food industry, affecting United State's economy a big time. The

[1] E-mail address: farooqui.2@osu.edu.

main objective of this review is to discuss the significance of biogenic amines-mediated signaling in relation to olfactory learning and memory in honeybees and shed the light on a potential molecular mechanism underlying CCD.

Keywords: Honeybees, *Apis mellifera*, Colony collapse disorder, Reactive oxygen species, Olfactory learning and memory, Acetylcholine, Nicotine, Octopamine, Acetylcholine mimics, Octopamine mimics, Neonicotinoids, Formamidines.

1. Introduction

Honeybees belong to the insect order Hymenoptera. The Western (European) honeybee *Apis mellifera* (genus *Apis* and specie *mellifera*) is the most commonly adaptable species and best known among all insects. Worker honeybees need to forage for a wide diversity of pollen and nectar to raise a healthy brood in the hive and maintain strong immune systems. Honeybees use their proboscis to extract the nectar from flowers and hairy bodies to gather pollen. They are adapted for feeding on flower nectar (as an energy source) and pollen (as source of protein and other nutrients). Many insects consume nectar, whilst honeybees refine and concentrate nectar to make honey, and consume it as a food during winter. Not all insect-dependent pollination is provided by honeybees, but they are the most economically valuable pollinators of agricultural crops worldwide. Therefore, honeybees have been extensively used for commercial pollination of crops and flowering plants (such as many orchard fruits, agricultural food, vegetables, and nuts). Furthermore, honeybee products (honey, propolis, royal jelly, and bee venom) are known to have therapeutic uses. Therefore, honeybees are beneficial insects for humans due to food supply through pollination as well as providing honey and other products that may have potential therapeutic importance [1], [2]. Propolis, one of the major hive products of honeybees, is used for clinical purposes as well as in many products including toothpastes, mouth washes and skin creams due to its antibacterial, antifungal and antiviral properties [3]-[6]. Bee venom, released from the stingers of honeybees *Apis mellifera*, contains many biologically active components, and has therapeutic uses in traditional medicine as well as in animal models of variety of diseases [7]-[9]. Royal jelly, synthesized in the hypopharyngeal and mandibular glands of young nurse worker bees *Apis mellifera*, is specifically used for feeding queen destined larvae in the colony, serving as the principal food for the queen honeybee. Royal jelly is known to have a variety of biological activities such as antitumor, antihypercholesterolemic, anti-inflammatory, antihypertensive, immuno-modulatory, and anti-osteoporosis [10]-[15]. Therefore, royal jelly is used worldwide for many years as a dietary supplement to offer various health benefits to humans.

Honeybees are currently disappearing from their hives, and the term used for this phenomenon is Colony Collapse Disorder (CCD). In CCD, dead bodies are not found in or around the hive, suggesting that honeybees do not return to the hive after they leave for foraging. The underlying molecular mechanism for CCD is unknown. Sudden losses of bees have been reported in past with similar symptoms as being experienced in CCD-mediated colonies [16]-[39], except that in past losses dead bodies were found in or around the hive. It is still not yet clear whether CCD is caused by the same or new factors contributed to bee losses. Although several reports have been recently published on CCD, but they do not deal with molecular mechanism associated with CCD.

In this review, I have briefly discussed factors affecting honeybee's biology (development, life span, labor specialization and diseases), CCD, biogenic amines-mediated signaling and its disruption by biogenic amines-based pesticides (neonicotinoids and formamidines), and biogenic amines-mediated olfactory learning and memory. Finally I have made an attempt to link pesticides with disruption of biogenic amines-mediated olfactory learning and memory and empty hives in CCD, suggesting that pesticide exposure to honeybees results in excessive production of ROS-mediated oxidative stress in their brain that causes olfactory dysfunction in honeybees, which may be a prominent reason of their failure in returning to the hive.

2. Factors Affecting Honeybee's Development, Life Span and Labor Specialization

Bees are social insects. As many as 40,000 honeybees live as a colony in a nest/hive, containing a single queen, thousands of workers bees (sterile females) and hundreds of male bees called drones. There are four major stages in honeybee development: egg, larva, pupa and adult. Unfertilized eggs develop into drones, whereas fertilized eggs develop into either workers or queens depending on the quantity and quality of the food given to the developing bee during the first three days of larval life. These nutritional factors trigger different larval feeding rates, different levels of developmental regulatory hormones, resulting in different developmental programs. Juvenile hormone is considered as a major modulator of behavioral development in the honeybee [40]. The developmental switch depends not on inherited genetic differences between queens and workers but on the differential expression of sets of genes involved with larval fate. The development time and the quality of the emerged adult are dependent on temperature, nutrition, and the race [41].

All three castes in honeybees have specialized functions. Queens are fed by workers, and are the primary if not sole egg layers (reproductive). They live generally 1-3 years. Drones do not collect nectar or pollen. Their sole function is to mate with the queen, live an average of 21-32 days during the spring to mid-summer period, and can survive up to 90 days during late summer and autumn. As the winter approaches, drones are usually expelled from the nest, so few or no drones survive the winter. In contrast, after emergence, workers progress through a defined series of duties both inside and outside the hive. They initially function as nurse bees, whose tasks include a variety of housekeeping duties such as brood cell cleaning, feeding larvae, producing beeswax and processing honey. From this stage, the maturing adult progresses through one to a few other task specializations to guard duties. Finally, a worker commences its life as a forager that includes collecting nectar to make honey and pollinate crops. These behavioral changes are accompanied by regular shifts in the activity of exocrine glands. Workers retain flexibility to accelerate, retard or even reverse this behavioral development depending on environment-based needs of the colony [41], [42]. A honeybee worker's life span varies from only a few days to months, depending primarily on seasonal factors, food availability, activities performed during lifetime, and race. Within its foraging lifetime, a worker must learn about the association of odors with important ecological events, which involves locating flowers in order to harvest nectar and pollen resources for colony survival. A worker bee's life span can vary from only a few days to almost a year, depending primarily on seasonal factors, food availability, activities performed during lifetime, and race. The flight distance accumulated by a forager has more of an influence on her life span than chronological age. Workers die after flying a total of ~ 800 km due to breakdown in the enzymatic mechanisms that metabolize carbohydrates into glycogen [43].

There are number of diseases that can influence managed honeybees *Apis mellifera* population [12]. These diseases include: (1) nosema caused by microsporidian, spore forming, unicellular parasites (*Nosema apis and Nosema ceranae*), (2) amoeba by a protozoan (*Malpighamoeba mellificae*), (3) acariosis by acarine tracheal mites (*Acarapis woodii*), (4) deformed wings produced by infestation with varroa parasitic mites during development (*Varroa destructor*, *Varroa jacobsoni*), (5) viral infections by (Acute bee paralysis virus, Israel acute paralysis virus and invertebrate iridescent virus type 6), and (6) fungal brood diseases such as Chalkbrood caused by the fungus *Ascosphaera apis*, a honeybee-specific disease that can persist in colonies for years, and Stonebrood by another fungus *Aspergillus flavus*, a rare facultative pathogen that affects hosts other than honeybees and can likely survive outside insect hosts. These diseases affect the life span of honeybees, but no means has yet been found to completely eliminate these diseases from them [44]. However, the routine inspection, proper care, and disease treatment may help in controlling bee diseases to some extent. In addition, other factors such as the environment and socio-economic factors alone or combined with other extrinsic/intrinsic factors may have impact on the honeybee colony health.

3. CCD

In past few years, CCD phenomenon in which managed populations of western honeybees experiencing a substantial loss in adult population of worker bee colonies has been reported throughout North America. Similar large scale colony losses have been reported in several other countries as well as USA, receiving worldwide press coverage. In CCD, remaining bee population in hives consists of a queen and only a small number of young workers (if any, but no dead bodies are found in or near the hive) with plenty of food resources (honey and pollen) left in colonies [2], [32]-[39]. These findings suggest that total loss of honeybees from their hives is not due to lack of food. Large-scale losses of honeybees are not unusual. In past (1960s-1970s), honeybee population has been wiped out due to extreme weather, pesticide exposure, and pest infestation. In some instances, the disease even disappeared within a short time without any treatment. The symptoms during large-scale colony losses (past vs. present) are significantly different [35], [38]. Then how could all of these instances governed by the same mechanism? Is colony collapse disorder governed by a different mechanism, which is influenced by combination of different factors? Researchers have been struggling for years to explain CCD. Here, I mainly focus on biogenic amines-mediated signaling, biogenic amines-based pesticides and their possible relation with CCD.

4. BIOGENIC AMINES-MEDIATED SIGNALING

The biogenic amines include the three catecholamines (dopamine, norepinephrine, and epinephrine), tyrosine metabolites (octopamine and tyramine), indoleamine (serotonin), imidazoleamine (histamine), and an ester of acetic acid and choline (acetylcholine).

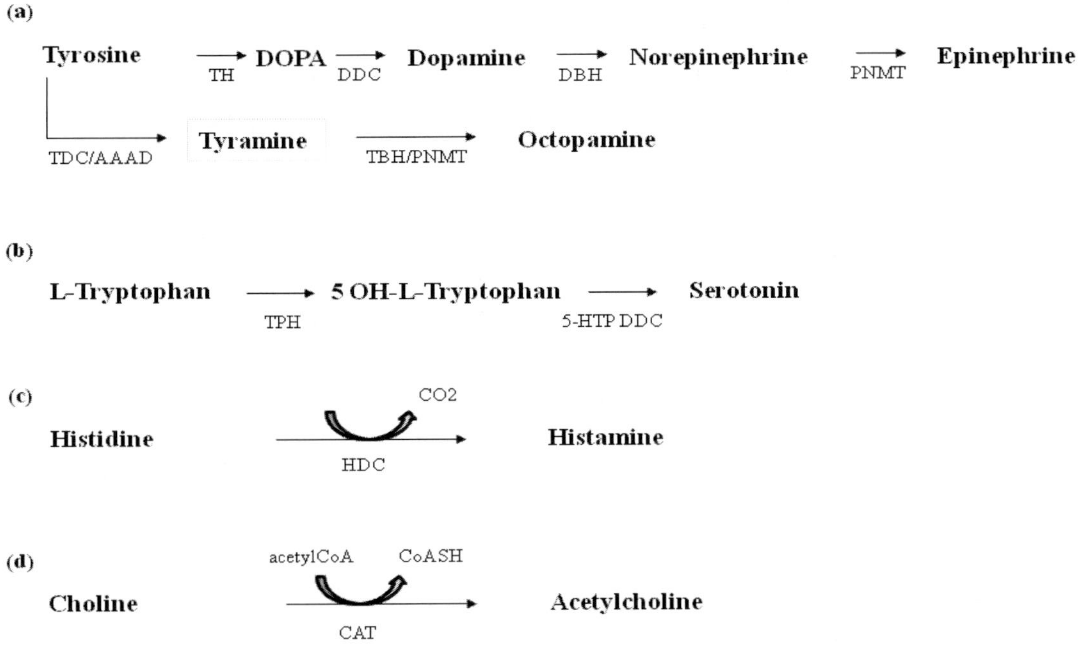

Figure 1. Biosynthetic pathways of dopamine, noepinephrine, epinephrine, serotonin. histamine and acetylcholine. (a) Catecholamines (dopamine, norepinephrine and epinephrine) are all synthesized in the same chemical pathway using an amino acid, tyrosine, as the precursor. Tyrosine is hydroxylated by tyrosine hydroxylase (TH) to dihydroxyphenylalanine (DOPA), which is decarboxylated to dopamine by DOPA decarboxylase (DDC). These reactions take place in catecholamine-secreting neurons or cells of the adrenal medulla. Norepinephrine is synthesized from dopamine by dopamine β-hydroxylase (DβH). Norepinephrine is then N-methylated to epinephrine by phenylethanolamine-N-methyl transferase. In invertebrates, tyramine decarboxylase (TDC) catalyzes conversion of tyrosine to tyramine, and tyramine-β-hydroxylase synthesizes octopamine from tyramine. In vertebrates these two reactions are catalyzed by aromatic amino acid decarboxylase (AAAD) and dopamine β-hydroxylase (DβH), respectively. (b) Serotonin (5-hydroxytryptamine) is synthesized from the amino acid tryptophan in the serotonergic neurons in the central nervous system by two enzymes: (1) tryptophan hydroxylase (TPH) and (2) amino acid decarboxylase (DDC). (c) Histamine is synthesized from the decarboxylation of the amino acid histidine, and this reaction is catalyzed by the L-histidine decarboxylase (HDC). (d) Acetylcholine is synthesized from choline and acetyl-CoA through the action of choline acetyltransferase (CAT) that facilitates the transfer of acetate to the choline, producing acetylcholine.

These biogenic amines are derived from amino acids: catecholamines from aromatic amino acid tyrosine, serotonin from tryptophan, and histamine from histidine, except acetylcholine that is synthesized from choline and acetyl-CoA (Figure 1). Some biogenic amines are preferentially synthesized (such as norepinephrine and epinephrine in vertebrates, and octopamine and tyramine in invertebrates). However, dopamine, serotonin, histamine and acetylcholine are synthesized in both vertebrates and invertebrates. After their release from neurons, biogenic amines bind and activate specific receptors that change intracellular concentration of second messengers, responsible for slow but long-lasting responses. They can act as neurotransmitters, neuromodulators, or neurohormones [45]. They mediate and/or regulate many functions, including endocrine secretion, cognitive function, aggression, emotional states, motivation, reward circuitry, and learning and memory, in the brain [45]-[47]. Therefore, any disruption in biogenic amines-mediated signaling may alter the levels of second messengers, resulting in behavioral disorders (such as olfactory dysfunction) that may affect honeybee's navigation ultimately leading to CCD.

5. Biogenic Amines-Based Pesticides

In agriculture, pesticides have been widely used throughout the world to protect food crops and to control disease-carrying pests. Depending on their type and/or dose used, pesticides can either alter insect behavior or completely destroy them. Pesticides that directly interact with biogenic amine receptors and interfere with biogenic amine-mediated signaling are discussed below.

5.1. Neonicotinoids

Neonicotinoids (systemic insecticides) are nicotine-based pesiticides [48]. These pesticides are similar to nicotine in the structure therefore they selectively act on the postsynaptic nicotinic acetylcholine receptors as agonists in the insect brain They are also referred as acetylcholine mimics. Their interaction with postsynaptic nicotinic acetylcholine receptors results in partial paralysis, and extended stimulation of these receptors leads to the death of insects. This class of pesticides includes compounds such as imidacloprid, acetamiprid, dinotefuran, nitenpyram, clothianidin, thiacloprid, and thiamethoxam (Figure 2). Agonist recognition by the nicotinic receptor involves cation-pi interaction for nicotinoids in mammals, and possibly a cationic subsite for interaction with the nitro or cyano substituent of neonicotinoids in insects [49], implicating a topological divergence of the agonist-binding sites in nicotinic acetylcholine receptors in insect and vertebrate systems. Nicotine based pesticide's binding with nicotinic acetylcholine receptors results in fast cholinergic synaptic transmission [50], [51]. Nicotinic receptor agonist, imidacloprid, has been shown to elicit 43% of the maximum acetylcholine-induced currents in the cultured antennal lobe neurons from adult honeybee brains. Epibatidine, an alkaloid found on the skin of neotropical poisonous frog *Epipedobates tricolor,* acts as an agonist of nicotinic acetylcholine receptors. Both nicotine and epibatidine (Figure 2) have been used in structural characterization of the agonist binding domain [52]. Neonicotinoid insecticides interfere with the transmission of

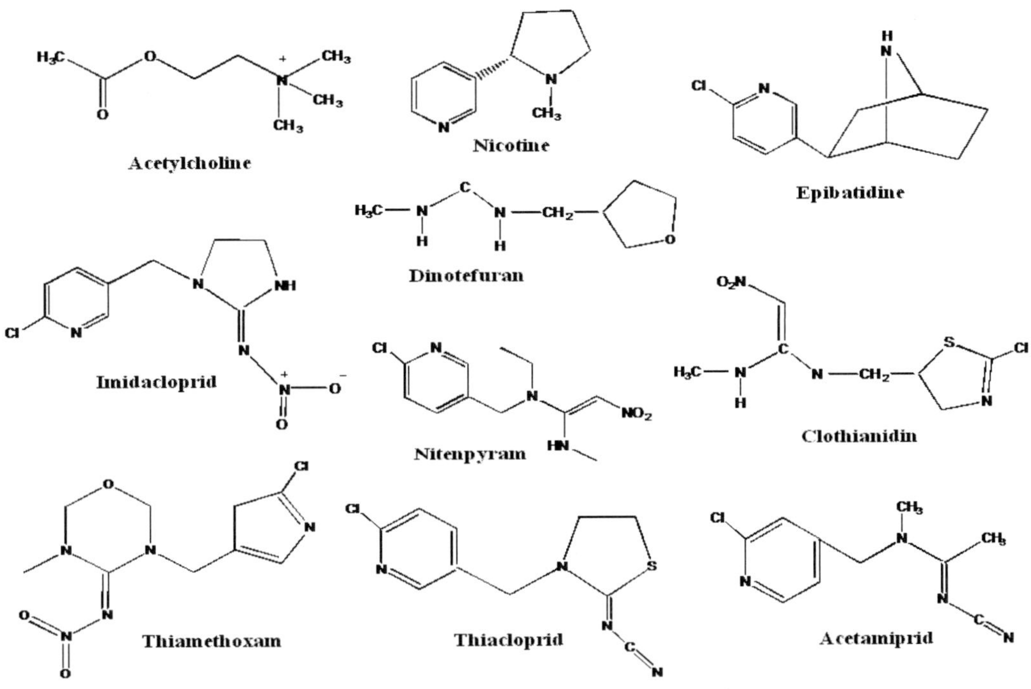

Figure 2. Chemical stuctures of acetylcholine, nicotine, epibatidine, and neonicotinoids such as imidacloprid, dinotefuran, clothianidin, nitenpyram, acetamiprid, thiacloprid, and thiamethoxam.

neural messages in insects much more efficiently than in mammals, therefore they possess selective toxicity to insects over vertebrates. A nitro or a cyano group in neonicotinoids is postulated to contribute directly to their selectivity [49], [52], [53].

The nitro-substituted compounds are clothiandin, dinotefuran, imidacloprid, thiamethoxam, and nitenpyram (Figure 2). The reported LD_{50} values of nitro-substituted compounds are 18 ng/bee for imidacloprid, 22 ng/bee for clothianidin, 30 ng/bee for thiamethoxam, 75 ng/bee for dinotefuran, and 138 ng/bee for nitenpyram [54]. In contrast, cyano-substituted neonicotinoids (Figure 2) have LD_{50} values of 7.1 μg/bee and 14.6 μg/bee for acetamiprid and thiacloprid, respectively [54]. These findings suggest that nitro-substituted compounds are more toxic to honeybees than cyano-substituted neonicotinoids. The reduced toxity of cyano-substituted neonicotinoids may also be due to their increased metabolism.

The subunit composition of a receptor determines functional and pharmacological properties of that receptor. The genome sequencing has revealed different number of nicotinic acetylcholine receptors subunits in different insect species: 10 in fruit fly *Drosophila melanogaster* [55], 10 in mosquito *Anopheles gambiae* [56], 11 in honeybee *Apis mellifera* [50], 12 in silk worm *Bombyx mori* [57], and 12 in red flour beetle *Tribolium castaneum* [58]. The presence of nicotinic acetylcholine receptors families, comprising several subunit encoding genes, provides a molecular basis for broad functional diversity [59]. Furthermore, the amino acid clusters between loop B and C of insect nicotinic acetylcholine receptor-α subunit are considered as essential amino acid residues for agonist binding [59]. Some important amino acid residues in insect nicotinic acetylcholine receptor-α and β -subunits (such as important residues in loop C, the region loop B to the N-terminus and loop B-C interval of insect α subunit, and important residues in loop D, E and F of insect β subunit) contribute to neonicotinoid's selectivity [60]. Collectively, drug-binding sites of nicotine based pesiticides are localized at subunit interfaces of the acetylcholine nicotinic receptor pentameric structure, and any mutation in selective amino acid residues in insect nicotinic acetylcholine receptor-α and β-subunits may contribute to insecticidal resistance of neonicotinoids.

5.2. Amidine Compounds

The formamidines, a group of acaricidal compounds, are chemically characterized by the presence of a phenylamidine moiety [61,62]. This class of insecticides includes commercially marketed compounds, including amitraz (N-(2,4-dimethylphenyl)-N'-[(2,4-dimethylphenyl)-imino]methyl-N-methylinethanimid-amide), chlordimeform (N'(4-chloro-O-tolyl)-N dimethylformamidine, CDM) and its degradable product N-desmethyl chlordimeform (DCDM) (Figure 3). Formamidines mimic the action of octopamine in insects, therefore increased cyclic adenosine monophosphate (cAMP) levels in brain and peripheral tissues result in tremors and convulsions. These compounds have been reported to mimic actions of octopamine at the locust neuromuscular junction [63]. At lower doses, formamidines suppress pest mating, reproduction, and feeding behavior [64]-[66]. Both *N*-demethylchlordimeform and octopamine increase the phosphorylation of proteins that are also phosphorylated by exogenous cAMP-dependent protein kinase in the two-spotted spider mite [67], confirming the agonistic effects of formamidines through octopamine-sensitive adenylate cyclase. Upon feeding, formamidines induce motor excitation, reduce phototaxis and impair olfactory learning in *Drosophila* [68].

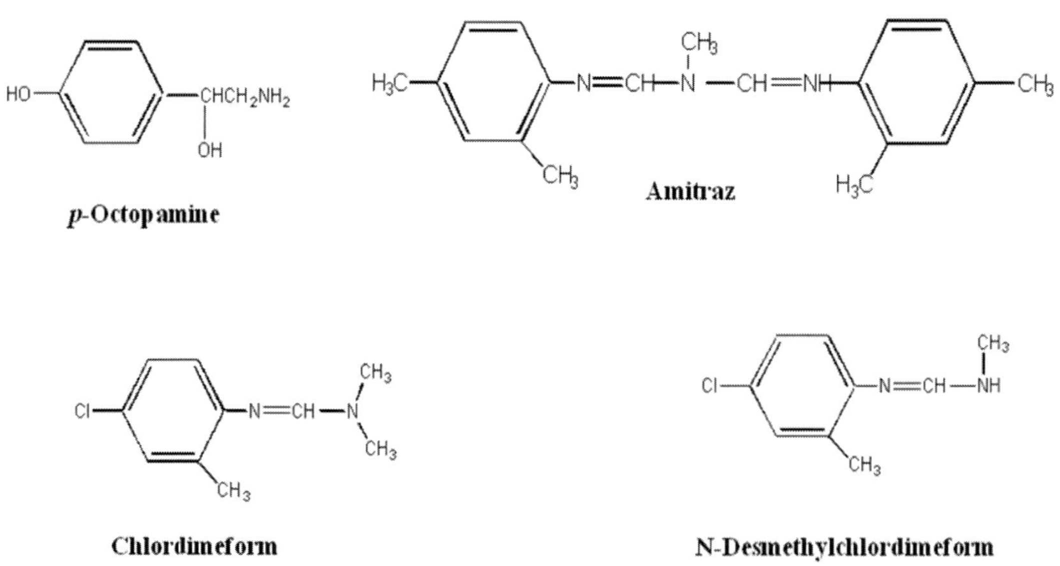

Figure 3. Chemical structures of *p*-octopamine, amitraz, and formamidine pesticides such as chlordimeform, and *N*-desmethylchlordimeform.

Formamidines, such as amitraz, has been used to control several agricultural pests, including parasitic mites in honeybees [69]-[71]. In general, formamidines exert their effect in the CNS of target species by interacting with octopamine receptors [63],[68], and inhibiting monoamine oxidase activity [72], [73], leading to alteration in the behavioral plasticity. Collective evidence suggests that octopamine is involved in modulating various behaviors and peripheral and sense organs through octopaminergic receptors, enabling insects to respond correctly to external stimuli [74]. Furthermore, the sensory system inputs are linked to motor-response systems through learning process, therefore changes in sensory system may alter insect behavior.

6. Biogenic Amines-Mediated Learning and Memory

Navigation of honeybees requires learning of the directions and distances of their travels between nest and food sources. Honeybees relate all navigational decisions to the origin of their flight path, and apply a navigational motor routine to bring them back to their hives by recognizing the landmark and multiple associations between landmarks and respective return heading and distance to their hives, suggesting that honeybees navigate according to a map-like organization of spatial memory [75]. Honeybees have sophisticated sensory systems, including well developed learning and memory capacities, therefore they serve as a valid model to study the underlying mechanism of learning and memory. Olfactory memory plays an important role in many aspects of honeybee behavior, including recognition of nestmates, foraging (food localization), food preferences, hive location, and various other behaviors [75]-[77].

In the honeybee *Apis mellifera*, an identified ventral unpaired medial (VUMmx1) cell, an octopaminergic neuron, produces an associative link between the unconditioned stimulus (US) and the conditioned stimulus (CS) pathways in the antennal lobe [78]. The VUM interneuron releases octopamine into most if not all glomeruli of the antennal lobe, innervating all three neuropils (antennal lobe, mushroom body, and lateral Protocerebrum), which are involved in the olfactory pathway. Electrical stimulation of VUM neuron substitutes for the US (e.g. sucrose) in an associative olfactory learning paradigm, supporting its role in olfactory learning and memory [78]. Local injection of octopamine into defined areas of honeybee brains increases their learning ability as well as recall, suggesting that octopamine is involved in the regulation of learning and memory [79]. How the VUM interneuron does exactly couple the CS and US pathways is not clear except that this neuron has a reinforcing property. RNA interference-mediated silencing of octopamine receptor gene as well as octopamine receptor antagonism by mianserin in the antennal lobe of the honeybee brain disrupt octopamine receptor function, which impairs olfactory conditioning and retrieval without disturbing odor discrimination [80], implicating that disruption in normal octopamine-mediated signaling in the antennal lobe frequently interferes with many forms of odor memory. Dopamine also exhibits profound effects on many physiological processes, such as emotion, cognition, motor control, and olfactory aversive learning, whereas octopamine modulates locomotion, grooming, conditioned courtship, and olfactory appetitive learning in insects [81], [82]. However, the role of serotonin in olfactory learning is not clear in insects.

In the insect brain, acetylcholine is the most abundant excitatory neurotransmitter, which binds to most numerously found acetylcholine receptors. Acetylcholine also functions as a neuromodulator in the peripheral and in the central nervous systems. It is involved with synaptic plasticity, specifically in learning and memory. Acetylcholine can bind and activate two types of receptors: (1) nicotinic acetylcholine receptors and muscarinic acetylcholine receptors.

In insects, nicotinic acetylcholine receptors are targets for neonicotinoids. However, limited data are available on the role of these receptors on insect cognitive processes. In honeybees, nicotine injections in brain have been reported to improve their short term memory [83], whereas mecamylamine (nicotine acetylcholine receptor antagonist) injections inhibit olfactory learning or recall, depending upon the site of injection [84]. Recently, Gauthier (2010) has suggested the existence of at least two subtypes of nicotinic acetylcholine receptors in the honeybee brain: (1) α-bungarotoxin (nicotinic antagonist)-sensitive receptors, and (2) α-bungarotoxin-insensitive receptors. The α-bungarotoxin-sensitive receptors are necessary for the formation of long-term memory, whereas α-bungarotoxin-insensitive receptors are involved in one-trial acquisition and in one-trial recall processes [85]. According to Gauthier's hypothesis, multiple-trial associative learning triggers activation of the α-bungarotoxin-sensitive nicotinic acetylcholine receptors that activate intracellular events, leading to long term memory formation. Methyl Parathion, an acetylcholinesterase inhibitor of the organophosphate family, enhances recall of learned tasks in the visual and olfactory domains, but does not affect the acquisition phase in either domain, suggesting enhancement in cholinergic transmission promotes recall in honeybees [86]. Collective findings confirm that octopaminergic and cholinergic systems are involved in mediating learning and memory in honeybees.

7. Molecular Mechanism Underlying Olfactory Dysfunction in Honeybees: A Possible Link with CCD

In CCD phenomenon, honeybees *Apis mellifera* adult workers disappear from the hive except queen remains inside the hive. No dead bodies of worker honeybees found in or around the hive suggest their death occurs in the field. Here I discuss the effect of reactive oxygen species (ROS)-mediated oxidative damage to the brain cell in relation to olfactory dysfunction. I hypothesize that (ROS)-mediated oxidative damage is responsible for olfactory dysfunction in honeybees, which may be associated with CCD.

7.1. ROS-Mediated Oxidative Stress

ROS include superoxide anions, hydroxyl, alkoxyl, and peroxyl radicals, and hydrogen peroxide. The major sources of ROS are the mitochondrial respiratory chain, an uncontrolled arachidonic acid cascade, and nicotinamide adenine dinucleotide phosphate-oxidase, NADPH oxidase activity [87]. Oxidative stress occurs when there is either an excess of free radicals or a decrease in antioxidant (e.g. glutathione) levels, or both [88]. Presence of iron potentiates ROS production and exacerbates oxidative stress [89]. Free radicals cause oxidative damage by attacking polyunsaturated fatty acids in neural cell membranes and generate metabolic waste products that interfere with cell-cell communication, disturb DNA/RNA and protein synthesis, lower energy levels, and vitally impede biochemical processes. The source of their devastating actions is the oxygen molecule's unpaired electron, which makes it unstable and electrically charged [90]. The oxygen free radical then attacks the nearest stable molecule to capture the needed electron to gain its stability, and then leaves nearest molecule as a free radical, beginning a chain reaction. Thus, once a process is started, it can cascade, finally resulting in the destruction of a living cell. Free radicals are generated during lipid peroxidation, which involves oxidative conversion of polyunsaturated fatty acids to lipid hydroperoxides. Lipid peroxidation is a chain reaction that is driven by ROS, and consists chain initiation, propogation and termination steps [91]. Lipid peroxidation utilizes molecular oxygen and produce ROS, which include superoxide anion ($O_2^{\cdot}$) and hydrogen peroxide (H_2O_2). Superoxide is rapidly converted to H_2O_2 by superoxide dismutase (SOD), and in turn, H_2O_2 is converted to H_2O by catalase [92]. In the presence of metal ions (such as Fe^{2+} and Cu^{2+}), H_2O_2 can be further converted to hydroxyl radical ($^{\cdot}OH$) through the Fenton reaction. Hydroxyl radicals can attack polyunsaturated fatty acids in membrane phospholipids forming the peroxyl radical ($ROO^{\cdot}$) and then propagate the chain reaction of lipid peroxidation [93].

To increase life span, there should be changes either in the regulation of mitochondrial genes, decreasing ROS production or elevated expression of antioxidant genes. Overexpression of antioxidative genes that encode CuZn-superoxide dismutase, SOD, Mn-SOD, and catalase enzymes has been shown to extend *Drosophila* life span [94-96]. These flies exhibit a delayed response to physical performance and lower oxidative damage, supporting the free radical theory of aging. The expression of antioxidant genes generally decreases with age in queen, but not in worker honeybees [97]. We have reported age-related olfactory dysfunction in worker honeybees housed in an indoor flight room maintained at 25-30°C on a 16/8-hour light-dark cycle, while controlling for other sources of mortality such as unfavorable weather, predators, or amount of time spent on flight. A marked deficit in olfactory learning occurring in aged workers compared to young workers, suggests that learning and memory processes in the brain appear to be vulnerable with aging [98]. In addition to ROS, reactive nitrogen species (RNS) may interact with catecholamines, producing oxidants and free radicals that are likely to trigger toxic effects, including damage to cellular proteins, lipids, and DNA.

7.2. ROS-Mediated Olfactory Dysfunction

Very little experimental data are available on ROS-mediated oxidative stress in the honeybee brain. To determine whether honeybees can be used as a model system for oxidative stress, and whether or not ROS-associated alterations in brain can modulate olfactory response of honeybees, we have injected ferrous ammonium citrate (FAC) into the antennal lobes of worker honeybees to produce OH• radical, and evaluated the effects of FAC on their olfactory learning and memory [99]. We have found that iron-induced inhibitory effect on olfactory learning and memory is dose-and time-dependent. Injections of reduced glutathione into the antennal lobes before FAC application block ROS-mediated inhibitory effect, implicating that glutathione protects brain against oxidative stress by modulating the redox state of specific thiol residues of target proteins. Similarly, injections of iron chelator (VK-28) prior to FAC overcome ROS-mediated inhibitory effect on olfactory learning and memory, implicating reduction in iron prevents initiation of Fenton reaction and thereby reduces oxidative stress. These findings clearly indicate that iron-induced oxidative stress in the honeybee brain leads to olfactory dysfunction [99].

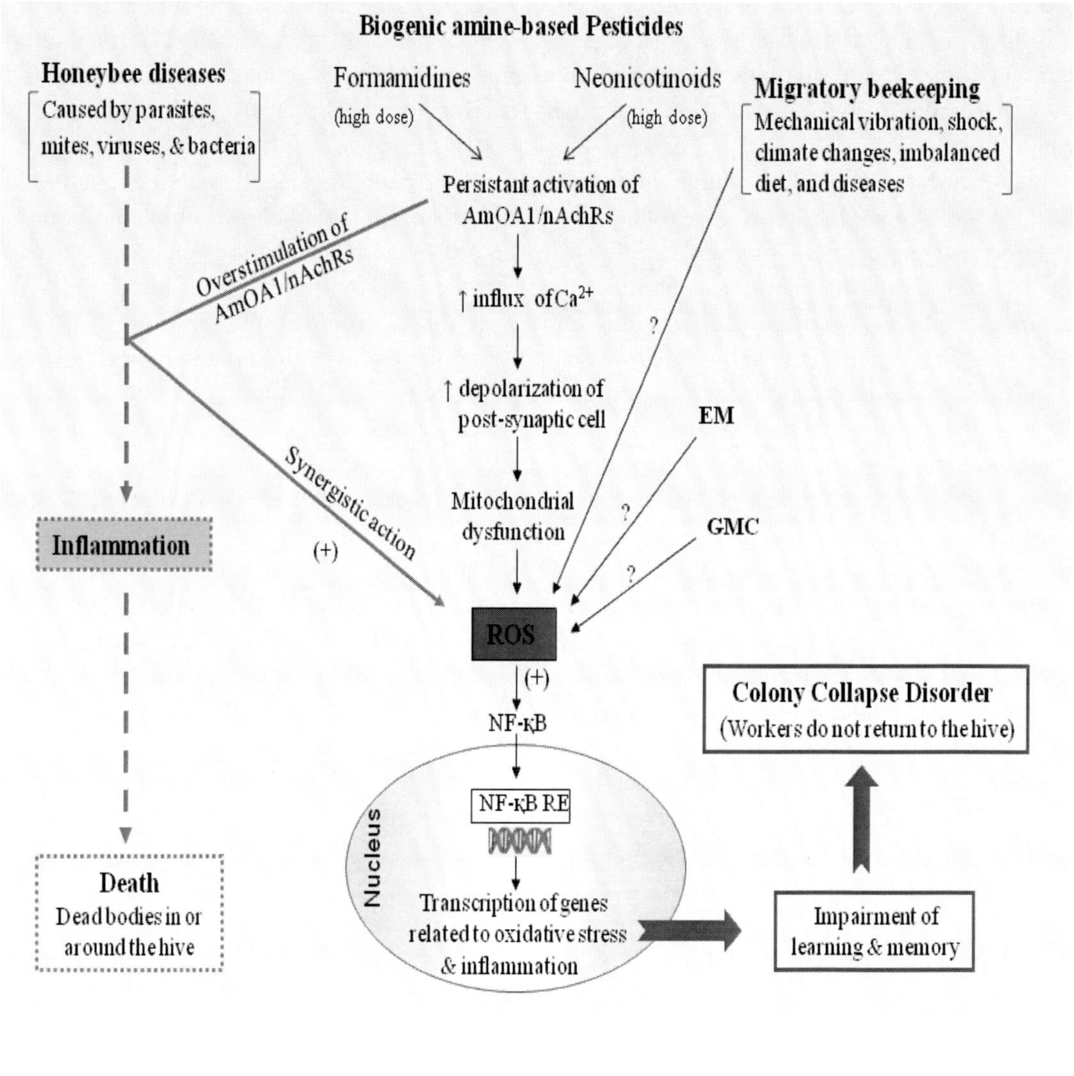

Figure 4. Hypothetical events involved in ROS-mediated olfactory dysfunction in honeybees resulting in colony collapse disorder. Generation of ROS increases oxidative stress that damages proteins, DNA/RNA and lipids causing olfactory dysfunction. GMC, genetically-modified crop; EM, electromagnetic radiation; nAchR, nicotinic acetylcholine receptor; AmOA1, octopamine receptor in *Apis mellifera* brain; ?, connection with biogenic amines signaling is notknown; (+) designates stimulation. Dashed arrows indicate diseases-mediated mechanism that may result in the death of honeybees in or around the hive. Solid arrows represent disruption in biogenic amines-mediated signaling that may lead to colony collapse disorder in which honeybees die in the field.

7.3. ROS-Mediated Olfactory Dysfunction: Implication to CCD

Although, several theories have been reported for their possible contribution to CCD, but what is common among these theories? Is CCD a multifactorial phenomenon based on synergistic effects? Here, I discuss the possible molecular mechanism focused on biogenic amines-based pesticides theory in relation to olfactory dysfunction and CCD.

It has been reported that calcium (Ca^{2+}) plays an important role in the formation of protein-dependent olfactory long-term memory in honeybees [100]. Ca^{2+} is necessary for mediating synaptic plasticity and regulating gene expression correlated with long-term memory formation. However, excessive Ca^{2+} influx alters intracellular Ca^{2+} homeostasis that can participate in the mechanisms generating potentially lethal plasma membrane injury and mitochondrial dysfunction. According to my hypothesis, the persistant activation of nicotine acetylcholine receptors or octopamine receptors with high doses of pesticides results in extensive influx of Ca^{2+} in the honeybee brain that is responsible for activation of ROS-generating enzymes (such as phospholipase A_2, cyclooxygenase, and lipoxygenase), resulting in production of arachidonic acid (ARA) and its oxygenated metabolites from neural membrane glycerophospholipids. Non-enzymic oxidation of ARA produces ROS that activates NF-κB, a key regulator of neuronal death (Figure 4). ROS interacts with NF-κB subunits in the cytoplasm, which promote NF-κB translocation from the cytoplasm to the nucleus. In the nucleus, NF-κB promotes the transcription of genes related to oxidative stress and inflammation, encoding number of proteins and enzymes related to oxidative stress and inflammation (Figure 4). Furthermore, enzymic oxidation of ARA produces eicosanoids that promote inflammation (not shown in Figure 4). These mediators contribute to oxidative stress and induce inflammation that may harm cholinergic neurons/octopaminergic neurons, which result in olfactory dysfunction, ultimately leading to CCD (Figure 4). These pesticides in conjunction with honeybee diseases and/or other factors may synergistically lead to failure in immune system and disorientation, resulting in CCD.

CONCLUSION

To satisfy demands of agriculture and complementary alternative medicinal science, it is necessary to understand the molecular mechanism underlying CCD. This devastating phenomenon does not affect queen bee and larvae except workers that disappear from the bee hive. Several factors have been attributed to this syndrome, without explaining the molecular mechanism. In my opinion, ROS-mediated oxidative stress suppresses immune system as well as impairs learning and memory of foraging worker honeybees, and that may be a molecular mechanism underlying CCD. ROS-mediated oxidative stress may be a main reason why worker honeybees fail to return to the hive and die in the field due to impairment in their learning and memory.

CCD is most likely to be a multifactorial phenomenon. The potential molecular mechanism underlying this phenomenon is ROS-mediated oxidative stress that is produced by the disturbance in cellular prooxidant and antioxidant ratio. Since honeybees are known to have a reduced number of genes that express resistance to toxic substances and diseases compared to fruit flies and ants, therefore, it is necessary to improve genetic stocks of honeybees. In laboratory experiments, honeybees have been shown either fail to discriminate between a known and an unknown odorant or face a significant impairment in olfactory learning or memory performance after exposure to some of these pesticides. Therefore, intrinsic toxicity of pesticides should always be taken into consideration when evaluating the risk of exposure of these compounds to honeybees in fields. Major concern is that some of these pesticides may even be more toxic due to synergistic effect if honeybees get exposed to several of them at the same time. Furthermore, sublethal doses of pesticides in conjunction with honeybee diseases and other factors may act synergistically, and result in the failure of immune system, disorientation, and olfactory dysfunction, leadig to CCD. Therefore the use of these pesticides for crop protection should be banned immediately. Although, Bt toxins including Cry1Ab have not shown any adverse effects on honeybees yet, but further research is still needed to evaluate the exact effect of Bt toxins on cognitive functioning in honeybees. To protect honeybee colonies from CCD phenomenon, it is good to adopt organic farming instead of facing worse decline of colonies in future due to occurrence of synergic effects between Bt biopesticides and other factors.

Overall, the disruption of biogenic amines-mediated signaling by pesticides impair leaning and memory of foragers, which may be a reason for their failure in returning to the hive.

REFERENCES

[1] Cooper E. L. Colony Collapse Disorder May Affect Complementary and Alternative Medicine. *Evid. Based Complement. Alternat. Med.* 2007, 4(3):275-277.

[2] Vanengelsdorp D., Meixner M. D. A historical review of managed honey bee populations in Europe and the United States and the factors that may affect them. *J. Invertebr. Pathol.* 2010, 103 Suppl. 1:S80-S95.

[3] Bankova V. Recent trends and important developments in propolis research. *Evid. Based Complement. Alternat. Med.* 2005, 2:29-32.

[4] Lotfy M. Biological Activity of Bee Propolis in Health and Disease. *Asian Pac. J. Cancer Prev.* 2006, 7:22-31.

[5] Sforcin J. M. Propolis and the immune system: a review. *J. Ethnopharmacol.* 2007, 113:1-14.

[6] Farooqui, T., Farooqui, A. A. Molecular mechanism underlying the therapeutic activities of propolis: a critical review. *Curr Nutri Food Sci.* 2010, 6:186-199.

[7] Gajski G, Garaj-Vrhovac V. Genotoxic potential of bee venom (Apis Mellifera) on human peripheral blood lymphocytes in vitro using single cell gel electrophoresis assay. *J. Environ. Sci. Health A Tox Hazard Subst Environ Eng.* 2008, 43(11):1279-1287.

[8] Park J. H, Kim K. H., Kim S. J., Lee W. R., Lee KG, Par K. K. Bee venom protects hepatocytes from tumor necrosis factor-alpha and actinomycin D. *Arch Pharm Res.* 2010, 33(2):215-223.

[9] Huh J. E., Baek Y. H., Lee M. H., Choi D. Y., Park D. S., Lee J. D. Bee venom inhibits tumor angiogenesis and metastasis by inhibiting tyrosine phosphorylation of VEGFR-2 in LLC-tumor-bearing mice. *Cancer. Lett.* 2010, 292(1):98-110.

[10] Bincoletto C., Eberlin S., Figueiredo C. A., Luengo M. B., Queiroz M. L. Effects produced by Royal Jelly on haematopoiesis: relation with host resistance against Ehrlich ascites tumour challenge. *Int. Immunopharmacol.* 2005, 5(4):679-688.

[11] Vittek J. Effect of royal jelly on serum lipids in experimental animals and humans with atherosclerosis. *Experientia.* 1995, 51(9-10):927-935.

[12] Fujii A., Kobayashi S., Kuboyama N., Furukawa Y., Kaneko Y., Ishihama S., Yamamoto H., Tamura T. Augmentation of wound healing by royal jelly (RJ) in streptozotocin-diabetic rats. *Jpn. J. Pharmacol.* 1990, 53(3):331-337.

[13] Matsui T., Yukiyoshi A., Doi S., Sugimoto H., Yamada H., Matsumoto K. Gastrointestinal enzyme production of bioactive peptides from royal jelly protein and their antihypertensive ability in SHR. *J. Nutr. Biochem.* 2002, 13(2):80-86.

[14] Erem C., Deger O., Ovali E., Barlak Y. The effects of royal jelly on autoimmunity in Graves' disease. *Endocrine.* 2006, 30(2):175-183.

[15] Hidaka S., Okamoto Y., Uchiyama S., Nakatsuma A., Hashimoto K., Ohnishi S. T., Yamaguchi M. Royal jelly prevents osteoporosis in rats: beneficial effects in ovariectomy model and in bone tissue culture model. *Evid Based Complement Alternat Med.* 2006, 3(3):339-348.

[16] Carr E. G. An unusual disease of honey bees. *J. Econ. Entomol.* 1919, 11:347-351.

[17] Bullamore G. W. Nosema apis and Acarapis (Tarsonemus) woodi in relation to Isle of Wight disease. *Parasitology* 1922, 14:53-62.

[18] Foote H. L. The mystery of the disappearing bees. *Am. Bee J.* 1966, 106:126-127.

[19] Roberge F. The case of the disappearing honeybees. National Wildlife 1978, 16:34-35.

[20] Kauffeld N. M. Disappearing disease...a longevity problem? *Agricultural Research.* 1973, 22:14.

[21] Williams J. L., Kauffeld N. M. Winter conditions in commercial colonies in Louisiana. *Am. Bee J.* 1974, 114:219-221.

[22] Oertel E. *Many bee colonies die of an unknown cause. Am. Bee J.* 1965, 105:48-49.

[23] Mraz C. Disappearing disease south of the border. *Gleanings in Bee Culture.* 1977, 105:198.

[24] Kulinčević J. M., Rothenbuhler W. C., Rinderer T. E. *Disappearing disease: III. Acomparison of seven different stocks of the honey bee (*Apis mellifera*).* The Ohio State University, Research Bulletin 1160. Ohio Agricultural Research and Development Center, Wooster, Ohio, 1984, 21 p.

[25] Thurber F. F. Disappearing-yes, disease-no! *Gleanings in Bee Culture.* 1976, 104:260-261.

[26] Kauffeld N. M., Everitt J. H., Taylor E. A. Honey bee problems in the Rio Grande Valley of Texas. *Am. Bee J.* 1976, 116:220-222, 232.

[27] Olley K. Those disappearing bees. *Am. Bee J.* 1976, 116:520-521.

[28] Kulinčević J. M., Rothenbuhler W. C., Rinderer T. E. Disappearing disease. Part I. Effects of certain protein sources given to honey-bee colonies in Florida. *Am. Bee J.* 1982, 122:189-191.

[29] Aikin R. C. *Bees evaporated, a new malady.* Gleanings in Bee Culture 1897, 25:479-480.

[30] Finley J., Camazine S., Frazier M. The epidemic of honey bee colony losses during the 1995-1996 season. *Am. Bee J.* 1996, 136:805-808.

[31] Faucon J. P., Mathieu L., Ribiere M., Martel A. C., Drajnudel P., Zeggane S., Aurieres C., Aubert M. F. A. Honey bee winter mortality in France in 1999 and 2000. *Bee World* 2002, 83:14-23.

[32] CCD Working Group (2006) Colony Collapse Disorder (CCD). Available: http://maarec.cas.psu.edu/pressReleases/FallDwindleUpdate0107.pdf. Accessed 2010 Jan 15

[33] vanEngelsdorp D., Underwood R., Caron D., Hayes Jr. J. An estimate of managed colony losses in the winter of 2006–2007: a report commissioned by the Apiary Inspectors of America. *Am. Bee J.* 2007, 147:599-603.

[34] Underwood R., vanEngelsdorp D. Colony Collapse Disorder: have we seen this before? *Bee Cult.* 2007, 35:13-18.

[35] Cox-Foster D. L., Conlan S , Holmes E. C., Palacios G., Evans J. D., Moran N. A., Quan P. L., Briese T., Hornig M., Geiser D. M., Martinson V., vanEngelsdorp D., Kalkstein A. L., Drysdale A., Hui J., Zhai J., Cui L., Hutchison S. K., Simons J. F., Egholm M., Pettis J. S., Lipkin W. I. A metagenomic survey of microbes in honey bee colony collapse disorder. *Science* 2007, 318:283-287.

[36] vanEngelsdorp D., Hayes Jr. J., Underwood R. M., Pettis J. A Survey of Honey Bee Colony Losses in the U.S., Fall 2007 to Spring 2008. *PLoS ONE 3*, e4071.

[37] Debnam S., Westervelt D., Bromenshenk J., Oliver R. Colony collapse disorder: symptoms change with seasons and are different with various locations. *Bee Culture.* 2008, 137:30-32.

[38] Vanengelsdorp D., Evans J. D., Saegerman C., Mullin C., Haubruge E., Nguyen B. K., Frazier M., Frazier J., Cox-Foster D., Chen Y., Underwood R., Tarpy D. R., Pettis J. Colony Collapse Disorder. *A Descriptive Study PLoS One.* 2009, 4(8):e6481.

[39] Bromenshenk J. J. Colony collapse disorder (CCD) is alive and well. *Bee Culture.* 2010, 138:51.

[40] Fahrbach S. E, Robinson G. E. Juvenile hormone, behavioral maturation, and brain structure in the honey bee. *Dev Neurosci.* 1996, 18(1-2):102-114.

[41] Winston, M. L. *The biology of the honey bee*. Harvard University Press, Cambridge, Mass, 1987.

[42] Robinson G. E, Page R. E. Jr, Strambi C., Strambi A. Hormonal and genetic control of behavioral integration in honey bee colonies. *Science.* 1989, 246(4926):109-112.

[43] Neukirch A. Dependence of the lifespan of honeybee (*Apis mellifera*) upon flight performance and energy consumption. *J. Comp. Physiol.* 1982, 146:35-40.

[44] Shimanuki H. Bacteria. In: Flottum, K. (Ed.), *Honey Bee Pests, Predators, and Disease*, third ed. pp. 33-54, Medina, Ohio, 1997.

[45] Blenau W., Baumann A. Molecular and pharmacological properties of insect biogenic amine receptors: lessons from Drosophila melanogaster and Apis mellifera. *Arch Insect. Biochem. Physiol.* 2001, 48(1):13-38.

[46] Farooqui T. Octopamine-mediated neuromodulation of insect senses. *Neurochem Res.* 2007, 32:1511-1529.

[47] Roeder T., Seifert M., Kahler C., Gewecke M. Tyramine and octopamine: antagonistic modulators of behavior and metabolism. *Arch Insect. Biochem. Physiol.* 2003, 54(1):1-13.

[48] Elbert A., Haas M., Springer B., Thielert W., Nauen R. Applied aspects of neonicotinoid uses in crop protection. *Pest Manag. Sci.* 2008, 64:1099-1105.

[49] Tomizawa M., Casida J. E. Neonicotinoid insecticide toxicity: mechanisms of selective action. *Annu Rev. Pharmacol. Toxicol.* 2005, 45:247-268.

[50] Jones A. K., Raymond-Delpech V., Thany S. H., Gauthier M., Sattelle D. B. The nicotinic acetylcholine receptor gene family of the honey bee, *Apis mellifera. Genome Res.* 2006, 16:1422-1430.

[51] Barbara G. S., Grunewald B., Paute S., Gauthier M., Raymond-Delpech V. Study of nicotinic acetylcholine receptors on cultured antennal lobe neurones from adult honeybee brains. *Invert Neurosci.* 2008, 8:19-29.

[52] Tomizawa M., Talley T. T., Maltby D., Durkin K. A., Medzihradszky K. F., Burlingame A. L., Taylor P., Casida J. E. Mapping the elusive neonicotinoid binding site. *Proc. Natl. Acad. Sci USA.* 2007, 104:9075-9080.

[53] Decourtye A., Devillers J., Cluzeau S., Charreton M., Pham-Delègue M. H. Effects of imidacloprid and deltamethrin on associative learning in honeybees under semi-field and laboratory conditions. *Ecotoxicol. Environ. Saf.* 2004, 57(3):410-419.

[54] Iwasa T., Motoyama N., Ambrose J. T., Michael Roe R. M. R. Mechanism for the differential toxicity of neonicotinoid insecticides in the honey bee, *Apis mellifera. Crop Protection.* 2003, 23:371-378.

[55] Adams M. D., Celniker S. E., Holt R. A. The genome sequence of *Drosophila melanogaster. Science* 2000, 287:2185-2195.

[56] Jones A. K., Grauso, M., Sattelle D. B. The nicotinic acetylcholine receptor gene family of the malaria mosquito, *Anopheles Gambiae. Genomics.* 2005, 85(2):176-87.

[57] Shao Y. M., Dong K., Zhang C. X. The nicotinic acetylcholine receptor gene family of the silkworm, *Bombyx mori. BMC Genomics,* 2007, 8:324.

[58] Jones A. K., Sattelle D. B. The cys-loop ligand-gated ion channel gene superfamily of the red flour beetle, *Tribolium castaneum. BMC Genomics.* 2007, 108:327.

[59] Jones A. K., Sattelle D. B. *Diversity of Insect Nicotinic Acetylcholine Receptor Subunits.* In: Insect Nicotinic Acetylcholine Receptors. Series: Advances in experimental medicine and biology , v. 683, pp. 25-43, S. H. Thany (ed.), Landes Bioscience and Springer Science+ Business Media, New York, Austin, 2010.

[60] Liu Z., Yao X., Zhang Y. Insect nicotinic acetylcholine receptors (nAChRs): Important amino acid residues contributing to neonicotinoid insecticides selectivity and resistance. *African. J. Biotechnol.* 2008, 7:4935-4939.

[61] Hollingworth R. M. Chemistry, biological activity, and uses of formamidine pesticides. *Environ Health Perspect.* 1976, 14:57-69.

[62] Beeman R. W. Recent advances in mode of action of insecticides. *Annu Rev Entomol.* 1982, 27:253-281.

[63] Evans P. D., Gee J. D. Actions of formamidine pesticides on octopamine receptors. *Nature,* 1980, 287:60-62.

[64] Phillips J. R. Bollworm control with chlorphenamidine. *Arkansas Farm Res.*1971, 4:9.

[65] Knowles C. W. Structure-activity relationship among amidine acaracides and insecticides. *In*: Insecticide Mode of Action. J. R. Coats (ed.), Academic Press, New York, 1982.

[66] O'Brien P. J., Elzen G.W., Vinson S. B. Toxicity of azinphosmethyl and chlordimeform to parasitoid *Bracon mellitor* (Hymenoptera: Braconidae): Lethal and reproductive effects. *Environ Entomol.* 1985, 14:891-894.

[67] Ismail S. M. M., Matsumura F. Influence of pesticides and neuroactive amines on cAMP levels of two-spotted spider mite (Acari: Tetranychidae). *Insect Biochem.* 1989, 19:715-722.

[68] Dudai Y., Buxbaum J., Corfas G., Orgad S., Segal D., Sher B., Uzzan A., Zvi S. Defective cAMP metabolism and defective memory in Drosophila. *Acta Biochim. Biophys. Hung.* 1986, 21(3):177-192.

[69] Liac A. Y., Daveybc R. B., Millerbc R. J., Georgec J. E. Detection and characterization of amitraz resistance in the southern Cattle Tick, *Boophilus microplus* (Acari: Ixodidae). *J. Med. Entomol.* 2004, 41:293-200.

[70] Elzen P. J., Baxter J. R., Westervelt D., Causey D., Randall C., Cutts L., Wilson W. T. Acaricide rotation plan for control of *Varroa*. *Am. Bee J.* 2001, 141:369-371.

[71] Floris I., Satta A., Garau V. L., Melis M., Cabras P., Aloul N. Effectiveness, persistance, and residue of amitraz plastic strips in the apiary control of *Varror destructor*. *Apidologie* 2001, 32:577–585.

[72] Atkinson P. W., Binnington K. C., Roulston W. J. High monoamine oxidase ativity in the tick *Boophyzlus microplus,* and inhibition by chlordimeform and related pesticides. *Aust. J. Entomol.* 1974, 3(3):207210.

[73] Schuntner C. A., Thompson P. G. Metabolism of ^{14}C-amitraz in larvae of *Boophilus microplus*. *Aust. J. Biol. Sci.* 1978, 31:141-148.

[74] Roeder T. Tyramine and octopamine: Ruling behavior and metabolism. *Annu. Rev. Entomol.* 2005, 50:447-477.

[75] Menzel R., Greggers U., Smith A., Berger S., Brandt R., Brunke S., Bundrock G., Hülse S., Plümpe T., Schaupp F., Schüttler E., Stach S., Stindt J., Stollhoff N., Watzl S. Honey bees navigate according to a map-like spatial memory. *Proc. Natl. Acad. Sci. USA*. 2005, 102:3040-3045.

[76] Menzel R., Müller U. Learning and memory in honeybees: From behavior to neural substrates. *Annu. Rev. Neurosci.*1996, 19:379-404.

[77] Giurfa M. Cognitive neuroethology: dissecting non-elemental learning in a honeybee brain. *Curr. Opin. Neurobiol.* 2003, 13:726-735.

[78] Hammer M. An identified neuron mediates the unconditioned stimulus in associative learning in honeybees. *Nature* 1993, 366:59/-63.

[79] Hammer M., Menzel R. Multiple sites of associative odor learning as revealed by local brain microinjection of octopamine in honeybees.*Learn Mem.* 1998, 5:146-156.

[80] Farooqui T., Robinson K., Vaessin H., Smith B. H. Modulation of early olfactory processing by an octopaminergic reinforcement pathway in the honeybee. *J. Neurosci.* 2003, 23:5370-5380.

[81] Han K.-A., Millar N. S., Davis R. L. A novel octopamine receptor with preferential expression in *Drosophila* mushroom bodies. *J. Neurosci.* 1998, 18:3650-8.

[82] Schwäerzel M., Monastirioti M., Scholz H., Friggi-Grelin F., Birman S., Heisenberg M. Dopamine and octopamine differentiate between aversive and appetitive olfactory memories in *Drosophila. J. Neurosci.* 2003, 23:10495-502.

[83] Thany S. H. Gauthier M. Nicotine injected into the antennal lobes induces a rapid modulation of sucrose threshold and improves short-term memory in the honeybee *Apis. Mellifera. Brain Research.* 2005, 1039:216–219.

[84] Lozano V. C., Bonnard E., Gauthier M., Richard D. Mecamylamine-induced impairment of acquisition and retrieval of olfactory conditioning in the honeybee. *Behavioural. Brain Research.* 1996, 81:215-222.

[85] Gauthier M. State of the art on insect nicotinic acetylcholine receptor function in learning and memory. *Adv. Exp. Med. Biol.* 2010, 683:97-115.

[86] Guez D., Zhu H., Zhang S. W., Srinivasan M. V. Enhanced cholinergic transmission promotes recall in honeybees. *J. Insect. Physiol.* 2010, 56(9):1341-1348.

[87] Sun G. Y., Horrocks L. A., Farooqui A. A. The roles of NADPH oxidase and phospholipases A2 in oxidative and inflammatory responses in neurodegenerative diseases. *J. Neurochem*, 2007, 103(1):1-16.

[88] Halliwell B. Oxidative stress and neurodegeneration: where are we now? *J. Neurochem.* 2006, 97(6):1634-1658.

[89] Liu R., Liu W., Doctrow S. R., Baudry M. Iron toxicity in organotypic cultures of hippocampal slices: Role of reactive oxygen species. *J. Neurochem.*2003, 85:492-502.

[90] Halliwell B., Gutteridge J. M. C. Free radicals in biology and medicine. Oxford (UK): Oxford University Press, 1999.

[91] Porter N. A., Caldwell S. E., Mills K. A. Mechanisms of free radical oxidation of unsaturated lipids. *Lipids* 1995, 30:277-290.

[92] Farooqui A. A., Ong W. Y., Horrocks L. A. *Neurochemical Aspects of Excitotoxicity.* Springer, New York, 2008.

[93] Farooqui T., Farooqui A. A. Oxidative stress and inflammation in the pathogenesis of Parkinson's disease. *Parkinson Disease* 2011, 2011. Article ID 247467, 9 pages. doi:10.4061/2011/247467

[94] Orr W. C., Sohal R. S. Extension of life-span by overexpression of superoxide dismutase and catalase in *Drosophila melanogaster*. *Science* 1994, 263:1128-30.

[95] Parkes T. L., Elia A. J., Dickinson D., Hilliker A. J., Phillips J. P., Boulianne G. L. Extension of *Drosophila* lifespan by overexpression of human SOD1 in motorneurons. *Nat Genet.* 1998, 19:171-4.Parkes and others 1998,

[96] Sun J., Molitor J., Tower J. Effects of simultaneous over-expression of Cu/ZnSOD and MnSOD on *Drosophila melanogaster* life span. *Mech Ageing Dev.* 2004, 125:341-349.

[97] Corona M., Hughes K. A., Weaver D. B., Robinson G. E. Gene expression patterns associated with queen honey bee longevity. *Mech Ageing Dev.*2005, 126(11):1230-1238.

[98] Farooqui T. Octopamine-mediated neuronal plasticity in honeybees: implications for olfactory dysfunction in humans. *Neuroscientist* 2007, 13:304-322.

[99] Farooqui T. Iron-induced oxidative stress modulates olfactory learning and memory in honeybees. *Behavioral. Neurosci.* 2008, 122:433-447.

[100] Perisse E., Raymond-Delpech V., Néant I., Matsumoto Y., Leclerc C., Moreau M., Sandoz J. C. Early calcium increase triggers the formation of olfactory long-term memory in honeybees. *BMC Biol.* 2009, 7:30.

In: Bees
Editor: Richard M. Florio

ISBN: 978-1-61324-825-6

Chapter 8

□ “¤… –¤ŒÕ–¤› Ã ¤ —“ □Ÿ¤ Œ¤‘ □Ã¤‹ □ÕÕ→ □Ã ¤ □“‹ “—“ □ÕÕ→ □Ã ¤ ‘ ¤□Œ“ “Ÿ –¤Œ□—Œ’ □“› ¤ “ ÿ —“¤« □¤¤ÕÚ

David Guez*
The Australian National University,
Research School of Biology,
Australia

□æ¨Æ ¨

Honeybee learning studies often utilize the well known proboscis extension reflex (PER) technique, a method that is known to be influenced by a honeybee's age. In this study, we investigated whether the PER technique may be linked to some of the discrepancies that have been previously noted between associative and non-associate learning performance in honeybees. Whilst associative learning has been shown to improve with age, non associative learning, such as habituation, has shown an apparent decline in performance with age. Here we investigated changes in the sugar elicited PER threshold using 6 different sugars to evaluate the relative value of a given sugar solution as a bee ages. Our results revealed an interesting switch in sugar response between the ages of 7 and 8 days when sucrose or fructose was presented. We found that the sucrose threshold decreased suddenly between 7 and 8 days of age and response to fructose ceased at day 8. The same pattern of results was found when comparing the fructose response's of pollen and nectar foragers. Further results obtained with nectar and pollen foragers suggest that this was not due to a cessation of the perception of fructose.

Our sucrose results explain the apparent discrepancies that have been observed between associative and non associative learning performance in honeybees. We suggest that in associative learning the perceived value of the sugar reward increases with age and so learning only appears to improve. Conversely, in non associative learning the stimulus value increases with age, and subsequently habituation of the response is harder to achieve. We discuss the importance of perspective in interpreting behavioural data, especially when conducting age related experiments in honeybees.

*E-mail address: david.guez@mac.com

–□› Õ05.45-a, 52.35.Mw, 96.50.Fm.

’ ªß©±Æ Ê□ ª₩ ªÆ, reinforcement, sugar threshold, associative learning, nonassociative learning, sucrose, fructose.

ÔÚ ¨Æº´ ¨±

The age related division of labour in honeybee workers has been well documented. Honeybee workers start their adult life as cleaners and hive maintenance workers, developing to become in turn nurses, food manipulators and hive guards before venturing outside to start their final lifeÕs journey as foragers [24]. Arguably, it is at this foraging stage that the honeybee needs to rely most on its capacity for learning and memory. For example, foragers need to learn the location and characteristic of food sources (ªÚÚthe color, shape and odors of particular flowers) as well as the time of the day at which these particular food sources are available. It therefore could be expected that if any variation of learning performance is to be uncovered, learning performance should be highest in older bees, or at least in forager bees.

It has been shown that learning performance does appear to improve with age using the olfactory conditioning of the proboscis extension reflex (PER) [12], [21], [20]. However, in previous work [8] we showed that honeybee performance in the habituation of the PER (a form of learning that is considered by most, but not all [6], as a non associative form of learning), seems to worsen with age. The number of trials needed to achieve habituation increased with age with a notable switch occurring between the ages of 7 and 8 days after emergence.

We therefore undertook to shed light on these apparent discrepancies and based our approach on the following methodological observations. In olfactory conditioning of the PER experiments, the olfactive cues or conditional stimuli are kept identical across ages, as are the unconditional stimulus used (in this case, sugar stimulation) and their temporal relationships. Therefore, an apparent improvement in learning performance could have a non cognitive origin and may be the result of perceptual change. For example, if the honeybees failed to detect either the conditional (CS) or unconditional stimuli (US) they would display an apparent learning deficit that was unrelated to their cognitive competency, but would point toward diminished perceptual capacities. In the case of habituation of the PER experiments, non reinforced monotonous antennal stimulation by a sugar solution leads to the suppression of the PER. Once again the physical quality of the sugar solution used and the frequency of the antennal stimulation are kept stable across ages. Therefore, an increase in the number of stimulations needed to achieve habituation may not be a reflection of a non associative learning deficit but may instead be linked to changes in the perceptual capacity of the bees. For example, if the sugar stimulation is perceived as being sweeter with age, it would increase in the number of stimulations necessary to achieve habituation of the PER but would not necessarily reflect any changes with regards to the underlying neurobiology of habituation or cognitive abilities. Therefore, performance changes may not always reflect of changes in learning ªÆª, they may also be a reflection of perceptual changes.

ǾÚ ¤® ªÆ ª ¨ÔÊ¤™¥¨ ± ± ÿ ± ªßæªª Õ' øÆÃ Æ ±¥ ©¨ ◘ ª

ǾǾÚ ¨Æº´ ¨ ±

The olfactory conditioning of the PER and the habituation of the PER both use sugar stimulations as US. An age dependant decrease in the honeybee's sugar threshold would differentially impact their displayed learning performance in these two paradigms. An age dependant decrease of the sugar threshold would induce an apparent age dependant increase in the learning performance in the olfactory conditioning of the PER, since the reward value of the US increases in function of the decreasing the sugar threshold. On the contrary, an increase in the value attributed to the sugar stimulation would be expected to increase the number of trials needed to achieve habituation, leading to an apparent decrease in learning performance for this task. For our first experiment we therefore tested the hypothesis of a change in sugar threshold as the underlying driver for the apparent age dependant changes in both associative and non-associative learning performance.

ǾǾÚ " ø¨ª Æþ¥ ø º " ª¨ ±º

To obtain worker honeybees of known age a single frame of capped brood was removed from a hive and incubated in darkness at 31°C (80% humidity). Emerged adult insects were transferred to a honey frame and collected randomly at desired times using a pair of tweezers. Each queenless box (580 × 485 × 100 mm) contained up to 150 bees. On the day of the experiment, bees were placed on ice until immobile and mounted in thin-walled aluminium tubes (7 mm inner diameter) using a thin strip of fabric-reinforced tape (GAFFA). After mounting the bees were fed a 60% (w/w) sugar solution. After feeding, the bees were placed back in the incubator and submitted to a 4-h fast before testing. All experiments were performed at 25°C at the same time of the day.

ǾǾǾÚ ‹ ª¨ª Æ ø¨ ± ± Õ' øÆÃ Æ ±¥

Honeybees of known age were assayed with accenting concentration of sucrose to determine their sugar threshold, defined as the lower concentration of sugar for which PER was elicited. The concentration of the sucrose solutions tested were 0.1%, 0.3%, 1%, 3%, 10%, 30% and 60% (w/w). A score was attributed to each concentration from 1 for 0.1% to 7 for 60%, and 8 if none of the concentration of sucrose had elicited a PER. Sugar stimulations were interspersed with water stimulations in order to control for sensory sensitization resulting from antennal stimulation with sucrose [18].

ǾǾǾÚ Õ'ø¨ ¨ ø¥◘ ø¥ß

The Systat 12.0 package was used for data analysis. Data sets were first examined with an overall Kruskal-Wallis test and if significative ($p < 0.05$), followed by pairwise comparisons using a Mann-Whitney test, values of $p < 0.05$ were considered significant.

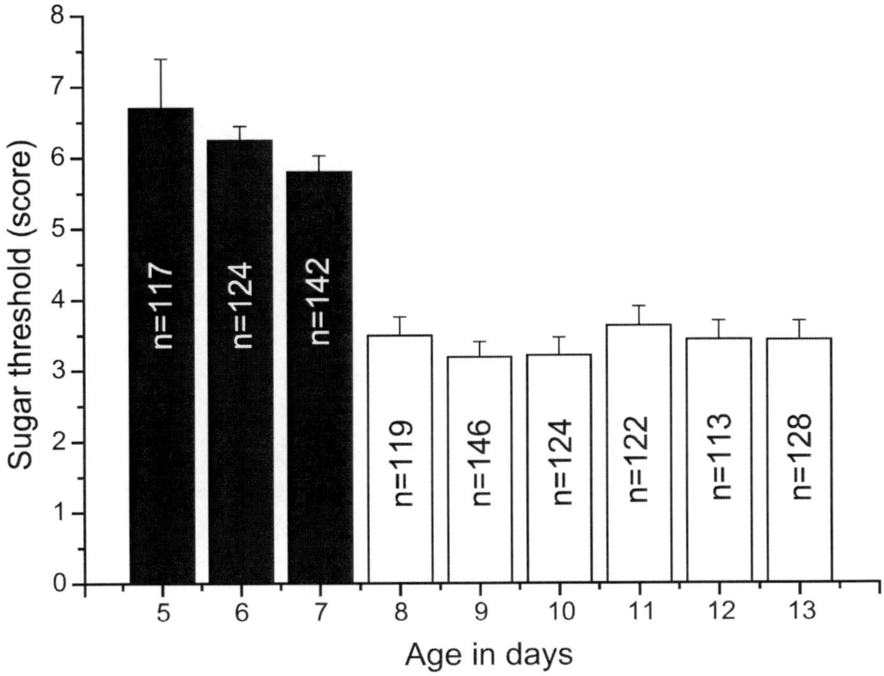

Figure 1. Mean score index ($\pm$SE) of sucrose threshold in function of honeybee ages. Different grey levels indicate statistically significant differences ($p < 0.05$). Overall Kruskal-Wallis test $p < 0.001$.

ÓÏÚ Œ ´¥ ø º ‹ ´ ±

Figure 1 shows that the sucrose threshold in honeybees changed with age. Young bees (5 to 7 days old) exhibited a higher sucrose threshold than did older bees (8 to 13 days old). Interestingly, a sudden switch in sucrose threshold was evident between the age of 7 and 8 days after emergence. These results fully explain the age dependant changes in habituation that we observed in our previous experimental work [8], [7] where it was demonstrated that younger bees, aged 4 to 7 days old, habituated faster than older bees (aged 8 to 10 days old). Here the sucrose threshold decreased suddenly between 7 and 8 days of age, so it could be said that the relative value of a given sucrose solution increased suddenly resulting in an increase in the number of trials needed to achieve habituation of the PER [8], [7]. Thus, the changes in sucrose threshold reported here are sufficient to explain the age related changes in the habituation of the PER that was previously observed in control individuals [8], [7].

Explaining an age related change in associative learning solely in terms of sucrose threshold may be more difficult based on the results shown in Figure 1. Maleszka et al. [12] and Ray et al. [20] have showed that performance in the olfactory conditioning of the PER improves with age, and consistent with this finding, here we found that sucrose threshold decreases with age. However, the sudden shift in the sucrose threshold that we detected between 7 and 8 days of age does not appear to translate into a sudden improvement in learning performance. In fact Maleszka et al. [12] found that a performance shift occurred between the ages of 4 and 5 days, and that performances before and after this time were stable. Thus, although the Ø´ø¥¨ø¨™ change of sucrose threshold is consistent with improved learning performance with age, in this paradigm, the timing of the sucrose threshold change that we demonstrate does not appear to map a performance shift in associative learning at these ages. One possible explanation is that there is another switch in sucrose threshold that occurs between the ages of 4 and 5 days. However, this could not be tested within the present experiment due to the already high sucrose threshold displayed by 5 days old bees that would have prevented the detection of any potential threshold switch occurring at this time. Nevertheless, if age dependant changes in sugar threshold in honeybees is a major factor in the observed age dependant evolution of associative performance, it could be predicted that if one uses a reward that is equally appetitive to all test bees regardless of age, associative performance should become identical and independent of age.

ÌÚ ¤® ªÆ ª ¨ŒÊÿ ± ªß ø ø Œ©øÆ

ÌŒÚ ¨Æº´ ¨±

After emergence honeybees essentially consume honey. It is therefore reasonable to assume that honey is equally appetitive to all honeybees regardless of their age. If indeed the main factor underlying the evolution of honeybee performance in the olfactory conditioning of the PER is perceptual and not cognitive, we should be able to condition bees aged 4 to 8 days old to the same performance level. However, if the main factor is cognitive maturation or due to the maturation of the olfactory system, the use of honey as a reward should not suppress the difference in performance that is found between the ages of 4 and 5 days. We decided to address this question in a second experiment. We also attempted to reproduce the result presented by Maleszka et al [12] as a “positive” control and therefore adopted the same learning protocol.

ÌŒÚ ” ø¨ª Æø¥ ø º ” ª¨ ±º

To obtain worker honeybees of known age a single frame of capped brood was removed from a hive and incubated in darkness at 31°C (80% humidity). Emerged adult insects were transferred to queenless box with free access to honey and collected randomly at desired times using a pair of tweezers. Each queenless box (580 × 485 × 100 mm) contained up to 150 bees. On the day of the experiment, bees were placed on ice until immobile and mounted in thin-walled aluminium tubes (7 mm inner diameter) using a thin strip of fabric-reinforced tape (GAFFA). The insects were fed 2 M sucrose solution via a syringe fitted with a 19-gauge needle and those that did not feed, as well as those that seemed sluggish, were discarded. After being fed, bees were arranged in a Perspex holder and kept overnight at room temperature.

Bees were trained to distinguish between a positive (reward-bearing, 1M sucrose solution or honey) odour and a negative (non reward-bearing, saturated NaCl solution) odour. The training procedure was as in Maleszka et al. [13], [12], but performed in one trial (see below). The training procedure was therefore a modified version of the well-known olfactory conditioning protocol of the PER, which usually involves training to a single odour

[10], [4] and the 3 trials procedure previously proposed by Maleszka et al. with a 'positive' and a 'negative' odour [13], [12].

The positive odour was natural lemon essence (4 μl/ml), dissolved in 1 M sucrose solution (or water when honey was the reward), which constituted the reward. The negative odour was natural vanilla essence (4 μl/ml), dissolved in saturated NaCl solution, which constituted the non reward solution [13]. The training protocol was as follows. A drop of the positive odour/reward solution or negative odour/salt solution, emerging from the tip of a syringe needle, was waved in front of the antennae for 5 seconds (without contact). This training was carried out only once, i.e. the bees were subjected to 1-trial learning.

Testing was always carried out 1 hour after the training to ensure it was well after the consolidation of the memory trace [9]. A test consisted of waving a drop of the positive odour solution in water and a drop of the negative odour solution in water in front of the antennae. Each drop was presented, in turn, for 5 seconds. The order of the presentation (positive odour followed by negative odour, or vice versa) was randomised for each test and each bee was tested only once. The test stimuli carried no sugar reward or salt detriment, but possessed the same odour concentration as the training stimuli.

The test results were scored as follows. (i) If the positive odour elicited a proboscis extension, but did not with the negative odour, the test trial was regarded as having yielded a correct response. (ii) If both odours elicited a proboscis extension, the response was regarded as incorrect. (iii) If the negative odour elicited a proboscis extension but there was none with the positive odour, the response was regarded as incorrect (this response rarely occurred). (iv) If neither odour elicited a proboscis extension, the response was discounted. (Although this last case could be indicative of a memory/learning problem, the absence of a proboscis extension to either odour could also have been due to lack of olfactory sensitivity. Because the two possible cases could not be disambiguated it would have been inappropriate to label either of these as correct or incorrect responses and so they were omitted.) The frequency of correct responses in the tests was calculated as $\frac{n_i}{n_i+n_{ii}+n_{iii}} \times 100$, where n_i, n_{ii} and n_{iii} denote the number of responses in categories (i), (ii) and (iii), respectively.

ÌÚÕÚÚ Ṏø¨ ¨ ø¥□ ø¥ß

The Systat 12.0 package was used for data analysis. Hypothesis testing on proportions were carried out for each age. The null hypothesis was the equality of the proportion against an alternative hypothesis which was that performance at 4 days of age for the 1M sucrose group was less than for the honey group, and for all others cases the performance of the 1M sucrose group was more than the performance of the honey group.

ÌÚÚ Œ ´¥ ø º ‹ ´ ±

Figure 2 presents the performance in the olfactory conditioning of the PER in function of age using two different reward types, 1M sucrose solution and honey. When using the 1M sucrose solution as a reward we were able to reproduce the results presented by Maleszka et al. [12] following one trial conditioning. In particular, we were able to observe a sharp improvement in performance between 4 and 5 days of age (Figure 2). Most importantly, the performance switch previously observed between 4 and 5 days of age when 1M sucrose solution was used was suppressed when honey was presented as the reward. In fact, the learning performance displayed by honeybees was found to be stable across ages, suggesting that indeed the value attributed by an individual to the reward is critical to the type of performance profile later observed. Indeed as honey is the base food for all adult workers, one could reasonably expect its relative value to be stable across all ages. These results suggest that the performance profile in the olfactory conditioning of the PER may be due to a variation of sugar threshold with age.

Surprisingly however, a higher rate of conditioning was observed when a 1M sucrose solution was used as the positive reward rather than honey (compare the resulting learning profile shown in Figure 2). Learning theories such as the Rescorla-Wagner model [22] can shed some light to this somehow puzzling result. In the Rescorla-Wagner model the variation of the associative strength of a cue I after each trial is given by the equation:

$$\Delta V_I = \alpha_I \beta(\lambda - \Sigma V), \tag{1}$$

where λ is the learning asymptote, α_I (with $0 \leqslant \alpha \leqslant 1$) is the learning rate (or salience) of the given cue (here an odor), β (with $0 \leqslant \beta \leqslant 1$) is the learning parameter associated with the given reinforcer (here honey or 1M sucrose solution), and ΣV is the sum over the associative strengths of all relevant cues present during the trial. From our honey results one can assert that the salience of our fragrant cue does not vary with age since performance was seen to be stable with age. Therefore, the result observed when 1M sucrose solution was used as the reward can not be explained in terms of the changing salience of the fragrant cue, since the same cues were used in both the honey and sucrose procedures. Assuming that the Rescorla–Wagner model [22] is valid, one could conclude from these results that the cause of the change was not due to the maturation of the olfactory system and the capacity of individual bees to perceive, or the intensity with which a specific odor is perceived. Most likely the difference lies in the associability of the reinforcer (or β parameter), or in other words, in the capacity of the reinforcer to be predicted by a given type of cue. Since floral scents in natural settings are generally not predictive of the discovery of honey, it is reasonable to consider that such a US would not be readily associated with a floral scent. On the contrary a 1M sucrose solution is closer in term of concentration to naturally occurring nectar [2] and therefore may be expected to be more readily associated with floral scents .

Another possible explanation is what is known as the US pre-exposure effect, whereby pre-exposure to an unsignaled US leads to a subsequent retardation of acquisition during CS-US pairings. (This effect was first described in vertebrates but has also been found to exist in honeybees [1].) In the present experiment, the honeybees were kept in an incubator with free access to honey and were therefore pre-exposed to it. Therefore, it is reasonable to expect that their preexposure to honey may have affected the ability of honey to enter into a new CS-US association. In contrast, the honeybees used in this experiment did not have a previous exposure to a 1M sucrose solution and so displayed higher acquisition rates. Regardless if the US preexposure effect or a difference in the associability of the reinforcer are at the origin of our results, it is possible to conclude that the evolution of honeybee performance in the olfactory conditioning of the PER is at least partly due to changes in sugar perception.

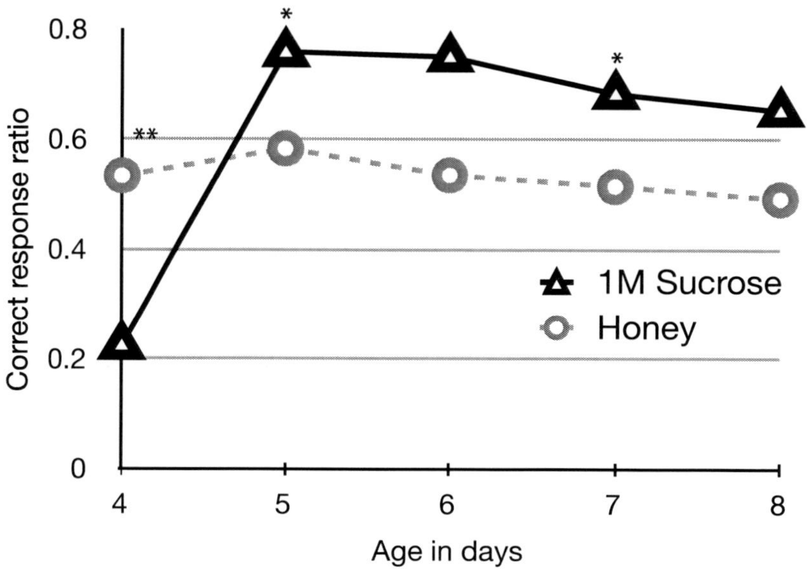

Figure 2. Effect of reward type (honey ™Ú1M sucrose solution) on honeybee performance in olfactory conditioning of the PER. The numbers of individuals for each time point were as follows: 1M sucrose solution: 40, 29, 16, 41, 23; honey: 45, 54, 43, 68, 59. 1M sucrose ™ honey with null hypothesis H_0=equality of proportions, ($^{**}p \leq 0.001$, $^{*}p < 0.05$): 4 days $p = 0.001$ (alternative hypothesis H_1=1M sucrose less than honey), 5 days $p = 0.048$ (alternative hypothesis H_1=1M sucrose more than honey), 6 days $p = 0.061$ (alternative hypothesis H_1=1M sucrose more than honey), 7 days $p = 0.04$ (alternative hypothesis H_1=1M sucrose more than honey) and 8 days $p = 0.092$ (alternative hypothesis H_1=1M sucrose more than honey).

ÏÚ ¤® ªÆ ª ¨ÌÊÕ´ øÆ-ªÆª ¨± ø º ‹ ™ ± ± ' øæ±Æ

ÏÚÚ ¨Æº´ ¨±

So far all the results presented here have involved honeybees of known ages reared in queenless group conditions in an incubator as previously described. It has been argued in previous work that these conditions lead to a default developmental path without the influence of brood or queen pheromones [12], [8]. In the light of our previous results and other published work [17] we wondered if sugar perception profile could be used to associate the default developmental age path with normal hive functions such as foraging activity. To this end we endeavoured to establish patterns in sugar preference.

In selecting the sugars to be tested we considered both honey and nectar compositions. In nectar and honey the most frequent sugars are the hexoses fructose and glucose and the disaccharides sucrose and maltose [2], [15], [19]. All of theses sugars are known to be perceived as 'sweet' by honeybees [5]. Some other sugars such as xylose appear to be neutral to bees while galactose is known to be toxic to honeybees [5]. We therefore chose xylose (as a control), fructose, glucose, maltose and sucrose as our test sugars.

ÏŰÚ ” ø˙ª ÆÞ¥ ø º ” ª¨ ±º

To obtain worker honeybees of known age a single frame of capped brood was removed from a hive and incubated in darkness at 31°C (80% humidity). Emerged adult insects were transferred to a box with free access to honey and collected randomly at desired times using a pair of tweezers. Each queenless box (580 × 485 × 100 mm) contained up to 150 bees. Nectar foragers were collected on a feeder placed at around 10 m from the hives on the morning of the experiment, whereas pollen foragers were visually identified by the fact that they were carrying pollen and were captured at the entrance of the hive upon their return on the morning of the experiment.

On the day of the experiment, bees were placed on ice until immobile and mounted in thin-walled aluminium tubes (7 mm inner diameter) using a thin strip of fabric-reinforced tape (GAFFA). After mounting the bees were fed a 60% (w/w) sugar solution. After feeding, the bees were placed back in the incubator and submitted to a 4-h fast before testing. All experiments were performed at 25°C at the same time of the day. They were then subjected to the alternated presentation of water and sugar solutions before finally being tested with honey. For each antennal stimulation the presence or absence of proboscis extension reflex was recorded. All sugar solutions were 60 % w/w (excluding honey that was presented pure). We tested the sugars found in honey: fructose, glucose, maltose and sucrose. We also tested xylose (to serve as a control). Each experiment was repeated at least two times.

ÏŰÚ Œ ´¥ ø º ‹ ´ ±

The result obtained in experiment 3 are presented in Figure 3. Although we expected the xylose response to be identical to that for water the response that was recorded was actually lower. It would appear that xylose is perceived and acts as an anti appetitive, at least in the case of caged bees (Fisher exact test: 7 days old $p = 0.008$, 8 days old $p = 0.044$). The proportion of PER was generally higher in caged bees that it was for forager bees (except for honey and xylose), which may reflect a higher motivational state in the caged bees compared with the foragers. Although one may expect the same pattern of response for honey the differences between the groups (i.e caged ™ forager bees), this may have been masked by a ceiling effect whereby despite a higher motivational state for the caged bees no larger response could be recorded because a maximum response rate had already been reached. Interestingly, the relative response rates for fructose, glucose, maltose, sucrose and honey between 7 and 8 day old bees appeared to be qualitatively identical to the one observed between pollen foragers and nectar foragers. This suggests a possible developmental correspondence between 7 days old bees and pollen foragers on the one hand, and 8 days old bees and nectar forager on the other. This possibility is made more evident if we take a look at the case of fructose.

In the case of fructose the response rate appeared to be much higher for 7 days old bees than 8 day old bees and the same pattern seems to be true in the case of pollen foragers versus nectar foragers. This hypothesis was statistically verified (given the null hypothesis

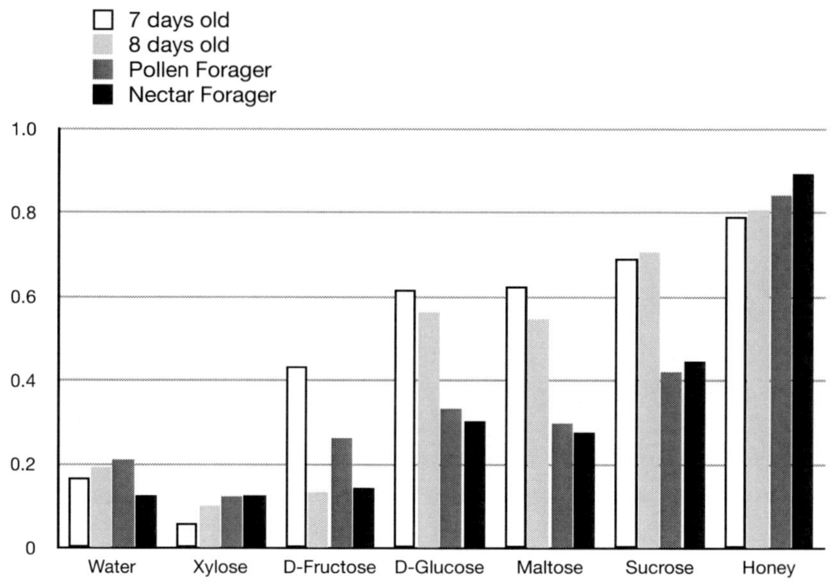

Figure 3. Proportion of honeybees responding to water and different type of sugars (60% w/w): xylose, D-fructose, D-glucose, maltose, sucrose and honey. The number of bees tested were: 7 days old n=120, 8 days old n=120, pollen forager n=57 and nectar forager n=112.

H_0= the proportions are equal, and the alternative hypothesis H_1=the proportion of 7 day old bees and pollen foragers responding were higher than 8 days old bees and nectar foragers respectively). If p is the probability for H_0 to be true we found that for the caged bees $p < 0.001$ and for the forager bees $p = 0.028$. This clearly indicates that indeed the proportion of responses observed at 7 days of age or in pollen foragers is significatively higher than those displayed at 8 days of age or by nectar foragers, respectively.

One possible interpretation of this result is that there is a perceptive switch between the ages of 7 and 8 days (in caged conditions) during which honeybees stop perceiving fructose and thereby stop displaying the PER following antennal stimulation by this monosaccharide. It would also appear that the same perceptive switch occurs during the transition from pollen to nectar forager. However, we also observed that when we replaced our sucrose feeders by fructose feeders, foragers did not stop visiting the feeders and still rapidly emptied them. In contrast, visits would stop if we replaced the sucrose feeders with water feeders. It is therefore highly probable that we were not observing a switch in the perception of fructose but instead observed a suppression of the PER to this sugar between the age of 7 and 8 days, which corresponds to the transition from pollen to nectar forager. Nevertheless, since it has been found that pollen foragers are generally more responsive to sucrose than nectar foragers [17], [23] this interpretation may be lacking as 7 days old bees were found to be less responsive to sucrose than 8 days old bees (see experiment 1).

On average honey contain roughly 38% of fructose, 31% of glucose, 2% of sucrose and 3% of maltose [15], [19]. Furthermore, flowers visited by honeybee have a high hexose content (fructose + glucose) compared with their sucrose content, with 83% of them having a sugar ratio ($\frac{sucrose}{glucose+fructose}$) smaller than 50%, and 44% smaller than 10% [2]. In addition, plants that rely 100% on honeybee for pollination such as the almond tree (◘ ß °ø¥ ± ´) [11] provide a nectar that exclusively contains glucose and fructose in terms of sugar content. Thus, it is unclear as to why such a switch in the PER to fructose, one of the major sugar components of honey, may occur in honeybees. However it is interesting to note that out of 765 species of plants pollinated by honeybees none were found to provide only fructose in their nectar [2].

› ± ¥ ±

The changes observed across ages in habituation performance[8] can be fully explained in terms of changes in sucrose threshold, making it unnecessary to hypothesise any change in cognitive function. However, one can not fully explain the changes observed across ages in honeybee associative performance [12]. Although qualitative sucrose threshold changes were observed in an expected way (ŰÚa decrease, which is consistent with an apparent improvement in learning performance as the relative value of the reward is increased), no significant performance change were observed between the ages of 7 and 8 days. Such a change in performance would be expected to account for the sudden drop in sucrose threshold detected between these ages. Nevertheless, one possible explanation might lie in the existence of a ceiling effect whereby despite a sudden drop in sucrose threshold (meaning an increase in reward value) no performance improvement is detected because the performance level was already close to maximum.

Our honey experiment suggest that the changes observed with age in the performance recorded in the olfactory paradigm is essentially due to sugar perception and not to cognitive development linked to brain maturation. In fact, studies that have investigated brain maturation in honeybees have found that this change occurred very rapidly and predated any observed changes in associative learning [16], [25], [14]. It would therefore be surprising if the changes observed here were directly related to changes in honeybee associative learning. More recent studies have shown that newly emerged honeybees can display significant performance in associative learning as long as their recorded sugar threshold is low enough (less than 10% (w/v)) [3]. These results taken together with those presented here definitely suggest that the main obstacle to adequate associative learning performance in honeybee is sugar perception and therefore the associated value of the reward used.

We have previously argued that our caged procedure results in a default developmental pathway that is removed from the developmental influence of the queen pheromone and the brood pheromone [8]. We have also previously reported that in our hive the youngest foragers observed were 7 days old [12] and we often observed 8 days old bees foraging [7]. These observations taken together with the fact that younger foragers are pollen foragers and our sugar "perception" test result (see experiment 3), suggest that our caged 7 days old bees are in fact developmentally equivalent to pollen foragers whereas the caged 8 days old bee are developmentally equivalent to nectar foragers. If this is the case it would mean that the division of labour in honeybees is developmentally determined and is not the result of experience.

The results presented here also strongly suggest that the changes in honeybee performance with age that have been observed both in associative [20], [21], [12] and non associative learning [8] are not the result of cognitive maturation but simply reflect a change in sugar perception. They also highlight that if we are to study changes in learning performance across different age groups it is important to ensure that the unconditional stimulus used is equivalent for all subgroups studied.

□ µ ±©¥ª º ª ¨

I would like to thank Ryszard Maleszka for providing advice and laboratory space and Paul Helliwell for his valuable technical support in some of the experiments. The experiments were performed in the all weather bee facilities of the Australian National University.

Œª ª Æ ª

[1] C. I. Abramson and M. E. Bitterman. The us-preexposure effect in honeybees. □ ø¥ ¥ªøÆ ø º □ª ø™±Æ 14(4):374–379, 1986.

[2] H. G. Baker and I. Baker. Floral nectar sugar constituents in relation to pollinator type. In C. E. Jones and R. J. Little, editors, ÿ ø ºæ±±µ ± ª® ªÆ ª ¨ø¥ ±¥ ø¨± æ±¥± ß, pages 117–141. Van Nostrand Reinhold Company, N.Y., 1983.

[3] A. Behrends and R. Scheiner. Evidence for associative learning in newly emmerged honey bees (apis mellifera). □ ø¥ ± ¨± , 12(2):249–255, 2009.

[4] M. E. Bitterman, R. Menzel, A. Fietz, and S. Schafer. Classical conditioning of proboscis extension in honeybees (apis mellifera). ÷±´Æø¥± › ± øÆß¨™ – ß ±¥± ß, 97(2):107–119, 1983.

[5] K. Von Frich. □ªª Û¨ ª Æ™ ± Ù ª ø¥ ª ª ø º ¥ø ´ø ª. Cornell University Press, Ithaca, N.Y., 1950.

[6] D Guez. A bio-logical theory of animal learning. □ ±¥± ø¥Ã ª±Æß, 4(2):148–158, 2009.

[7] D. Guez, L. P. Belzunces, and R. Maleszka. Effects of imidacloprid metabolites on habituation in honeybees suggest the existence of two subtypes of nicotinic receptors differentially expressed during adult development. – øÆ ø ±¥± ß □ ± ª ¨Æß ø º □ª ø™±Æ 75(1):217–222, 2003.

[8] D. Guez, S. Suchail, M. Gauthier, R. Maleszka, and L. P. Belzunces. Contrasting effects of imidacloprid on habituation in 7- and 8-day-old honeybees (apis mellifera). “ª´Æææ ±¥± ß ± ‘ ªøÆ ø º ” ª ±Æß, 76(2):183–191, 2001.

[9] D Guez, H Zhu, S Zhang, and M Srinivasan. Enhanced cholinergic transmission promotes recall in honeybees. ÷±´Æø¥± ª ¨– ß ±¥± ß, 56:1341–1348, 2010.

[10] M. Kuwabara. Bildung des bedingten reflexes von pavlovs typus bei der honigbiene, apis mellifica. ÷±´Æø¥± ¨ ª ◘ø ´¥ß ± Õ ª ªÚÿ ±µµø º ± À ™Æ ¨ßÚ13:458–464, 1957.

[11] I. London-Shafir, S. Shafir, and D. Eisikowitch. Amygdalin in almond nectar and pollen - facts and possible roles. –¥ø ¨Õß ·ª ø¨ ø º ¤™±¥ ¨± , 238:87–95, 2003.

[12] R. Maleszka and P. Helliwell. Effect of juvenile hormone on short-term olfactory memory in young honeybees (apis mellifera). ÿ ±Æ ± ª ø º ◘ª ø™±Æ40:403–408, 2001.

[13] R. Maleszka, P. Helliwell, and R. Kucharski. Pharmacological interference with glutamate re-uptake impairs long-term memory in the honeybee, apis mellifera. ◘ªÛ ø™±´Æ¥◘Æ Œ ªøÆ , 115(1):49–53, 2000.

[14] C. Masson and R. Arnold. Organization and plasticity of the olfactory system of the honeybee, apis mellifera. In R. Menzel, editor, “ª´Ææ ±¥± ß ø º ◘ª ø™±Æ± ÿ ± ªßæªª , pages 280–295. Springer Verlag, Berlin, 1987.

[15] R. Mateo and F. Bosh-Reig. Sugar profiles of spanish unifloral honeys. ◘±±º › ª Û ¨Æß, 60(1):33–41, 1997.

[16] S. Morgan, V. M. Butz-Huryn, S. J. Downes, and A. R. Mercer. The effects of queenlessness on the maturation of the honeybee olfactory system. ◘ª ø™±´Æ¥◘Æ ŒÛ ªøÆ , 91:115–126, 1998.

[17] T. Pankiw and R. E. Page. Response thresholds to sucrose predict foraging division of labor in honeybees. ◘ª ø™±Æ¥¤ ±¥± ß ø º Õ± ±æ ±¥± ß, 47:265–267, 2000.

[18] T. Pankiw, K. D. Waddington, and R. E. Page. Modulation of sucrose response thresholds in honey bees (apis mellifera l.): influence of genotype, feeding, and foraging experience. ÷±´Æø¥± › ± øÆ ¨™ – ß ±¥± ß◘, 187:293–301, 2001.

[19] P. Y. Qiu, H. B. Ding, Y. K. Tang, and R. J. Xu. Determination of chemical composition of commercial honey by near-infrared spectroscopy. ÷±´Æø¥± ◘ Æ ´¥´Æ¥ø º ◘±±º › ª ¨Æß, 47:2760–2765, 1999.

[20] S. Ray and B. Ferneyhough. The effects of age on olfactory learning and memory in the honey bee apis mellifera. “ª´ÆÆ ±Æ, 8(3):789–793, 1997.

[21] S. Ray and B. Ferneyhough. Behavioral development and olfactory learning in the honeybee (apis mellifera). ‹ ª™¥± ª ¨ø¥– ß ±æ ±¥± ß, 34:21–27, 1999.

[22] R.A. Rescorla and A.R. Wagner. Classical conditioning II. In A. Black and W.F. Prokasy, editors, ◘ ¨ ª±ß ± ø™±™ø ± º ¨± Ê øÆø¨± ¨ ª ª ª ¨™ ª ± Æ ±Æª ª ¨ø º ± Æ ±Æª ª ¨, pages 64–99. New York: Appleton-Century-Crofts, 1972.

[23] R. Scheiner, R. E. Page, and J. Erber. Sucrose responsiveness and behavioral plasticity in honey bees (apis mellifera). ▫ º±¥± ª, 35:133–142, 2004.

[24] M. L. Winston. Ã ª▫ ±¥± ß ± Ã ª ÿ ± ªß▫ªª. Harvard University Press, Cambridge, 1987.

[25] G. S. Withers, Fahrbach S. E., and G. E. Robinson. Selective neuroanatomical plasticity and division of labour in the honey bee. “ø¨´Æ, 364:238–240, 1993.

In: Bees
Editor: Richard M. Florio

ISBN: 978-1-61324-825-6

Chapter 9

ACOUSTIC NOISES GENERATED BY INSECT COLONIES

E. K. Eskov and V. A. Toboev
Russian State Agrarian Correspondence University,
Balashikha, Moscow, Russia

INTRODUCTION

Formation and functioning of aggregations of humans, animals, birds, and also development of many geophysical processes is attended with generation of a quasistationary acoustic noise, described by different mathematical methods in accordance with its temporal, spectral and spatial structure. In the majority of cases the noise is regarded as interference, but not at all seldom it is used as a source of information. For example, by the structure of acoustic noise it is possible in general features to determine the mood of human crowds [1, 2], the state of families or clusters of social insects [3—7], while by the radio emission of stars and other cosmic formations a study is made of the physical processes ongoing there [8].

For analysis of nonstationary acoustic processes, use is traditionally made of spectral-correlation analysis. However, it is ineffective if the temporal scale of the evolving nonstationarity is smaller than the duration of the analyzed noise process, because in this case there is averaging of the spectrum of fluctuation power over the entire time interval of observation of the possible detected signal. As a rule, spectral analysis is conducted on a data sample of large extension [5, 7]. This in a significant measure excludes the contribution of separate local regions of nonstationarity into the resulting Fourier spectrum of the signal. But in many analyzed time series it is the local changes in the properties of investigated noises on small time scale that can contain useful information. In addition upon a statistical approach to analysis of acoustic processes the dynamic entity that gives birth to them is, as a rule, not traced.

In recent time still broader application is found by special spectral methods oriented onto analysis of stationary processes. Of such methods the greatest distribution has been gained by window Fourier analysis [9—11], wavelet analysis [12, 13] and polyspectral analysis [14].

Application of algorithms with sliding windows permits substantially increasing the resolving power of analysis in the time domain with retention of a high enough resolution in the frequency domain. But this is coupled with a significant increase of the computation volume and accordingly with an increase of time expenditures in analysis [15, 16]. Besides, three-dimensional (frequency-amplitude-time) surface images are too complex for formal recognition.

The goal of the present work included development of a noninvasive method of investigation of nonstationary acoustic noises, not introducing uncontrolled error and ensuring simplification of interpretation on the basis of visualization and statistical analysis. The method has been approved in studying the connections between the changes in the physiological state of bees and the components of sounds generated by them, giving rise to the sound noise (background) of a bee family. The presence of these connections was established earlier by methods of spectral analysis [4, 7].

EXPERIMENTAL

The method of recurrent integration was used to reveal the main tendencies in the change of amplitude values in the isolated regions of analyzed noise by their diminution or augmentation. This permitted finding statistically homogeneous fragments, comparing them by the value of the correlation function and detecting intrinsic (i.e. not researcher-imposed) diagnostic characters. They were isolated by the sequence of ranked amplitudes of relative fluctuations (SRARF) and trends obtained by the procedure of optimal linear smoothing (POLS) by the minimal relative error. The length of fragmentation regions and the correlation dependence between trends of different regions were determined by the generalized Pearson correlation function (GPCF).

The use of a ranked amplitude sequence permits expressing quantitatively the relative fluctuations in terms of some "universal" set of reduced (fitting) parameters entering the approximate analytical expression for SRARF. They are required for quantitative comparison of arbitrary sections of acoustic noise. Increasing the number of fitting parameters, one can when necessary reveal quantitative differences between two noise sections. If these differences are inessential, one can always determine a confidence interval fitting into which two compared sections become "indistinguishable." A generalized mean function (GMF) expressed in terms of higher and fractional moments permits predicting the possible behavior of the analyzed random noise. Since SRSRF and GMF are free of any model (a priori) notions about the nature of acoustic noise, then these characteristics can be applied to analysis of nonstationary noises. In the present work an attempt is made to apply the indicated approach to determination of the physiological state of bees by the acoustic noise they generate (acoustic background of the bee family).

Experimental bees were kept in standard single-casing twelve-frame hives. For recording sound noise, use was made of a microphone probe (model 4182 of Bruel&Kjaer, Denmark) with a condenser mike (sensitivity 3.16 mV/Pa) and a portable digital sound analyzer SVAN 912M (Poland). The analyzer was also provided with a condenser mike of sensitivity 50 mV/Pa. Acoustical pickups had a linear frequency-amplitude characteristic in the analyzed range 20—1000 Hz. The nonlinear distortion coefficient (in a confidence channel at a frequency of 1 kHz) did not exceed 0.01%. The analyzed acoustic noise of bee families represented a set of unidimensional series of digital data recorded at a discretization frequency of 44 100 Hz. Processing of the recorded signal was conducted with the use of mathematical packages Matlab and MathCAD.

The acoustic noise generated by insects was analyzed by a noninvasive method by the temporal structure of its short fragments. Depending on the chosen algorithm of analysis of noise fragments we isolated sections having the same structure. This permitted us to reveal regularities in the elected set of isolated noise sections. By frequently repeated fragments we revealed statistically homogeneous (background) sections of noise, and also its hidden periodicities and uncharacteristic (anomalous) peculiarities.

RESULTS

Substantiation of noninvasive method of analysis of quasistationary acoustic processes. The proposed approach to analysis of acoustic noises can be united into a unified algorithm. It includes in itself three main stages and a sequence of main steps necessary for conducting the analysis.

Stage 1. Isolation of peculiarities and choice of optimal length of fragments (sections) of noise. Formally for finding the length of the partition section it is required to build a detector that permits choosing one of the two hypotheses: H_0 — the chosen section corresponds to background (the most frequently repeating sections with the same characters); H_1 — there exists a time moment τ such that the explored section of noise corresponds to background (synchronized and/or statistically homogeneous sections of noise) at $t < \tau$ and uncharacteristic (marginal) peculiarities at $t > \tau$.

For choosing the optimal length τ_{opt} of noise partition sections at the first stage, necessary is realization of the following steps:

S1. In the analyzed noise, exarticulated is a test interval $[t_k,\ t_k + \tau\ ]_{+}$ in which $t_k + \tau < T$, while $t_k = \kappa\Delta T$, at $\kappa = 0,1,2,3, \ldots$. The values τ and Δτ are determined by the characteristic time intervals exerting definitive influence on the dynamics of change of the explored noise. The value Δτ, corresponding to the interval between the reference points of the digitized signal, is determined by the Kotelnikov theorem and depends on the discretization frequency. As regards the amount of characteristic temporal intervals τ, it depends on the structure of the explored noise process.

S2. In the chosen noise section of duration τ, determined is a random sequence n_j $(j = 1, \ldots N)$, which represents the difference of the values of the initial data and their arithmetic mean. Acoustic noise is regarded as a set of unidimensional series of digital data recorded with a certain discretization frequency. Proceeding from this, a random sequence n_j represents in itself a time series in equidistant points x_j $(j = 1, \ldots \mathrm{N})$.

S3. Recurrent integration is conducted by the trapezium rule [17]:

$$J_j = J_{j-1} + x_j - x_{j-1}\ R_j , \qquad (1)$$

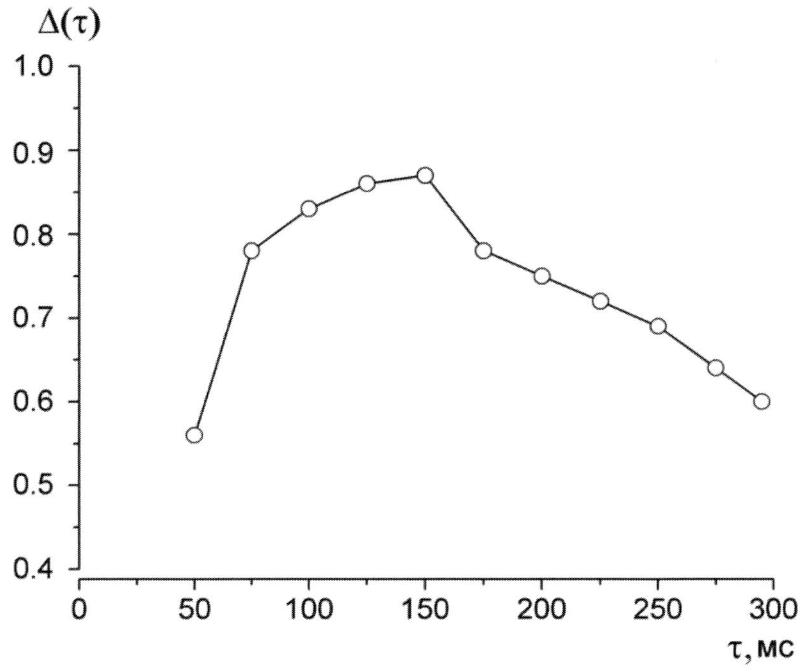

Figure 1. Choice of durations of noise partition into statistically homogeneous sections by the maximal value of correlation A.

where

$$R_j = 0.5 \; n_j + n_{j-1} \quad (2)$$

S4. The obtained sequence J_j is averaged by the minimal value of relative error with the use of POLS with a Gaussian kernel. This permits separating relative fluctuations from the trend containing only low-frequency components. The value of the optimal smoothing window ω is found from the relation:

$$\delta(\omega) = \min \left| \frac{\sqrt{\frac{1}{N}\sum_{j=1}^{N}(J_j - Js(x_j, J_j, \omega))}}{\frac{1}{N}\sum_{j=1}^{N} J_j} \right|, \quad (3)$$

where J_j is integrated sequence; ω is smoothing window value,

$$Js(x_j, J_j, \omega) = \frac{\sum_{i=1}^{N} K\left(\frac{x_j - x_i}{\omega}\right) J_j}{\sum_{i=1}^{N} K\left(\frac{x_j - x_i}{\omega}\right)}, \qquad (4)$$

$$K(u) = \exp(-u^2).$$

Expression (4) entering into expression (3) determines a smoothed function (trend), which is obtained with the use of a Gaussian smoothing kernel. Here and below by trends we imply smoothed sections of the temporal acoustic sequence which determine in essence an averaged function of mean values reflecting the behavior of this function on large times. The temporal dependence of a trend in a general case is determined by the initial noise and has a specific functional form for different sections.

S5.The section length τ_{opt} is determined by the maximal value of GPCF calculated between isolated trends for sections of the same duration. The GPCF $K_N(p; i, J)$ permits selecting statistically close trends determined by the maximal value of the correlation magnitude Δ (in more detail these notions are given in S12). Since in nonstationary acoustic noise the extent of correlation strongly depends on the length of chosen sections, then by function $\Delta(\tau)$ it is possible to determine the optimal length of the interval of partition of the noise process into short fragments having the same structure (Fig. 1).

Stage 2. Clusterization of isolated fragments by a statistically homogeneous character (character-wise description). Here analyzed are the properties of noise at every chosen section. Assuming that S_k — a noise section of duration τ_{opt} corresponds to the k-th fragment of nonstationary acoustic noise, the unified algorithm of processing all its fragments includes in itself the following steps:

S6. For the chosen section of noise S_k, found again is the integrated sequence and its smoothed mean value (i.e. the same calculations are repeated as in S2— S4 of the preceding stage).

S7. The obtained trend is subtracted from the integrated sequence, which permits one to determine the sought (already detrended) random sequence of distribution of relative fluctuations Y_k. In this sequence the newly calculated trend with the use of POLS gives a horizontal line practically coinciding with the abscissa axis.

S8. For extracting additional information Y_k is separated into two positive parts in accordance with the formula:

$$Y_k^{(\pm)} = \pm\frac{1}{2}\, Y_k \pm |Y_k| \qquad (5)$$

where Y_k is discrete sequence of relative fluctuations; $k$is fragment number.

S9. The obtained random sequences $Y_k^{(\pm)}$ are ranked (or ordered) in amplitudes. When the amplitudes obey the condition

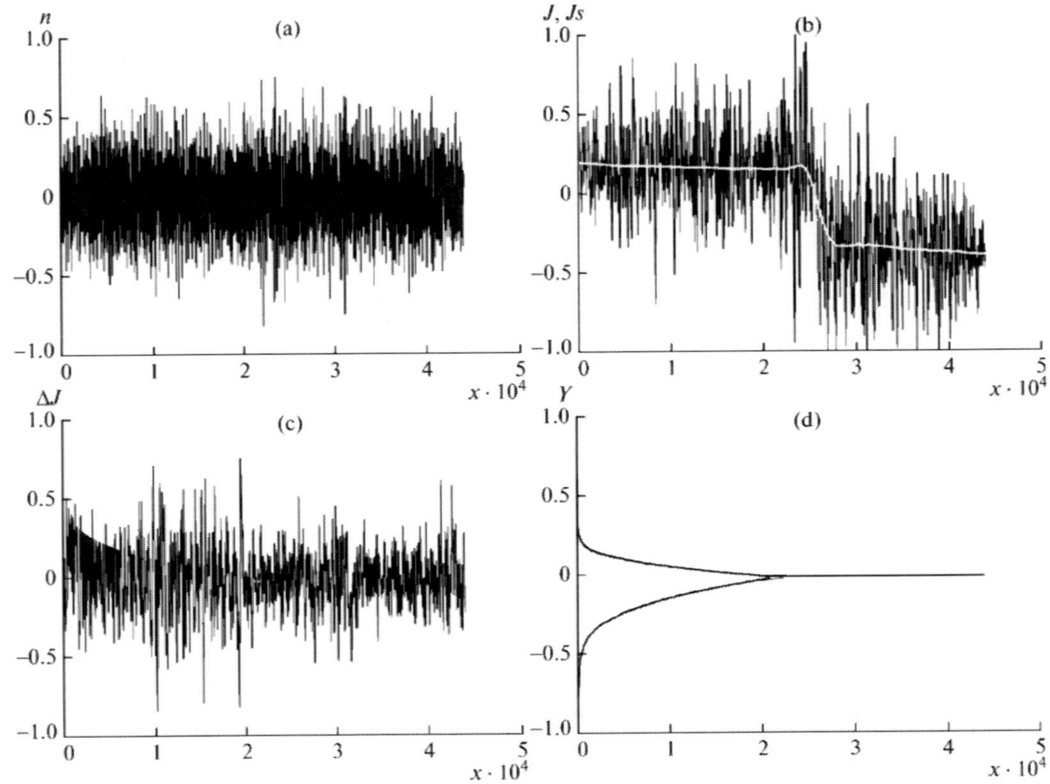

Figure 2. Main stages illustrating the general principle of obtaining relative fluctuations and SRARF: (a) normalized sequence of initial values (acoustic noise of duration 1 s recorded with a discretization fragment of44 100 Hz); (b) integrated sequence(y) and its smoothed values *(Js)* obtained with the aid of POLS; (c) relative fluctuations; (d) SRARF for relative fluctuations determined by dependences (6) or (7).

$y_1^k > y_2^k > ... > y_N^k$, then the ordered set of values { y_j^k } forms a SRARF.

The main steps illustrating the general principle of obtaining relative fluctuations and sequences of ranked amplitudes of these fluctuations for nonstationary acoustic noise of duration 1 s recorded with a discretization frequency 44 100 Hz are presented in Fig. 2.

S10. For quantitative comparison of relative fluctuations of different sections of the signal, SRARF envelopes are replaced with fitting functions $F\ t$ of the form:

$$F\ t\ = \sum_{i=1}^{m} C_i e^{\lambda_i t} \quad (6)$$

or

$$F\ t\ = \sum_{i=1}^{m} C_i t^{\alpha_i} \quad (7)$$

The choice between the hypotheses about an exponential or a power dependence of the SRARF envelope is performed by the minimal value of the standard deviation and closeness to unity of the Pearson correlation coefficient (PCC) calculated for SRARF and the corresponding fitting curve.

The calculated fitting parameters of function *F(t)* from formulae (6), (7) permit dividing the amplitudes of relative fluctuations of nonstationary noise y_j^k (j = 1, … N) into some optimal statistical groups with parameters (C_m, λ_m) or (C_m, α_m), which corresponds to reduced (abridged) description of the considered process. These parameters permit one to quantitatively compare arbitrary sections of acoustic noise and divide them in the time domain by character distinctions.

Stage 3. Study of the obtained regularities on the basis of clusterization of noise fragments and visualization of the found clusters by methods simplifying the analysis of the results obtained. For visualization and quantitative comparison of different sections of acoustic noise besides the above-indicated points it is proposed to use the statistics of fractional moments (SFM) [18].

S11. In the framework of SFM any sequence can be expressed with the use of GMF in the space of fractional moments:

$$F_N(p;i) = \left(\frac{1}{N} \sum_{k=1}^{N} \left| a_k^i - \langle a_k^i \rangle \right|^p \right)^{1/p}, \qquad (8)$$

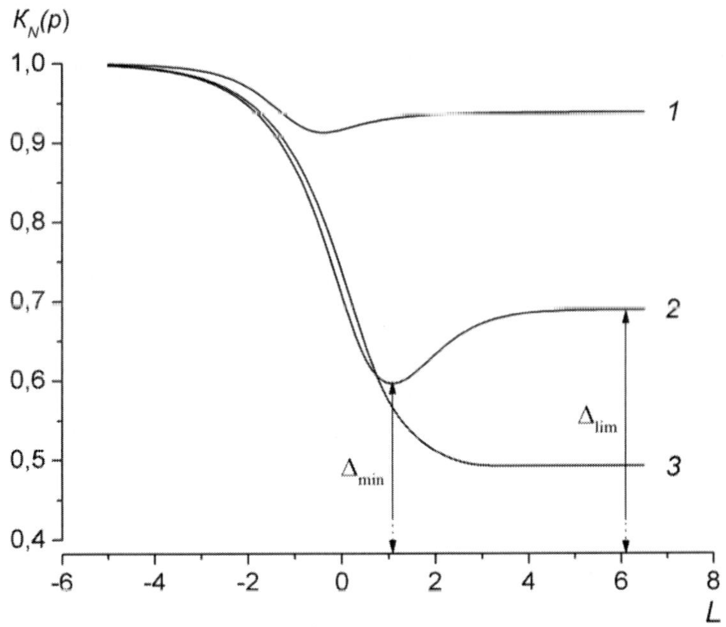

Figure 3. Dependences of GPCF on the order of moment p for different sections, at chosen value of interval duration т, L = ln (p). Correlations: *1*–strong; 2–medium; *3*– weak.

where p is the order of the moment, the value of which can take integer as well as fractional values ($0<p<\infty$); k is the current value of discrete readings of sequences

($k = 1, \ldots N$); $\{\alpha_k^i\}$ is the integral sequence of the i-th section or its smoothed trend. The use of GMF gives a possibility of finding the true correlation between two compared sequences in plotting these functions relative to each other. If such plotting gives a segment of a straight line, then this points to their statistical closeness. Such a comparison permits obtaining four parameters: (1) the slope (Tg) of the straight line, which in the case of complete coincidence is equal to unity or the coefficient of geometric similarity upon fractality of the compared sequences; (2) the cutoff (B) of the straight line, which in the ideal is equal to zero; (3) the value of relative error (5) of fitting, which in the ideal is close to zero; (4) the generally accepted PCC, in the ideal equal to unity.

S12. Besides that, on the basis of GFM one can expand the notion of the correlation coefficient, by introducing the generalized Pearson correlation function GPCF, which depends on the magnitude of the moment [17]:

$$K_N(p;i,j) = \frac{\left[\frac{1}{N}\sum_{k=1}^{N}\left[\left|n_k^i - \langle n_k^i\rangle\right|\left|n_k^j - \langle n_k^j\rangle\right|\right]^p\right]^{\frac{1}{p}}}{\left[\frac{1}{N}\sum_{k=1}^{N}\left|n_k^i - \langle n_k^i\rangle\right|^{2p}\right]^{\frac{1}{2p}}\cdot\left[\frac{1}{N}\sum_{k=1}^{N}\left|n_k^j - \langle n_k^j\rangle\right|^{2p}\right]^{\frac{1}{2p}}}, \tag{9}$$

where random sequences $\{\alpha_k^i\}$ and $\{\alpha_k^j\}$ correspond to isolated sections with numbers i and j. Function $K_N(p;\ i, j)$ at certain values $p > 0$ can have a minimum (Δ_{min}) and tend to its limit value at p □ □1 (Δ_{lim}). Typical curves of $K_N(p;\ i, j)$ calculated by formula (9) and accounting of the correlation dependence between different sections are shown in Fig. 3. Therewith three main types are possible for correlation peculiarities between compared sections:

[1] Statistically homogeneous sections are determined by value Δ_{min} at the point of minimum and attainment by GPCF of asymptotics at $p \gg 1$ equal or close to unity ($\Delta_{lim} \cong 1$).

[2] Behavior of GPCF with the value Δ_{min} and attainment of asymptotics not equal to unity corresponds to a decrease of the extent of correlation.

[3] Case when the GPCF minimum disappears, and it attains an asymptotics with a value Δ_{lim} that is substantially smaller than unity and corresponds to the case of weak correlation.

The generalized Pearson correlation function significantly expands the limits of applicability of the generally accepted Pearson correlation coefficient defined only at $p = 1$. The function $K_N(p;\ i, j)$ promotes finding a whole band of correlations localized between its minimal and limit values. Therefore in addition to the four parameters mentioned above, for evaluation of the statistical closeness of two noise sections it is expedient to add a new parameter quantitatively characterizing the width of the correlation band:

$$\Delta = \frac{1}{2}(\Delta_{min} + \Delta_{lim}). \quad (10)$$

The five found parameters permit obtaining a statistically homogeneous cluster, the values of which in the ideal form a family of homogeneous data in relation to some external character. Such a representation permits isolating statistically close sections especially important in analysis of acoustic noises.

The principal distinction of correlation parameters found with the aid of GPCF and GMF consists in that the former of them allows calculating correlations accounting of the mutual disposition of points of integrated sequences, while the second one, correlation irrespective of the mutual disposition of the aggregate of points of the pair of compared sequences. Therefore, for completeness of description, it is desirable to take into account these correlations simultaneously.

Thus, analysis of short fragments of unordered acoustic noise, describing relative fluctuations, permits revealing the regularities of change in its properties determined by the fragmentation algorithm. Finding of clusters (statistically homogeneous sections) allows one to determine in the time domain the background sections of the signal, hidden periodicities, uncharacteristic (marginal) peculiarities. In the frequency domain a possibility opens for comparing by the traditional spectroscopic method (by intensity, half-width) the dynamics of change in the main peaks of Fourier components of the signal and finding geometrically

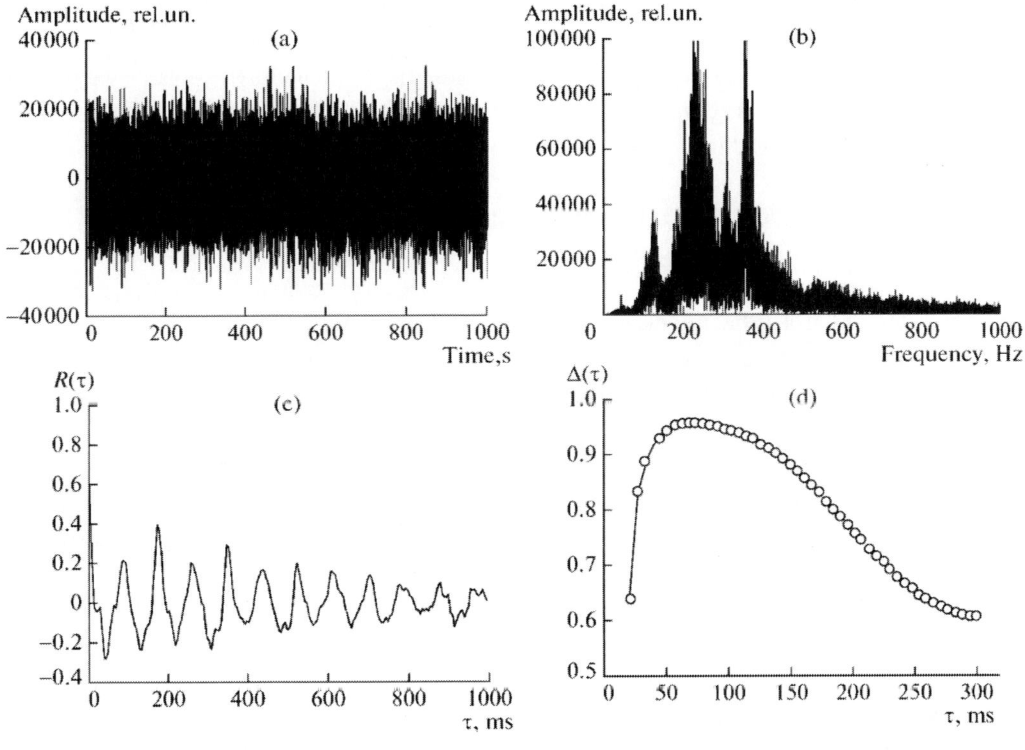

Figure 4. Acoustic noise of a bee family recorded in the period of reproduction of working bees: (a) section of sequence of sound fragments of duration 8 s; (b) spectrum of acoustic noise; (c) autocorrelation function; (d) dependence Δ (τ) and choice of optimal duration of noise partition.

similar (fractal) regions of spectrum for samples of different volume. By the dynamics of change in trends (smoothed curve) and SRA of relative deviations there opens a possibility of prognosticating the probable changes in the state of the explored non-stationary acoustic processes.

Reliability of acoustical diagnosis of the physiological state of bees. *Frequency-amplitude-time characteristic of acoustic noises generated by bees.* The overall acoustic background of a bee family is made up of many components which are generated by separate individuals. Their intensity, temporal and frequency structures depend on the physiological state of bees. As a result of superposition of own oscillations (microvibrations) of the thorax and flaps of the flight apparatus of separate individuals with time-varying amplitude and different periods, as well as interference taking place between them, the sum acoustic noise (Fig. 4a) and the spectrum reflecting it become nonstationary. From the total noise it is difficult to isolate stable informative characters, by which it would be possible to conduct identification of the physiological state of bee families. Presentation of the acoustic noise generated by them as a stationary quasidetermined random process permits isolating only some of its characteristic frequency regions [6].

The intense components of the acoustic background of a bee family reside in the frequency range 100—600 Hz (Fig. 4b). This range is generated by the sounds arising upon ventilation of the nest by means of wing flaps, generation of heat by microvibrations of the thoracic musculature, friction of bees against support substrates etc. [5, 8]. However the presence of correlation of the sounds of separate bees with the family background and the nonstationarity of the background itself do not allow one to differentiate the physiological state of bees immediately by the frequency-amplitude-time spectrum.

Obviously, it is more expedient to isolate from the total acoustic noise synchronized sections representing its short fragments that correspond to definite biological situations and bee states connected to them. The amplitude of the autocorrelation function of these fragments has a tendency to diminution with the growth of the correlation step. It possesses weakly damped periodical maxima and minima with an interval of 80—100 ms and amplitude from —0.3 to +0.4. (Fig. 4c).

Since autocorrelation does not tend to zero with step growth, then the nonrandom component isolated from the integrated sequence of fragments with an optimal value of the smoothing window represents a trend. The time structure of a smoothed curve in a general case is determined by the initial noise, the analytical expression of which originally can have a various functional form. Therefore smoothed sequences (trends) should be compared in terms of GPCF and GMF. Introduction of these additional parameters allows one to find the mutual correlation of points of compared sequences with engagement of SFM. The optimal length of noise fragmentation τ_{opt} is determined by the greatest value of parameter A, which accounts of the correlation dependence between different sections (Fig. 4d). The relative fluctuations obtained by subtraction from the integrated sequence of its smoothed curve are analyzed in terms of SRARF.

Analysis of the acoustic noise of a bee family is based on isolation of statistically homogeneous and often repeating fragments thereof, which represent the "true" total background for a definite physiological state of bees. The isolated sections of noise possess the same structure and do not contain sounds of separate bees, which represent local sections with unordered and uncharacteristic components. The fragments composed in this way are further used for identification of a multitude of physiological states of bee families, which

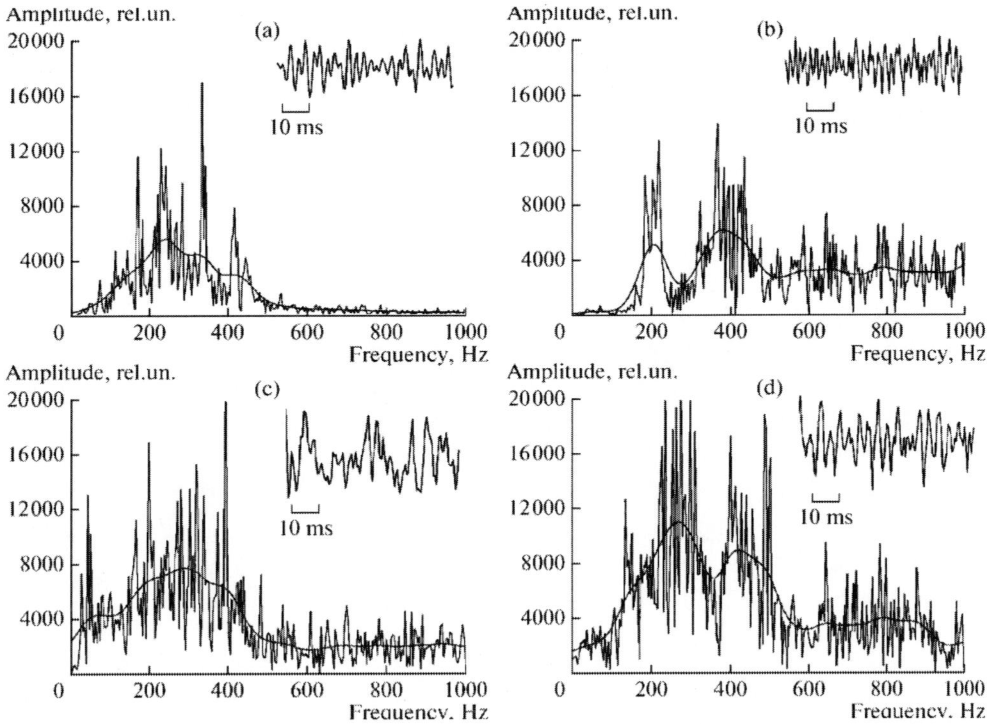

Figure 5. Frequency-amplitude spectra and oscillograms of sounds of bees being in different physiological states (duration of each signal 62.5 ms): (a) safely wintering family; (b) bees flapping their wings (ventilating); (c) reaction of wintering bees to a vibration pulse propagating over the hive; (d) reaction of bees to removal from the nest of the ovulating female (queen). Solid line, smoothed values of spectra (averaged spectral trend) at the optimal value of smoothing window, obtained with the aid of POLS.

permits forming their collective and statistically homogeneous "acoustic portraits." Sections with the same statistical structure of noise (clusters) are analyzed in the time and in the frequency domains.

Specifics of the temporal structure of bee acoustic noise. The temporal structure of acoustic noise has a complex form, which strongly restricts isolation therefrom of stable information characters and conduction of reliable identification of the states of bee families in different periods of the annual cycle of their life (Fig. 5). The most simple form, close to a sinusoid with amplitude changing in time, is possessed by the sound of separate bees aerating the nest (flapping their wings). This can be regarded as a stationary quasidetermined random process made noisy by random interference with mathematical expectation equal to zero. Therefore fluctuations of interference do not exert a statistically significant influence on the values of estimates of the Fourier spectrum. In the frequency spectrum of such sounds generated by aerating bees, distinctly isolated are intense components at frequencies 200—225 and 350—400 Hz (Fig. 5b). In many other cases the approach of harmonic approximation is low-effective (Fig. 5a,c,d).

The proposed algorithm of processing acoustic signals, expressed in SRA terms, presents in itself a unique possibility of comparing different sections of noise and determining their statistical closeness. Finding of statistically close sections in nonstationary acoustic noise permits one by its temporal structure to control the changes in the physiological state of bee families. Thus in a family safely wintering for five months, in the dynamics of fluctuations of acoustic noise presented by the dependence $Yi(Y_0)$ $(i = 1, 2, 3, 4)$ relative to the first month of wintering (December), one traced a strong correlation (Fig. 6a). Insignificant deviations from straight lines are connected with changes in the physiological state of the bee family in March and April in connection with appearance of brood. Therewith the maximal values of correlation coefficients A for all dependences reside in the interval 0.87—0.95. In smoothed frequency spectra, obtained by the BFT method after isolation of strongly correlated sections, one also traced insignificant shifts of the position of maxima and half-width of the intense components of spectra (Fig. 6b).

Detailing of analysis is achieved by way of quantitative comparison of fitting parameters of SRARF of sound noise of families in the course of wintering. The values of parameters of fitting with the use of a power dependence (formula 7) for them are presented in the table. The last two columns illustrate the quality of fitting of hypothesis (2) of specified SRARF. The relative error (5) is calculated as the ratio of standard deviation between SRARF and fitting function to the magnitude of the mean value of the fitted function. The PCC (ρ) calculated in a standard way [9] is confined in the interval [0,1].

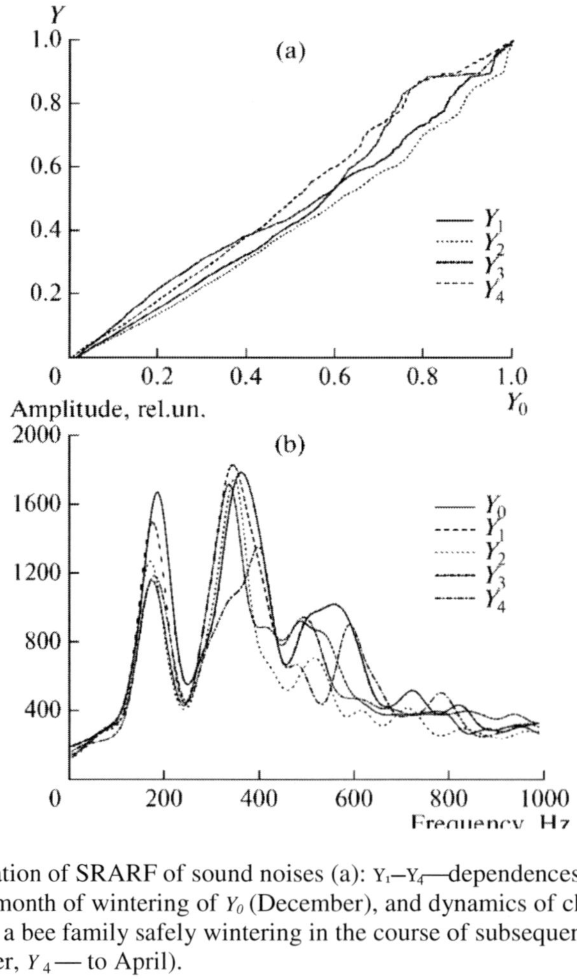

Figure 6. Mutual correlation of SRARF of sound noises (a): Y_1–Y_4—dependences plotted relative to correlations of the first month of wintering of Y_0 (December), and dynamics of change in smoothed frequency spectra (b) of a bee family safely wintering in the course of subsequent five months. (Y_0—corresponds to December, Y_4— to April).

Fitting parameters obtained for SRA describing nonzero relative fluctuations of the total sound noise of a family in the course of five months of wintering

Month	C1	α_1	C2	α_2	C3	α_3	δ	ρ_1
December	—0.003	—0.960	0.662	—0.272	—0.303	0.702	0.044	0.967
January	—0.005	—0.947	0.599	—0.265	—0.267	0.826	0.045	0.979
February	—0.001	—0.983	0.679	—0.264	—0.333	0.655	0.022	0.951
March	—0.001	—0.952	0.681	—0.281	—0.301	0.546	0.113	0.863
April	—0.001	—0.981	0.633	—0.291	—0.299	0.642	0.247	0.810

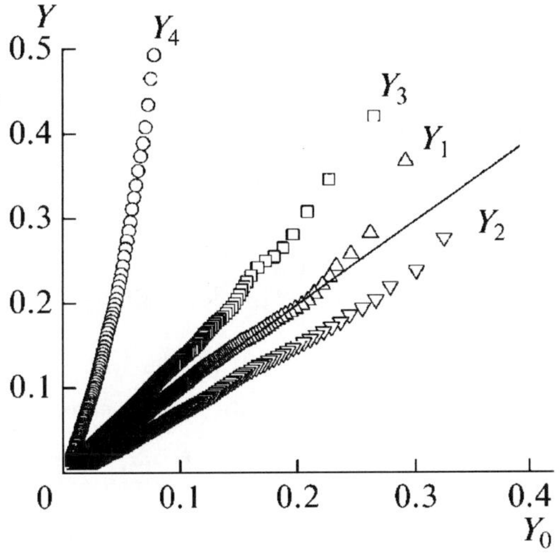

Figure 7. Different mutual correlation of SRARF for relative fluctuations of sound noise of five wintering bee families. All dependences plotted relative to correlations of the safely wintering family Y_0. Deviations from straight lines can be explained by a change of physiological states of corresponding families (families 2, 3, 4).

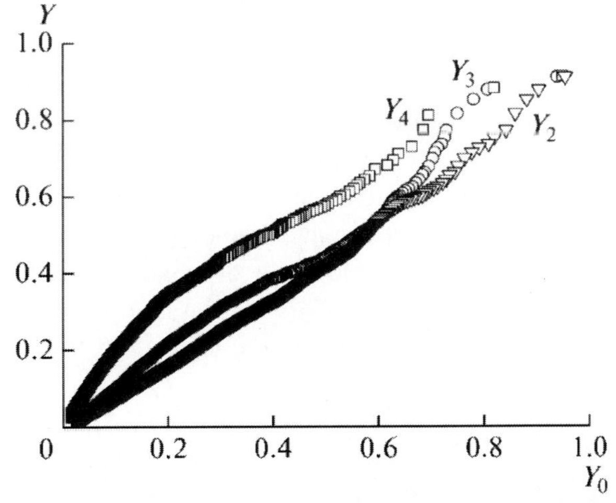

Figure 8. Change of mutual correlation of SRARF for relative fluctuations of sound noise of a family preparing to propagation, at 20 (Y_1), 8 (Y_2), 5 (Y_3) days before its separation and at the day of flight of separating bees (Y_4).

Analysis of SRARF with a power dependence (formula 7) of fitting parameters points to the possibility of prognosticating the changes in the physiological state of bees during wintering. For this it is necessary to determine SRARF at the initial stage of wintering. The extent of predictability is estimated by means of comparing the found fitting parameters for different families with standard (reference) values, the states of which are patently known.

Let us consider situations in which (Yo) is a SRA with known numerical characteristics (for example, SRARF isolated for the sound background of a safely wintering family or with a known initial physiological state), while (Y_n) reflects the explored sequence. In like cases by the dependences Yn (Yo) one can collate these two acoustic noises. In Fig. 7 such dependences are presented for five wintering families. For each of them the functions Y_n (Y_0) are different and therefore can be used for identification and quantitative differentiation by the physiological state. Thus the normal physiological state of bees in the period of wintering is characterized by the angle of the slope of the curve, the latter being close to 45 (Fig. 7, Y_1). Diminution of this angle is coupled with energization of bees (Fig. 7, Y_2), while augmentation, with shortage of feed (Fig. 7, Y_3, Y_4). These deviations reflect troubles in the course of wintering. As regards the reliability of analysis, it can be estimated by the relative value of the slope of the straight line and the magnitude of cutoff relative to SRARF related to a safely wintering family.

Investigation of the dynamics of the change of trends and SRARF permits isolating from bee family acoustic noise of stable information characters and conducting reliable identification of the states of bee families in different periods of their vital activity. For this it is enough to calculate the aggregate of fitting parameters of functions (6) or (7) and plot their dependence in relation to the controlled state. For example, along with analysis of changes in the state of families in the wintering period one can control their preparation to propagation, manifesting itself in separation of the family into two or more parts. The time of formation of a new family can be determined by the value of the slope of the compared curve, the latter approaching unity (Fig. 8).

Frequency characteristics of acoustic noise of bee families. Determination of frequency characteristics of noise without isolation of quasistationary sections does not permit one to reveal diagnostic characters (Fig. 5). This is connected with that the acoustic noise of a family is quite often modulated by low frequencies to 100 Hz, which hinders recognition of the main spectral components of noise. Using methods of analysis of statistically homogeneous sequences of sections of acoustic noise and window smoothing of Gabor [15], there is a possibility of localization of characteristic frequencies even for comparatively short noise fragments. This permits analyzing the dynamics of changes of the main peaks in the frequency domain by the traditional spectroscopic method (by intensity and half-width).

With a high probability the statistical closeness of isolated smoothed spectra along with using SRARF (5) and their fitting functions (6)—(7) can be estimated with the aid of GMF (8). Two sections of a spectrum are statistically close if GMFs for them plotted relative to each other in the space of fractional moments form a segment of a straight line. The quantitative characteristic of this closeness is estimated by five parameters: A, Tg, B, 5, PPC (Fig. 9).

Investigation of the degree of correlation of isolated statistically homogeneous sections of acoustic noise of different duration demonstrates fractality of Fourier spectra of these sections (Fig. 10). Thus GMFs of spectra for two time sections of sound noise the duration of which equals 62.5 and 125 ms upon plotting relative to each other lie with high accuracy on a segment of a straight line. This points to that smoothed spectra are fully similar to each other.

Apparently this similarity bears a common character, since the found regularity was observed for all intervals of statistically homogeneous sections chosen at an optimal value of the interval length t_{opt}.

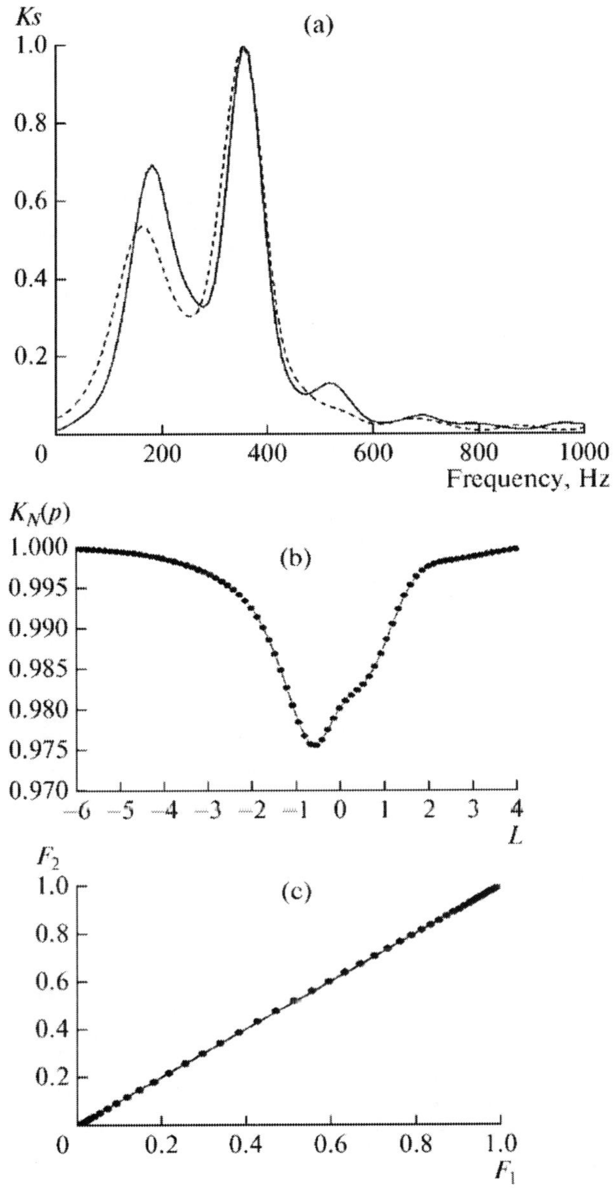

Figure 9. Comparison of normalized frequency spectra for two sections of acoustic noise of a bee family and quantitative values of its parameters characterizing the closeness of chosen spectra: (a) smoothed values of spectra obtained with the aid of POLS; (b) dependence of GPCF KN(p) on the order of moment p for these sections L = ln (p); (c) demonstration of statistical closeness of frequency spectra (depicted are GMF for smoothed spectra of two sections of background plotted relative to each other). Parameter values: Δ = 0.988; Tg = 1.006; B = 0.001; δ = 0.9%; PCC—0.999.

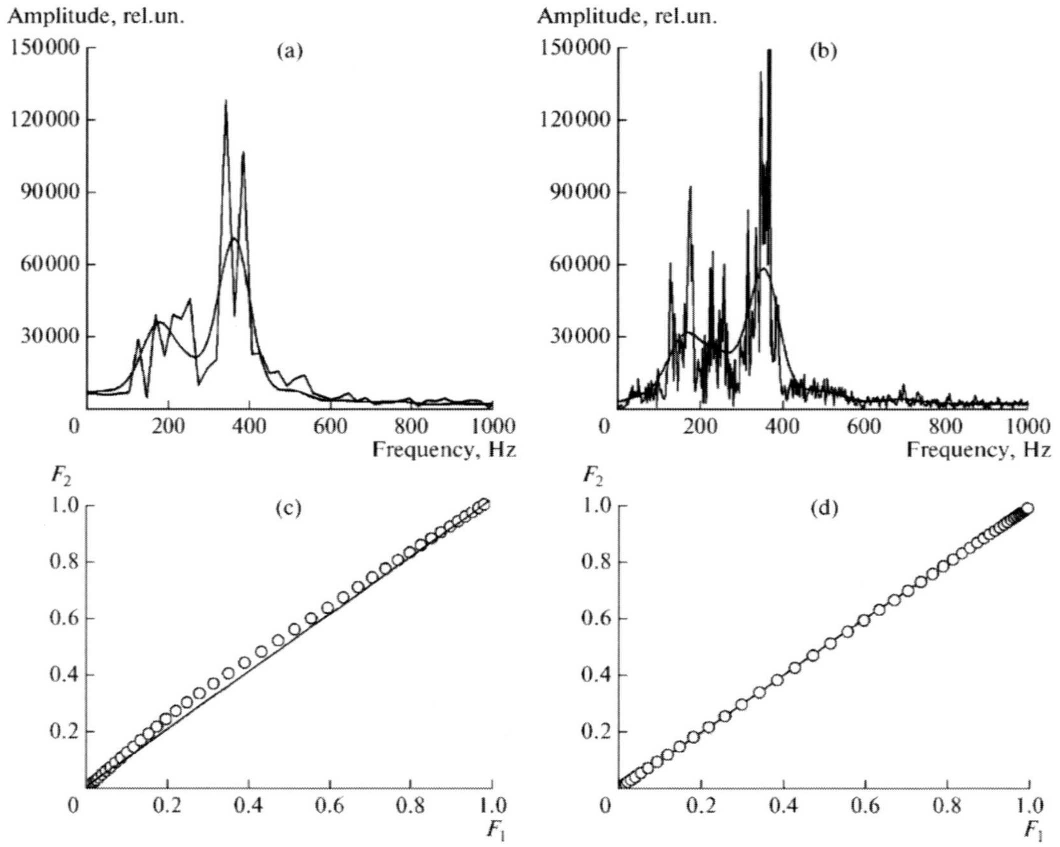

Figure 10. Comparison of Fourier spectra and their smoothed values obtained with the aid of POLS for temporal sections of sound noise of different duration: (a) 62.5 ms, (b) 125 ms, (c) and (d) correspond to GMFs plotted relative to each other for Fourier spectra and their smoothed values. Parameter values: (c)–Δ = 0.930; Tg = 1.022; B = 0.008; δ = 3.3%; PCC— 0.963; (d)– Δ = 0.984; Tg = 0.997; B = 0.008; δ = 0.3%; PCC—0.999.

CONCLUSION

The use of elements of structural modeling permits one by isolated characters to actualize compact description of nonstationary acoustic noise. On this basis there is a possibility of detection and classification of its inner structure by comparatively short fragments that give an idea of the explored process. Depending on the chosen algorithm of isolating noise fragments it is possible to isolate sections having the same structure, to reveal their anomalous (marginal) peculiarities. This permits revealing regularities in the obtained set of isolated sections of noise, which is confirmed by the established correspondence of changes in the physiological state of bees to the structure of sounds generated by them.

The generally accepted description of acoustic noise of a bee family appears quite simplified and does not correspond to its real characters depending on physiological state. POLS based on the use of a Gaussian kernel [17] permits one by the minimal value of relative error to reveal the optimal value of the smoothing window and obtain information on the physiological state of the bee family on the basis of analysis of the trend and detrended sequences reduced to SRARF. Having determined the universal function of distribution of relative fluctuations (formulae (6)—(7)), it is possible with a high accuracy to realize fitting of corresponding SRARF. This allows one to compare detrended sequences in terms of reduced fitting parameters entering into functions (6)—(7).

Optimal trends that remain after subtracting detrended sequences are quantitatively compared between themselves in the space of fractional moments [18] or else compared with the aid of GPCF. In particular, application of SRARF checked preliminarily for their statistical homogeneity, and also new quantitative characteristics based on GPCF and GMF, ensures high reliability of acoustical diagnosis of the physiological state of bee families in different periods of their annual cycle of life.

The developed algorithm solves the problem of clusterization of sections of acoustic sequence and, in essence, permits describing noise in the frequency-time domain. By choosing statistically homogeneous sections in the acoustic sequence by the GPCF and GMF criteria, it is possible to obtain a frequency partition of the signal describing its spectral structure. Since the discovered dynamic fractality of noises generated by bees has a high probability of correspondence to the indices characterizing the changes in their physiological state, then one can hope that the principle of fractality finds application in studying the mechanisms of coding of acoustic information, used in communications of animals of different levels of complexity.

REFERENCES

[1] V. Denda, *Noise as Information Source* (Mir, Moscow, 1993) [in Russian].

[2] *Psychology of Crowds* (Izd. KSP+, Moscow, 1998) [in Russian].

[3] E. K. Eskov, *Biofizika* 14 (1), 158 (1969).

[4] E. K. Eskov, *Zool. Zh.* 49 (2), 241(1970).

[5] E. K. Eskov, *Zool. Zh.* 51 (7), 1018 (1972).

[6] E. K. Eskov, *Zh. Obshch. Biol.* 34 (6), 861 (1973).

[7] E. K. Eskov, *Acoustic Signaling of Social Insects* (Nauka, Moscow, 1979) [in Russian].

[8] A. A. Potapov, *Fractals in Radiophysics and Radiolocation. Sample Topology* (Univ. Kniga, Moscow, 2005) [in Russian].

[9] J. Bendat and A. Pirsol, *Application of Correlation and Spectral Analysis* (Mir, Moscow, 1983) [in Russian].

[10] D. Brillinger, *Time Series. Data Processing and Theory* (Mir, Moscow, 1980) [in Russian].

[11] V. P. D'konov and I. V. Abramenkova, *MATLAB. Signal and Image Processing. Special Reference Book* (Piter, St. Petersburg, 2002) [in Russian].

[12] V. I. Vorob'ev and V. G. Gorbunin, *Theory and Practice of Wavelet Transform* (Piter, St. Petersburg, 1999) [in Russian].

[13] V. N. Vasil'ev and I. P. Gurov, *Computer Processing of Signals Applied to Interferometric Systems* (BKhV St. Petersburg, 1998) [in Russian].

[14] A. K. Novikov, *Polyspectral Analysis* (TsNII im. Kry-lova, St. Petersburg, 2002) [in Russian].

[15] S. L. Marpl, *Digital Spectral Analysis and Its Applications* (Mir, Moscow, 1990) [in Russian].
[16] A. B. Sergienko, *Digital Processing of Signals* (Piter, St. Petersburg, 2003) [in Russian].
[17] V. A. Toboev and R. R. Nigmatullin, *Nelineinyi Mir* 5 (4), 183 (2007).
[18] R. R. Nigmatullin, *J. Signal Process* 86**,** 2529 (2006).

In: Bees
Editor: Richard M. Florio

ISBN: 978-1-61324-825-6

Chapter 10

EXOGENOUS AND ENDOGENOUS FLUCTUATIONS OF THERMOREGULATORY ACTIVITY

E. K. Eskov and V. A. Toboev
Russian State Agrarian Correspondence University,
Balashikha, Moscow, Russia

INTRODUCTION

Certain cyclicity in thermoregulatory activity under constant environmental conditions is characteristic, to different degrees, of both poikilothermal and homoio-thermal organisms. In poikilotherms, a high variability of body temperature is related to changes in their loco-motor activity. When this activity decreases, the body temperature of a poikilotherm approaches the ambient temperature (Schmidt-Nielsen, 1982; Es'kov, 1992; Kipyatkov and Lopatina, 2003). The influence of loco-motor activity on body temperature in homoiothermal animals is lower than in poikilotherms (Ivanov, 1972; Aschoff and Wever, 1984; Es'kov, 1992), but variation in their body temperature obviously tends to increase with a decrease in body weight (Ivanov, 1972).

When at rest, endogenous rhythmic changes of body temperature in homoiothermal animals are maintained by a number of hierarchically interrelated oscillators differing in their ability to generate autonomous oscillations (Benson and Levies, 1976; Daan and BArde, 1978; Aschoff, 1984; Sharma and Shandrashekaran, 2005). In vertebrates, the role of the leading oscillator may be played by the suprachiasmatic nuclei of specialized cells in the central nervous system (Minnis and Pit-tendrigh, 1968). Unlike a single organism, consolidated aggregations of insects (families or isolated groups) have no coordinating center that could play the role of this oscillator (Es'kov, 1992, 1995, 2003). Every individual in a family of social insects has all the properties of an integral organism, but none of the individuals in social insect families can survive for long being out of touch with the family. Consolidation, interrelation, and interdependence of family members are especially high in bees of the genus *Apis* (Es'kov, 1995), which makes them a convenient model for studying the nature of periodic oscillations in complex systems.

The purpose of this study was to analyze the nature of rhythms in the thermoregulatory activity of honeybees *Apis mellifera* L., which stand out among social insects due to the perfect mechanisms of temperature regulation within the nest.

Materials and Methods

Temperature variability was studied in the nests and isolated aggregations of honeybees. The temperature was measured with electric thermometers and by means of thermovision scanning of nests in polycarbonate hives with removable front and side walls. The comb frames of the hive were hinged into a kind of ring book to be pulled apart for scanning. In this period, the open surfaces of the combs with bees were covered with polyethylene film to minimize the response of bees to the opening of their nest. Each scanning session was no longer than 5-6 s. The coefficient for converting radiance temperature into its actual value was 1.01.

Electric thermometers with platinum thermal resistor sensors HEL-705 (Honeywell, United States) were used in the computer system of digital temperature control. The system's software "questioned" each of the 15 sensors for 10 s. The sensors were either installed in the nest or inserted with a probe covered with protective net. The distance between neighboring sensors was 2.5 cm.

The test bee families were kept either outdoors, being exposed to the influence of changes in temperature, illumination, humidity, and other physical environmental factors, or in an unlit underground room at almost constant temperature (its daily variation did not exceed 0.1°C). Small groups of bees were kept in entomological cages placed for several days in a TS-80 dry-air thermostat (Russia) with a grounded metal case. The temperature in it was adjusted to a certain level, and relative humidity was no higher than 50-65%. To prevent daily oscillations of illuminance, thermostats in a shaded room were additionally covered with a thick black cloth. Hence, the maximum illuminance of the thermostat walls did not exceed 0.1 lx, being still lower inside it. For comparison, the threshold level of daily oscillations perceived by honeybees is around 0.2 lx (Mazokhin-Porshnyakov, 1965; Schriker, 1965).

The results of temperature measurements were processed (smoothed) by the fractional moment method, which allowed large-scale (high-amplitude) temperature fluctuations to be distinguished. Circadian periods were determined by regression analysis of weekly temperature fluctuations, and phase shifts were estimated from the difference between smoothed peaks of thermal activity.

Results and Discussion

Bees regulate the temperature in their nests. Its stability and absolute values depend primarily on the number of bees per nest, their physiological condition, and external temperature. The regulation of the temperature in the nest is based on using ethological and physiological mechanisms allowing the bees to change the rates of heat production and loss intensity within a broad range. Heat is generated by fibrillar muscles of the flight apparatus

(Es'kov, 1969), and heat loss is regulated by changing the aggregation density of bees. This density decreases in response to increasing external temperature, and vice versa. A high aggregation density minimizes heat loss by limiting air exchange with the environment. The same mechanism of thermoregulation is used by the groups of bees isolated from their family, which form aggregations that are stable in time and in the space they occupy.

Bee family. Temperature in different zones of a bee nest may vary considerably. Nests occupied by brood are least dependent on fluctuations of external temperature. The temperature at the periphery of the nest is highly variable, especially in zones occupied with empty combs and food supplies.

In different nest zones occupied by the brood, the dependence of temperature on external changes is to some extent different. Under the influence of external temperature increasing from 5°C to 35°C, the temperature in a family consisting of 20-30 thousand individuals increases, on average, from 34.1 to 36.0°C at the upper boundary of area occupied by the brood, from 34.5 to 35.8°C at its lower boundary, and from 34.9 to 35.6°C in the center of the family.

The circadian dynamics of the heating of nest zones occupied by the brood directly depend on fluctuations of external temperature as long as they are not greater than 34-35°C. At the external temperature reaching or exceeding 37°C, the direct correlation between it temperature and the temperature in the nest is disrupted. For instance, during periods with hot weather continuing for several days, when the external temperature early in the day rises from 22°C to 32°C, the temperature in the nest center occupied by brood increases, on average, from 35.7°C to 36.4°C. Further increase in the external temperature to 36°C by midday coincides with a decrease in the nest temperature by an average of 0.15°C. When the external air temperature approaches 40°C (after noon), temperature variation in the central part of the nest increases and its average value decreases to 35.3°C.

Independently of changes in the external temperature, air temperature in the space between combs fluctuates, decreasing periodically for a short time. The decrease may sometimes reach 5.5-7°C but is usually no greater than 0.5-1.3°C. These fluctuations follow each other at intervals of 7-16 minutes, independently of the time of day (Fig. 1).

The temporal pattern of large-scale fluctuations in thermoregulatory activity changes both during the day and during the annual cycle of the bee family. Differences in this pattern are distinct between the spring-summer and the autumn-winter periods, probably due to seasonal variation in the locomotor activity of bees. In zones of temperate and cold climate, this activity reaches a maximum at the end of spring and remains high until the beginning of autumn, whereas in the course of wintering it decreases to a minimum. However, the length of the periods of large-scale temperature fluctuations between the spring-summer and the autumn-winter period increases, on average, from 2.15 ± 0.10 to 2.4 ± 0.3 hours. The daily duration of these fluctuations changes within similar limits: 2.3 ± 01 hours during the daytime and 2.1 ± 0.1 hours during the nighttime in the spring-summer period vs. 2.5 ± 0.5 and 2.2 ± 0.3 hours, respectively, in the autumn-winter period (Fig. 2).

The periodicity of large-scale fluctuations of the thermoregulatory activity of bees shows no distinct dependence on changes in the external temperature. Its stabilization also has no effect on the established rhythm of temperature fluctuations in the nest. This was observed in the bee families that wintered outdoors and then were transferred in the hives to a room where

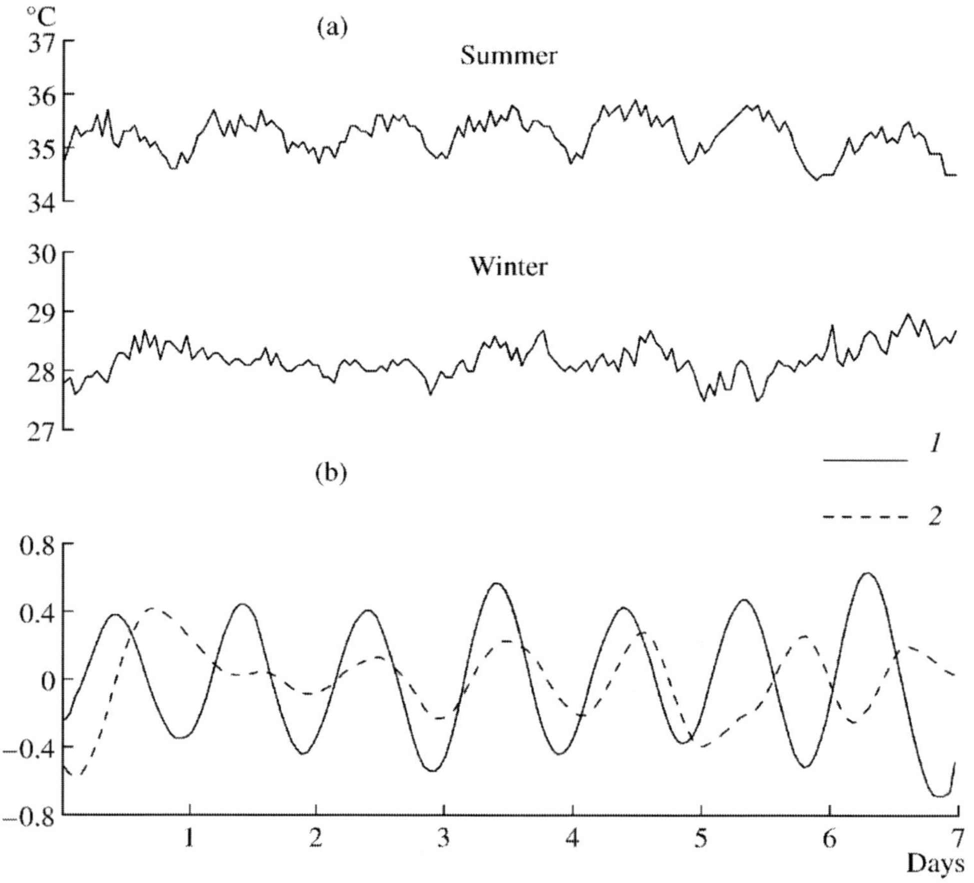

Figure 1. (a) Weekly dynamics of the warming-up of the heat center in different periods of the annual cycle of a bee family and (b) large-scale temperature fluctuations (1) in summer and (2) in winter (for Figs. 1 and 2).

the temperature was maintained at around ^.4°C. Six replications of this experiment showed that the adaptation of bees to the new environment was accompanied by a phase shift of the thermoregulatory activity peak. The magnitude of this shift initially varied from 0.5 to 1.5 hours and was directly dependent on the number of bees in the family. After one day, independently of the number of bees in the family, the shift from original phase decreased to 0.3 ± 0.2 hours. After 2-3 days, the rhythm of temperature activity was fully normalized and synchronized in all the families placed in the same room at distances of 20-60 cm from each other.

When two families with different numbers of bees are thermally isolated, the rhythms of large-scale fluctuations in the centers of their aggregation are often different. For instance, the duration of large-scale temperature fluctuations varied from 1.8 to 2.8 hours in one family and from 1.9 to 2.2 hours in the other family, with their sizes being 18000 and 10000 individuals, respectively. Both families wintered in the same hive, being separated with a solid wood partition 2 cm thick. The first family influenced the second after the partition was lifted 2-3 cm above the floor of the hive: after 4548 h, the second family synchronized its rhythm with that of the larger family.

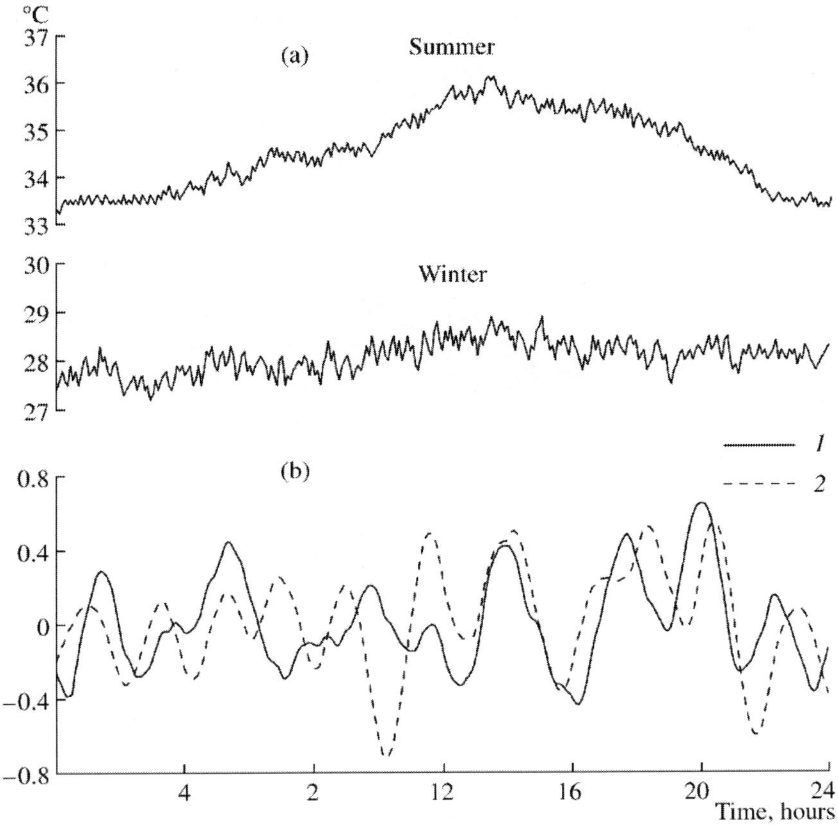

Figure 2. (a) Circadian dynamics of the warming-up of the heat center and (b) large-scale temperature fluctuations at varying external temperatures (15-24°C in summer, -13-4°C in winter).

In bees living under natural conditions, a short-term increase in the temperature of the nest may be stimulated by an increase in illuminance before sunrise. This is most probable during the spring-summer period, when the external temperature is favorable for flight. For instance, the increase of illuminance in the morning to 20-50 lx at 16-18°C was accompanied by an increase in the nest temperature by 0.7-0.8°C by the brood located in the part of the nest facing the entrance. On the next day, the external temperature decreased to 5-7°C, which prevented the bees from leaving the nest; hence, no temperature changes were observed in this area of the nest.

Under conditions favorable for flight, the productivity of the feeding territory has a certain effect on circa-dian fluctuations of temperature in the nest, which may increase by 1.8-2.3°C due to only to activation of bees involved in foraging. In this case, the temperature in the nest stabilizes at the level depending primarily on the external temperature, after the end of the period with conditions favorable for flight (Figs. 1, 2).

In the autumn-winter period, cooling stimulates the aggregation of bees around the zone of maximum heating (the heat center). Alternation of phases of relative immobility and active heat production leads to considerable destabilization of temperature in different zones of bee aggregation. The external temperature has maximum influence on the warming of the heat center at the beginning of wintering, changing its temperature within a relatively wide range (Fig. 3).

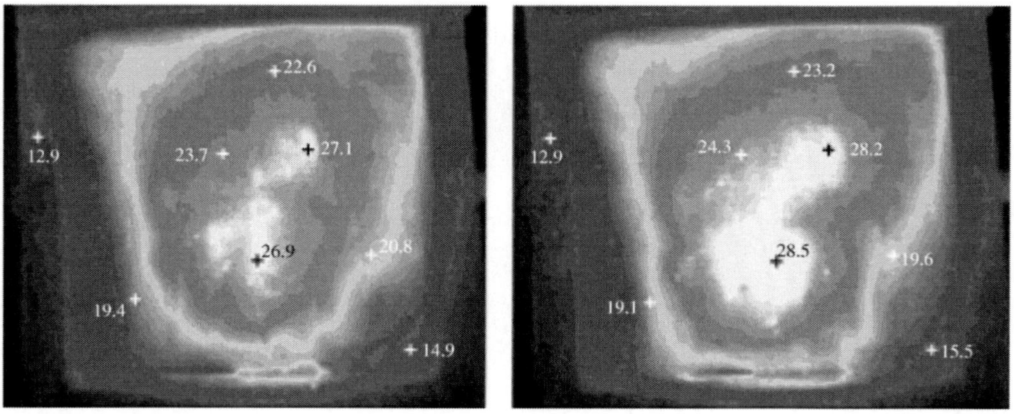

Figure 3. Thermograms of winter aggregations of bees scanned with a thermovision camera at intervals of 60 s at an external temperature of -2°C.

For instance, fluctuations of the heat center temperature in a family of 20 000 individuals may reach 2.5-2.7°C in November-December, compared to 1.0-1.6°C in February-March (Fig. 4). Correspondingly, the coefficient of correlation between the external temperature (varying from -17 to 11°C) and the temperature of the heat center decreases from -0.81 ± 0.10 to -0.13 ± 0.04.

The correlation between fluctuations of the external temperature and the nest temperature depends on the number of bees in the family. In relatively small fami- the onset of wintering, at the initial stages of adaptation lies (around 10000 individuals), a strong negative cor- to autumn cooling. Beginning from December, the correlation between these temperatures is observed only at relation coefficient decreases to -0.36 ± 0.12. In families consisting of almost twice as many individuals (around 18000), fluctuations of temperature in November-December reached 0.8-2.7°C, and the correlation coefficient was -0.84 ± 0.09. During the second part of wintering (beginning from January), bee families weakly responded by changes in thermoregulatory activity even to significant fluctuations of the external temperature (Fig. 4).

Isolated aggregations of bees. Bees separated from the family display certain aspects of group behavior, namely, aggregation and temperature regulation. The activity of aggregation depends on the number of bees and on the external temperature. Such bees begin to aggregate together when their number approaches 50, with the activity of aggregation increasing at greater numbers. In a group of about 500 isolated bees, 90% aggregation is observed at 24-27°C, and 100% aggregation, at 10°C. The greater the number of bees in an isolated group, the higher the temperature within their aggregation. At 24°C, the temperature in an aggregation of 50 bees is maintained at 28.7 ± 0.38°C; in an aggregation of 100 bees, at 29.5 ± 0.51; and in an aggregation of 500 bees, at 30.9 ± 0.42°C.

The quality of food also has a certain effect on the temperature in bee aggregations. For instance, this concerns the concentration of carbohydrates dissolved in water for feeding the bees. At 25°C, the temperature in an aggregation of around 500 bees feeding on 30% sucrose is maintained at 33.3 ± 0.39°C; when this solution is replaced with 60% sucrose, the temperature increases by an average of 0.3°C *($P \sim 0.95$)*.

Both the temperature in bee aggregations and the temperature in their nests are subject to fluctuations, but their range decreases in the course of adaptation to conditions of isolation. For instance, the temperature in the center of an aggregation of about 500 bees fluctuated by 1.6 ± 0.26°C during the first seven days and by only 0.8 ± 0.14°C during the following 15-30 days. In some cases, considerable short-term bursts of locomotor activity and rises in temperature to 41^3°C take place without any apparent cause.

In the course of adaptation to shading, temperature changes in bee aggregations develop circadian cyclicity manifested in temperature elevation during the daytime (Fig. 5). The temperature in the aggregation center reaches its maximum by noon and decreases to a minimum by midnight. The durations of the phases of temperature rise and fall are 8-12 and 11-13 hours, respectively. The periods of thermoregulatory activity reflected in by large-scale temperature fluctuations average 2.5 ± 0.2 hours.

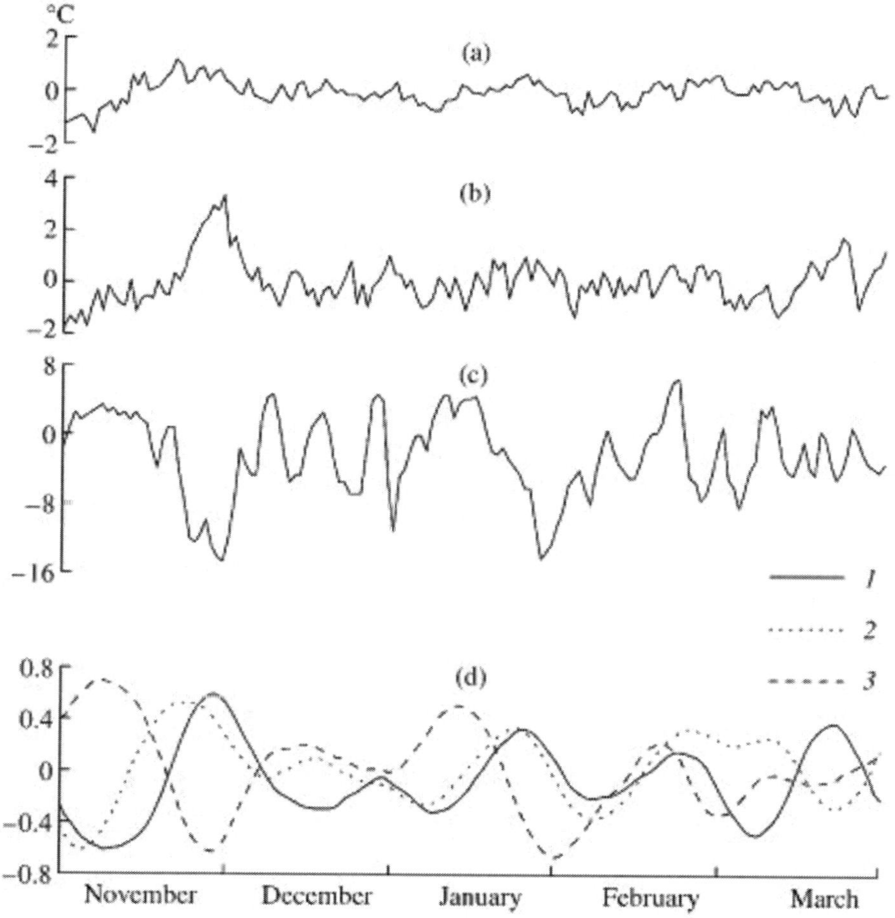

Figure 4. Fluctuations in the warming-up of heat centers in families consisting of about (a) 10 000 and (b) 18 000 individuals (c) at different external temperature and (d) normalized values of large-scale temperature fluctuations calculated with the fractional moment method: (1) temperature fluctuations in the heat center of family (a), (2) temperature fluctuations in the heat center of family (b), and (3) fluctuations of external temperature.

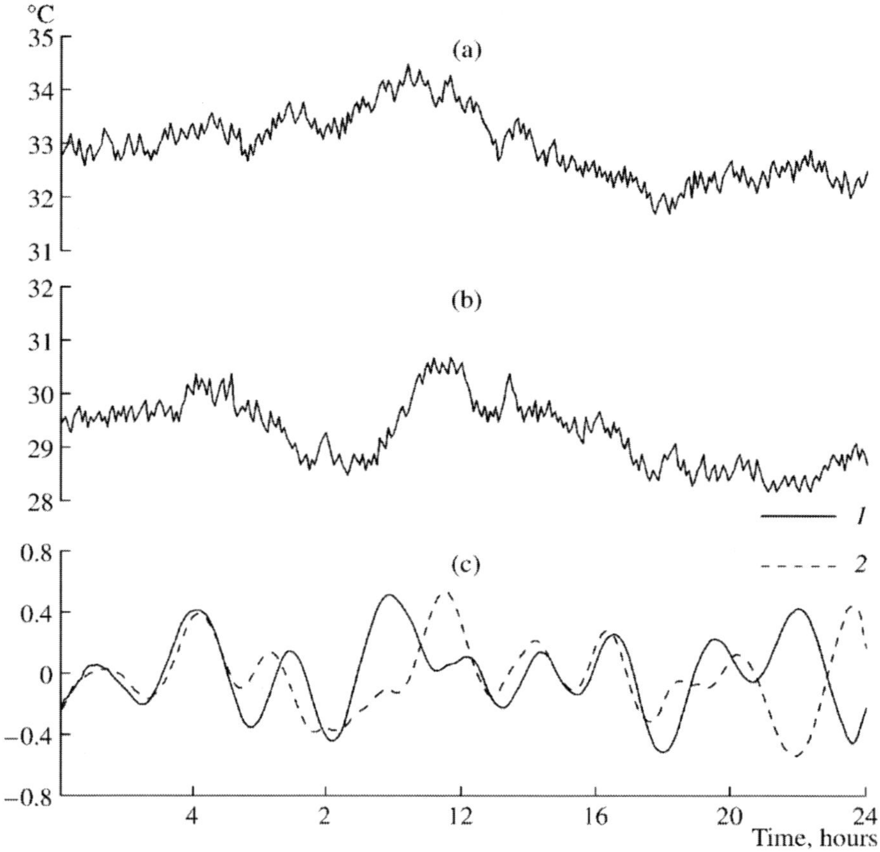

Figure 5. Circadian dynamics of the warming-up of heat centers in aggregations of approximately 500 worker bees and normalized values of large-scale temperature fluctuations at an external temperature of 4.2 ± 0.5°C: (a) bees from a family wintering outdoors, (b) bees from a family wintering indoors at illuminance of about 0.01 lx and temperature of 13°C; (7) bees from family a, *(2)* bees from family b.

Under conditions excluding the influence of fluctuations in the external temperature, air humidity, and illuminance, the range of circadian variation of heating in bee aggregations depends on the number of individuals and on the external temperature. At 23.0 ± 0.1°C, the maximum (daytime) and minimum (nighttime) temperatures in aggregations of 200-250 bees average 28.7 ± 0.31°C and 26.9 ± 0.29°C, respectively. When the number of bees in an aggregation is twice as great, the daytime temperature increases, on average, by 3.9°C, and the nighttime temperature, by 3.6°C *(P* - 0.99).

CONCLUSION

Adaptation to wintering in honeybees manifests itself in the decreasing dependence of their thermoreg-ulatory activity on fluctuations of external temperature. Thus, the functioning of thermoregulatory mechanisms becomes more autonomous in the course of cold adaptation. But considerable fluctuations of external temperature destabilize the warming of the heat center and reduce the probability of successful wintering. This effect is the stronger, the smaller the number of worker bees in the family. Larger families are more capable of adequately responding to changes in the external temperature. This is why the expansion of the honeybee range to zones with temperate and cold climates was accompanied by selection in favor of increase in the number of bees per family.

Short-period temperature rhythms used by the ther-moregulatory complex probably contribute to the preparedness of bees to recurrent changes in the external temperature and other ecological factors. Certain rhythms of thermoregulation may influence the coordination of behavior providing for optimal responses to sudden and significant changes in the environment. This allows aggregating bees to achieve certain autonomy, making them less dependent on the family in their functioning and helping them to save energy resources necessary for maintaining the required temperature.

In the course of evolution, bees acquired the program of behavior related to the circadian periodicity of changes in temperature and other environmental factors, which has improved their adaptation to these factors within natural ranges of their fluctuations. But conditions for biologically relevant realization of the genetic program for synchronizing the behavior with cyclic circadian changes in the environment are limited by the fact that the ranges of fluctuations in the intensity of these factors change considerably during the annual cycle of the bee family. This makes strict programming of behavior biologically inexpedient and stipulates the need for complementing the circadian program with individual experience. Individual experience allows bees to make thermoregulation in the nest increasingly independent of external conditions in the course of wintering. Misalignments between the circadian rhythms of heat production and heat emission that occur upon excessive external heating or cooling of a bee family result in desynchronization of the circadian system.

Circadian cyclicity of temperature regulation in bee families is retained in aggregations of worker bees isolated from them. The innate cyclicity of their behavior is apparently accounted for by a physiological mechanism allowing bees to reflect the circadian rhythms of environmental factors. This mechanism manifests itself in the fact that bees retain the circadian rhythms of their thermoregulatory activity even at constant external temperature and illuminance, which they never encounter under natural conditions.

The temporal structure of thermoregulatory processes in bee families and in aggregations of bees isolated from them depends on the number of bees per family or aggregation. However, bees from different families aggregated together and exposed at constant temperature and illuminance develop a single (averaged) rhythm of thermoregulatory activity. This activity is synchronized even in two different bee families kept under conditions allowing their thermal contact. In such cases, the resulting common rhythm is set by the family that is larger and, consequently, emits more heat.

REFERENCES

Aschoff, J., *Free-Running and Entrained Circadian Rhythms, in Biological Rythms*, Aschoff, J., Ed., New York: Plenum, 1981. Translated under the title Biologicheskie ritmy, Moscow: Mir, 1984, vol. 1, pp. 5^69.

Aschoff, J. and Wever, R., *Human circadian system, in Biological Rythms,* Aschoff, J., Ed., New York: Plenum, 1981. Translated under the title Biologicheskie ritmy, Moscow: Mir, 1984, vol. 1, pp. 362-388.

Benson, J.A. and Lewis, R.D., An Analysis of the Activity Rhythm of the Sand Beach Amphipod, *J. Comp. Physiol.,* 1976, vol. 105, pp. 339-352.

Daan, S. and Berde, C., Two Coupled Oscillators: Simulations of the Circadian Pacemaker in Mammalian Activity Rhythms, *J. Theor. Biol.*, 1978, vol. 70, pp. 297-313.

Es'kov, E.K., *The Acoustic Apparatus of Bees*, Biofizika,1969, vol. 14, no. 1, pp. 158-166.

Es'kov, E.K., *Etologiya medonosnoipchely* (Ethology of the European Honeybee), Moscow: Kolos, 1992.

Es'kov, E.K., *Ekologiya medonosnoi pchely* (Ecology of the European Honeybee), Ryazan': Russkoe slovo, 1995.

Es'kov, E.K., Individual and Social Adaptations of the European Honeybee to Wintering, *Usp. Sovrem. Biol.*, 2003, vol. 123, no. 4, pp. 383-390.

Ivanov, K.P., *Bioenergetika i temperaturnyi gomeostazis* (Bioenergetics and Temperature Homeostasis), Leningrad: Nauka, 1972.

Kipyatkov, V.E. and Lopatina, E.B., Temperature and Photo-periodic Control of Seasonal Development Cycles in Ants (Hymenoptera Formicidae): *Exogenously Heterodynamic Species, Entomol. Obozr.*, 2003, vol. 82, no. 4, pp. 801-819.

Mazokhin-Porshnyakov, G.A., *Zrenie nasekomykh* (Insect Vision), Moscow: Nauka, 1965.

Minnis, D.H. and Pittendrigh, C.S., Circadian Oscillation Controlling Hatching: Its Ontogeny during Embryogenesis of a Moth, *Science*, 1968, vol. 159, pp. 534-536.

Schmidt-Nielsen, K., Animal Physiology: Adaptation and Environment. Translated under the title Fiziologiya zhivot-nykh. *Prisposoblenie i sreda, Moscow: Mir*, 1982.

Schriker, B., Die Orientierung der Honigbiene in der Dan-merung, zugleeich ein Beitrag zur Frage der Ocellenfunktion bei Bienen, *Zeit. Verg. Physiol.*, 1965, vol. 49, no. 5, pp. 420-458.

Sharma, V.K. and Chandrashekaran, M.K. Zeitgebers (Time *1* Cues) for Biological Clocks, *Current Sci.,* 2005, vol. 89, pp. 1136-1146.

In: Bees
Editor: Richard M. Florio

ISBN: 978-1-61324-825-6

Chapter 11

THERMAL COMMUNICATION BETWEEN DEVELOPING AND ADULT HONEYBEES

E. K. Eskov and V. A. Toboev
Russian State Agrarian Correspondence University,
Balashikha, Moscow, Russia

INTRODUCTION

The body temperature of an individual honeybee, as in other poikilotherms, is highly dependent on the ambient temperature. However, bee clusters and families actively regulate their temperature [1—4]; the most precise control is maintained over the brood. Whatever the physiological state of the family, its size, and the ecological situation, the averaged brood temperature varies by fractions of a degree, and only in some extreme cases may deviate more than 1 C. Lacking any centralized mechanism of thermal control, bee aggregations successfully adjust heat production and heat loss in the brood zone through activities of adult worker bees.

Clearly, such regulatory activity requires some inborn program, but it is unknown how it is launched, what signals from the brood may urge the worker bees to provide heat or stop doing it. This work is an attempt to shed some light on these questions.

EXPERIMENTAL

Heat production by *Apis mellifera* larvae and pupae was measured with a microelectrocalorimeter (sensitivity 2.27 V/W, accuracy 1 μW). The developing insects were placed into the 25-cm3 chamber of the instrument in their capped wax cells, which were excised from brood combs taking care to avoid any mechanical influence on the organisms to be tested.

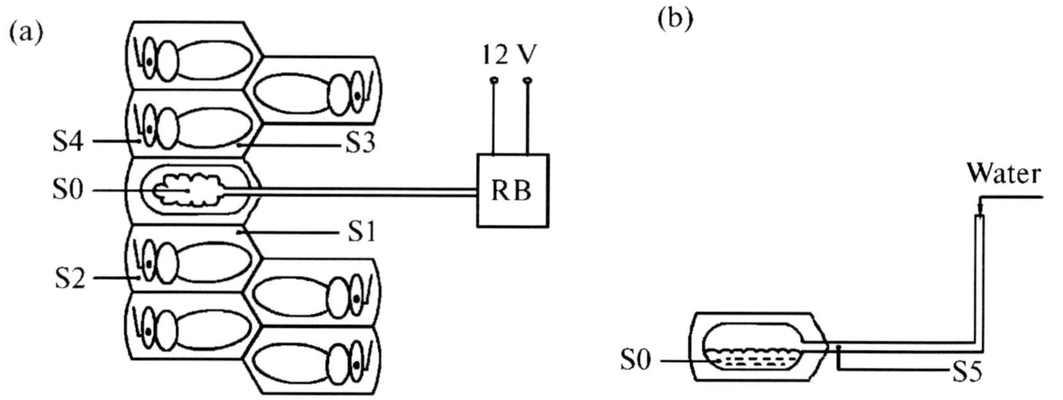

Figure 1. Scheme of assessing the effect of (a) heating or (b) cooling of a brood cell with a thermal dummy: RB, resistor block; S0, heater/cooler sensor; S1, S3, sensors near the bottom in neighbor cells; S2, S4, sensors under the caps; S5, water temperature sensor.

The brood-heating behavior was studied on honeybee families, each kept in a one-frame hive with removable sidewalls. The temperature of the comb cell surface and of the bee body was monitored by thermal imaging (IRTIS-2000, Russia; ThermaCam SC3000, USA), upon replacing the hive walls with IR-transpar-ent plastic film. The temperature of pupae in capped comb cells was measured with KTY21-6 semiconductor microsensors (Siemens, Germany) inserted through bottom perforations, adjusted along the cell wall at the head and the abdominal levels, and fixed with a thin layer of wax, which was also used to close the perforations.

The reaction of bees to heating and cooling of pupae was assessed using thermal dummies. A heated dummy was a 12-V DC heating element (Fig. 1a); a linear power increment was provided with a resistor block. A cooled dummy was a water-filled silicone vessel sized as a pupa (Fig. 1b). The dummies were inserted into brood cells from which pupae had been removed through the bottoms, and the holes were closed with wax.

The attitude of adults to dead brood was examined with worker-bee pupae in intact capped cells. Combs were taken out, selected areas were subjected to (sub)lethal UV irradiation (RDT-240 lamp, Russia) while the rest of brood cells were shielded with a metal screen [5], and the combs were put back in place. The temperature in irradiated and nonirradiated areas was monitored with microsensors.

RESULTS AND DISCUSSION

Depending on the ecological situation and the age composition of the honeybee family, the behavior of adult bees around the brood may vary significantly. At relatively low external temperature (15°C and lower) the bees aggregate on the brood combs and around, which attenuates the overall heat loss. As the temperature exceeds 22—25°C, the bees enter active locomotion and flight, so their number in the nest and density on the brood markedly decrease. When the ambient temperature becomes close to optimal for brood development (33—35°C), the bees start to cool the nest, fanning their wings to speed up air exchange and increase the supply of moisture.

Thermoregulatory activity of bees on brood. Not all bees that are on the brood combs take part in heating. Only separate individuals, having no outward distinctions from other worker bees, press close upon brood cells and heat them. The process is intermittent (Fig. 2). As the bee generates heat by shivering of the flight muscles, it is the thorax that heats the most.

Bees started to heat the brood cells with surface temperature of 31—32.6°C (mean 31.8°C). During a heating episode, which could last from 0.4 to 5.4 min, the temperature rose by 2.4 ± 0.1°C on average at 0.6 ± 0.3°C/min.

The temperature of the adult bee (dorsal side of thorax) before the heating episode averaged 36.7 ± 0.4°C; in 2.5 ± 0.15 min it increased to 37.4 ± 0.4°C. The total ranges recorded were 34.2—40.3°C before and 34.6—39.6°C after the heating episode. Note that the bees with initial temperature 38.5°C or higher actually cooled themselves while heating the cell.

Along with brood heating through the comb surface, bees also use empty cells: a bee enters the cell head first so that only the end of its abdomen remains outside. Such bees produce heat much longer than those on the comb surface, up to 24 min, which is about four times the maximal duration of the surface heating episode.

Warming response of pupae. Heating of the cell surface by an adult had different effects on the temperature inside the cell near the cap (pupa head, Fig. 1) and at the bottom. Before heating, the temperature under the cap exceeded that at the bottom; accordingly, the pupa head before heating was warmer than the abdomen by 0.6°C on average (range 0.3—0.8). In the course of heating the difference decreased, and when the bee left the heated cell, the

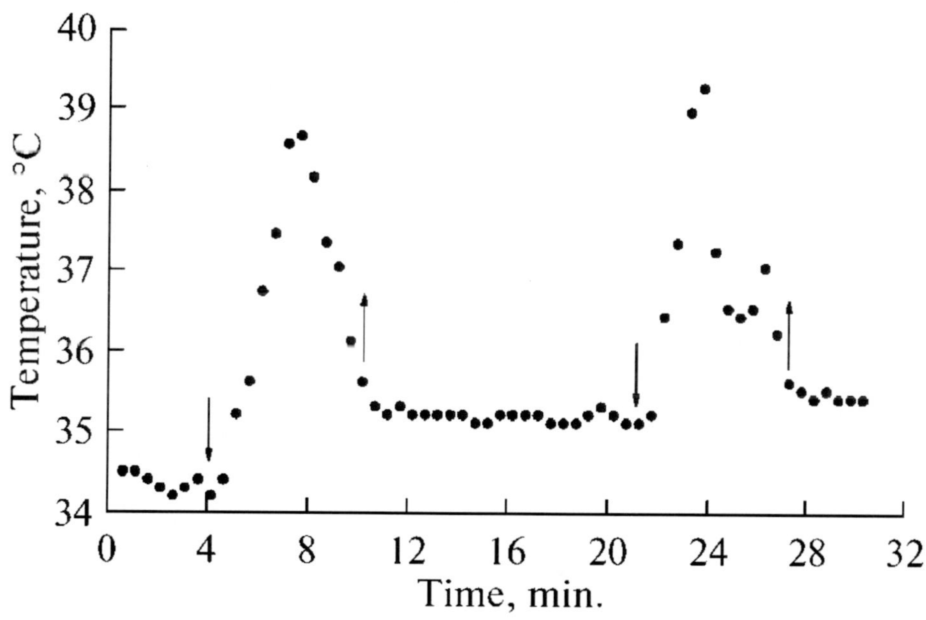

Figure 2. Surface temperature of brood cells with and without adult bees on them (direct readings of an IR imager). Downward arrow: a bee has pressed itself upon the cap and starts generating heat; upward arrow, the bee has left the heated cell.

bottom temperature exceeded the under-cap temperature by 1.1 ± 0.2°C. This "inversion" was due to that the pupa head warmed up at 1.0...1.1°C/min, the thorax at 1.3...1.4°C/min, and the abdomen at 1.5...1.6°C/min, most probably because metabolic processes in these parts of the body were intensified by external heating to different extents.

Temperature dependence of heat production by worker bee mid-age pupae (calorimetric data)

Air temperature, °C7	Heat production, mW	
	per pupa	per gram mass
29	0.23...0.26	1.85...2.01
34	0.29...0.33	2.24...2.33
37	0.51...0.59	4.49...4.54

Heat production by brood. The amount of heat produced by the brood itself depends on the stage of development and on individual mass. At the air temperature and humidity corresponding to those inside the nest, the highest heat production is observed in larvae, the lowest in pupae. Thus at 33—34°C a last-instar larva of 132...151 mg generates 0.96 ± 0.09 mW, which is three times more than a pupa does (table).

In all cases the amount of heat produced by the brood depends on external heating. Upon an increase of the ambient temperature by 8 C within the range admissible for bee development (29^37°C), the heat production by brood of varied age rises by a factor of 1.8...2.3 (table, P ~ 0.99). The age dependence of metabolic intensity is reflected in the dynamics of specific heat production in the brood. A body temperature shift by 1°C changes this index by 0.5 mW/g in larvae and by 0.3 mW/g in pupae.

Reaction of bees to dummy temperature shifts. When a dummy pupa inserted among the brood (Fig. 1a) was heated at 1.3°C/min, adult bees started to actively fan this brood area with their wings when the cell surface temperature reached 36...37°C (Fig. 3). The dummy

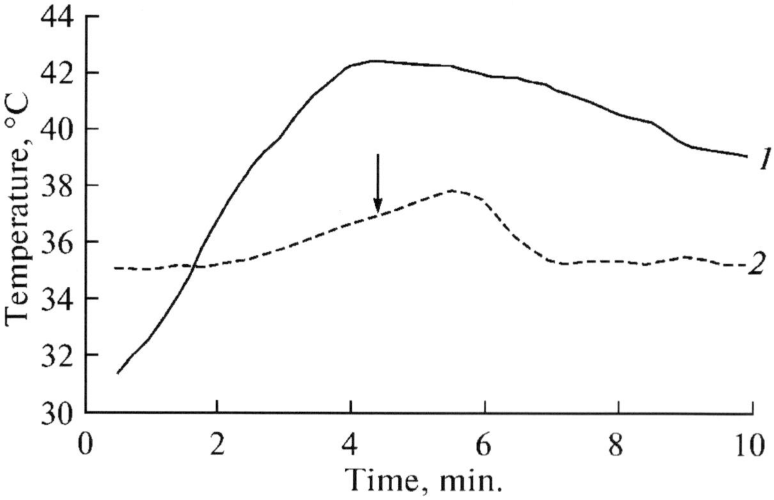

Figure 3. Dynamics of temperature for the heated dummy in a brood cell (1) and for the cell cap surface (2). Arrow marks the start of active cooling of the internally heated cell by adult bees.

temperature at that moment was maximally 43°C. Even though in the next 1.5—2 min the temperature of the dummy and of the cell declined, the bees continued cooling until the comb surface was brought down to 35…35.5°C.

Cooling of a dummy pupa (Fig. 1b) stimulated some of the adults near the cell to start warming it. Other bees remained indifferent. When the dummy cell temperature dropped to 32°C, the bee pressed itself to the cap (a typical heating posture). The thorax temperature in responsive bees ranged 36.6…39.8°C, whereas in indifferent bees it was 30.8…34.8°C.

Attitude of bees to irradiated brood. Adult bees obviously distinguished normal brood from those fatally injured by UV irradiation. From time to time they heated the areas of irradiated brood to 38— 39.5°C, while the unaffected areas were kept at the normal 34.6—35.4°C. After several such overheating episodes, the bees opened the cells and removed the dead pupae. Each such bee produced intense heat while pressing itself to the cap: in 2—3 min the temperature between the thorax and the cap was raised by 3— 4°C to reach 39.6°C.

Conclusion

Consolidated aggregations of honeybees are capable of precise temperature control ensuring proper development of the brood. Thereby they represent a decentralized system where thermoregulation is achieved via interplay of multiple subsystems (individuals), each of which operates on limited local information from its thermoreceptors [4].

Brood temperature control in bees is part of a complex instinct that developed in the course of their socialization, whereby a bee family has in essence become an evolutionary unit.

Numerous observations suggest that not all of the bees busy about the brood nest do heat (or cool) the brood. This is likely to mean that the inborn program of sustaining the brood temperature must first be somehow "enabled"—i.e., an adult bee must be made responsive by internal physiological changes—so that it can be launched into action by an external signal. The "program-enabling" factors that give rise to such special behavioral plasticity are clearly beyond the scope of the present communication. The stimulus to contact thermogenesis is apparently the decline of the brood cell surface temperature to a certain threshold. However, the stimulus as such is not sufficient for optimal temperature control: in the process, the adult and the pupa are engaged in "thermal communication," whereby heat emission by the pupa rises in response to external heating. This feedback may of course be reinforced by the olfactory modality, and particularly by enhanced release of carbon dioxide by the pupa, the sensor for which combines with the thermoreceptor in the bee antenna [6]. On the other hand, the absence of such response(s) permits the adults to detect dead(dying) pupae in capped cells; thereupon they use the same endothermic mechanism for obviously different purposes.

References

E. K. Eskov, *Ecology of the Honeybee* (Kolos, Moscow, 1 1992; Russkoe Slovo, Ryazan, 1995) [in Russian].

B. Heinrich, *The Hot-Blood Insects: Strategies and Mechanisms of Thermoregulation* (Springer, Berlin, 1993).

E. K. Eskov, *Usp. Sovr. Biol. 123* (4), 383 (2003).

E. K. Eskov and V. A. Toboev, Zh. Obshch. Biol. *70* (2), 1 110 (2009).

E. K. Eskov, Izv. RAN Ser. Biol. no. 6, 754 (1996).

V. Lacher, Z. Vergl. Physiol. *48* (6), 587 (1964).

In: Bees
Editor: Richard M. Florio

ISBN: 978-1-61324-825-6

Chapter 12

Electrocardiogram of the Honeybee

E. K. Eskov
Russian State Agrarian Correspondence University,
Balashikha, Moscow, Russia

Introduction

The amplitude and temporal structure of electrical oscillations related to the functioning of the heart (ECG) in insects varies over a wide range depending on their physiological state and ecological situation (Irishawa *et al.,* 1956; Tyshchenko, 1976; Eskov and Dolgov, 1986). In wasps and bumblebees, specific changes of ECG were observed under hypo- and hyperthermia (Eskov, 1998, 1999). This study deals with the effects of temperature on ECG in the honeybee, which differs from bumblebees and wasps in taxonomic position, level of social organization, and adaptations to the thermal factor.

Worker bees, drones, and queens of *Apis mellifera* L. were used in the study. The ECG was recorded with an N-338 analog recorder. The input resistance of the preamplifier for a range of up to 5 kHz was not less than 3 gΩ, and the noise level at the input was not less than 50 μV. The active electrolytically sharpened tungsten electrode was introduced into the cervical articulation, and the indifferent electrode was inserted into the abdominal region. The bees were preliminary immobilized by injecting 10% urethane solution into the abdominal cavity.

The experimental insects were placed in a thermo-stated shielded chamber. For controlling insect body temperature, the microsensor of an electronic thermometer (ST11-19 microthermistor) was placed 1-2 cm away from the insect. Hypothermia was usually limited to cooling up to a state of profound cold torpidity. In some experiments, exposure to hyper- and hypothermia reached lethal values.

Temperature optimum. Within the range of optimum temperatures (22-33°C), the ECG in worker bees, queens, and drones consists mainly of two-phase low-and high-amplitude oscillations. There is also a certain cyclicity in the ECG structure. More or less distinct cycles include from one to four periods of decaying electrical oscillations.

The frequency-amplitude structure of the ECG and the dependence of its changes on temperature differ in worker bees, queens, and drones. In queens, the high-amplitude oscillations at the optimum temperature exceed those in drones by a factor of 1.2 ($P \sim 0.9$) and those in worker bees by a factor of 2.6 ($P > 0.99$). An increase in temperature from the lower to upper limit of its optimum range stimulates an increase in the ECG amplitude in queens 1.3-fold ($P > 0.95$) and in worker bees and drones 1.1-fold ($P \sim 0.9$). This is accompanied by a 1.3-fold increase in the frequency of high-amplitude oscillations in all three castes ($P \approx 0.95$).

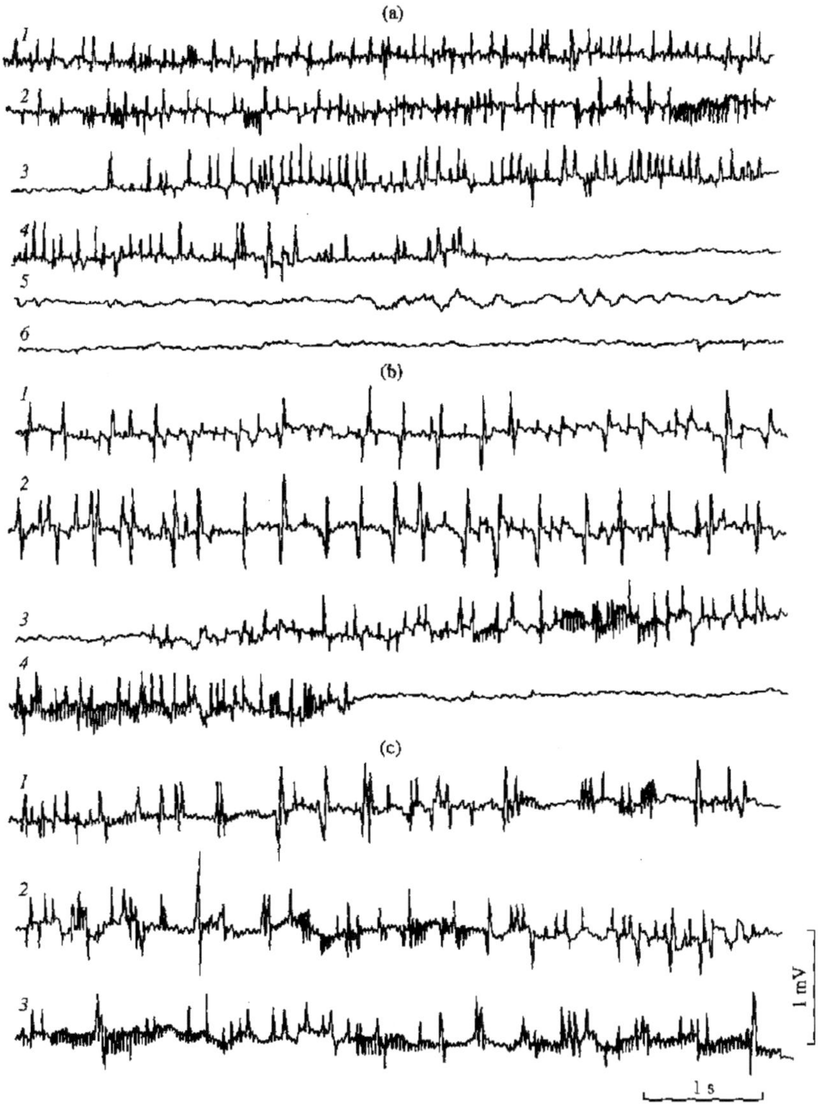

Figure 1. Changes in the amplitude-time structure of ECG under hyperthermia in (a) worker bees, (b) queens, and (c) drones, (a): *1*- 23°C; 2- 33°C; *3*- 45°C; *3, 4,* onset and progression of electrical oscillations after temporary cessation; 5, 10 min at 55°C; 6, 20 min at 55°C. (b): *1*- 23°C; 2- 33°C; 3,4- 50°C, onset and progression of electrical oscillations after temporary cessation. (c): 1- 23°C; 2- 33°C; 3- 40°C.

Hyperthermia. The increase in temperature beyond the upper limit of the optimum range initially stimulates an increase in the frequency of electrical oscillations. As the temperature approaches lethal values (differing in worker bees, queens, and drones), the generation of electrical oscillations repeatedly stops and begins again. Complete asystole (death from hyperthermia) occurs in worker bees at 55-60°C, in queens at 49-56°C, and in drones at 46-52°C.

Temporary cessations of the heart function begin in worker bees at 42 ± 0.9°C *(CV* = 16%), in queens at 41 ± 0.5°C *(CV* = 12%), and in drones at 39 ± 0.7°C (*CV*= 4%). This effect of hyperthermia occurs independently of the rate of heating, but the duration of unfavorable actions of identical temperatures that stimulate retardation of electrical activity noticeably vary. For example, in worker bees at 45°C the first cessation of electrical activity may take place within a range of 0.4 to 3.2 minutes from the beginning of heating to this temperature.

Changes in the frequency of electrical oscillations under the influence of increased temperature somewhat differ in worker bees (Fig. la), queens (Fig. lb), and drones (Fig 1c). Under the influence of temperature rise from 23 to 45-50°C, the frequency of high-amplitude oscillation increases in worker bees from 2.1 + 0.07 Hz (*CV*= 37%) to 4.3 ±0.11 (*CV* = 42%), in queens from 2.0 ± 0.21 (*CV* = 44%) to 7.8 ± 0.39 (*CV* = 29%), and in drones from 2.6 ± 0.14 (CV= 52%) to 4.1 ± 0.22 Hz (*CV* = 38%); the frequency of low-amplitude oscillations increases from 2.8 ± 0.18 Hz (*CV* = 49%) to 7.2 ±0.41 (CV=41%), from 2.5 ± 0.16 (CV=40%) to 7.9 + 0.48 (*CV*= 37%), and from 3.0 + 0.26 (CV- 63%) to 17.0 ± 0.93 Hz (CV= 26%), respectively.

The sublethal effects of hyperthermia are expressed as repeated temporary cessations of electrical activity. Their duration was 16.1 ± 0.73 s *(CV* = 19%) in worker bees, 7.2 ± 0.12 *(CV* = 9%) in queens, and 2.9 ± 0.44 s (CV = 69%) in drones. The periods of repeated activation of the heart function after cessation continue for 14.8 ± 1.96 s *(CV* = 43%) in worker

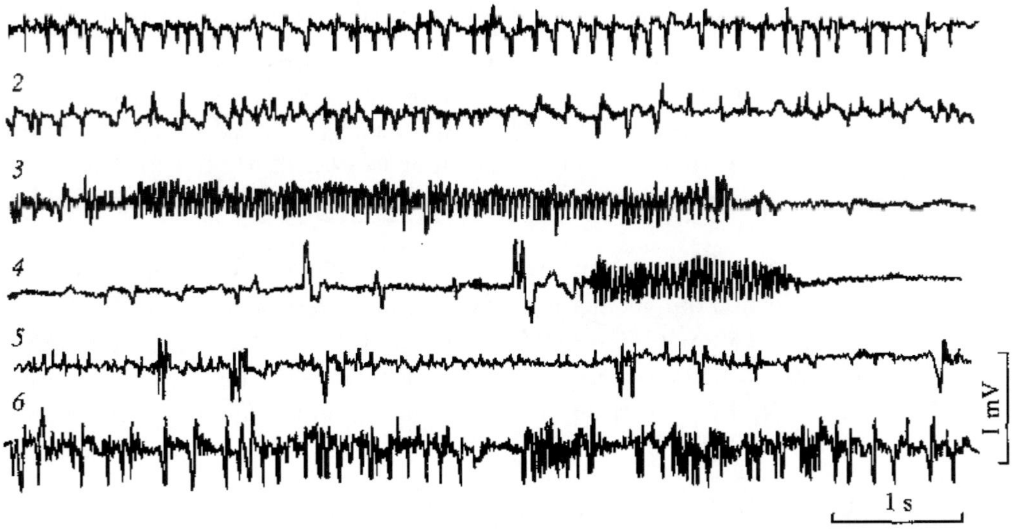

Figure 2. Changes in the ECG structure upon cooling of a worker bee at 5°C for *(1)* 2 min, (2) 8 min, (3) 10 min, and *(4)* 11 min to the state of cold torpidity and (5) 1 min and (6) 10 min after the beginning of its activation at 24°C.

bees, 4.1 + 0.37 (*CV* = 82%) in queens, and 1.8 ± 0.19 s (*CV* = 49%) in drones. The frequency of electrical oscillations increases, gradually reaching 13.1 ± 0.97 Hz (CV= 27%) in worker bees, 14.2 ± 0.93 (*CV* = 24%) in queens, and 28.0 ± 0.98 Hz (CV= 19%) in drones, and then decreases again. The amplitude of oscillations changes similarly to the frequency. Its maximum value reaches 600-900 *piV.* Prior to complete asystole, the frequency of electrical oscillations may remain at these levels, with the amplitude decreasing almost to zero (Fig. la, lanes *1, 2).*

Hypothermia. The structure of ECG significantly changes between the beginning of cooling and the complete cessation of electrical activity under the influence of cold. Similar to hyperthermia, this is preceded by temporary cessations of electrical activity. Upon cooling to +5°C, they begin after 5.4 + 2.2 min (CV- 76%), and complete asystole occurs after 9-24 min. The duration of the first cessations is 14 ± 2.1 s (*CV* = 82%). Before complete asystole, it decreases to 9 ± 1.6 s (*CV* = 68%). This is accompanied by a decrease in the periods of resumed electrical oscillations, from 13 ± 3.8 s (CV = 54%) to 4 ± 0.9 s (*CV*= 76%).

In the course of hypothermia, the amplitude and frequency of electrical oscillations change. The frequency of high- and low-amplitude oscillations changes in the same way, but the directions of changes in their amplitude are different (Fig. 2). In particular, in worker bees initially kept at 24 ± 1°C, the frequency of high-amplitude oscillations was 4.2 ± 0.24 Hz (*CV* = 49%); after 5 min of exposure at 0°C, it decreased to 3.6 ± 0.29 Hz (*CV* = 31%); after 10 min, it was 2.7 ± 0.21 Hz (CV=26%). The low-amplitude oscillations initially had a frequency of 5.8 ± 0.24 Hz (*CV*= 46%); after the aforementioned intervals of cooling, this frequency decreased to 5.6 ± 0.61 (*CV*= 68%) and 4.3 ± 0.32 Hz (*CV* = 34%), respectively. The high-amplitude oscillations were initially at a level of 647 + 11.2 jiV (*CV* = 29%); the low-amplitude oscillations, at 287 ± 6.4 jxV (*CV* = 41%). After 5 min of cooling, the amplitude of the former increased to 683 ± 6.9 jxV (*CV*= 11%); after 10 min, it was 698 ± 6.9 µV (*CV* = 12%). The amplitude of the latter decreased to 238 ± 7.9 (*CV* = 32%) and then tol87±8.8 µV(CV=47%).

The pattern of electrical oscillations changes to the greatest extent in the periods of activation of the heart function after cessation of heart function caused by cold torpidity. Each period in which the electrical activity is resumed begins with the generation of single electrical oscillations. Their frequency increases to 35 ± 4,6 Hz (*CV* = 53%) and then decreases by the end of the period of electrical activation to 19 ± 4.2 Hz (*CV* = 59%). This is accompanied by small changes in the amplitude, on average from 112 + 3 (*CV*- 37%) to 106 ± 3 µV (*CV*= 34%).

During cold torpidity, the generation of electrical oscillations ceases completely and is not resumed during the period of hypothermia. Warming at a temperature above the threshold of cold torpidity (in bees, 13.5 ± 0.1°C) stimulates the resumption of locomotor and electrical activity. Low-amplitude oscillations appear first, then follow single high-amplitude oscillations, which become more frequent within a few minutes. Normalization of ECG coincides with complete restoration of movements (Fig. 2). In particular, worker bees exposed at 0°C for 60 minutes began to recover electrical and motor activity within 6.4 + 2.1 minutes (CV= 74%). The structure of ECG became fully normalized within 18-26 minutes. Similar results were obtained for queens and drones.

Thus, the frequency-amplitude structure of electrical oscillations generated by heart functioning is highly similar in worker bees, queens, and drones. In all three castes, the ECG pattern includes both high- and low-amplitude oscillations. Their amplitude and frequency increase with an increase in temperature, which is obviously due to the higher rate of enzymatic processes related to heart functioning and the generation of electrical oscillations.

The response of worker bees, queens, and drones to the extreme effects of hyper- and hypothermia manifests itself in repeated temporary cessations of heart functioning. They precede complete asystole, which occurs as a result of heat shock or cold torpidity, similarly to the responses of paper wasps (Eskov, 1998) and bumblebees (Eskov, 1999). The species- or cast-specific differences in these responses are related to insect tolerance for warming and cooling. This tolerance correlates with different threshold values of hyper-and hypothermia stimulating temporary cessation and then complete arrest of the heart functioning.

In the honeybee, the cast specificity of the ECG manifests itself as differences in amplitude, which is maximum in worker bees and minimum in queens. These casts differ also in threshold temperatures causing temporary cessations of heart functioning and in the time course of this process, which precedes complete asystole. The threshold temperature causing cessations of heart functioning during hyperthermia in worker bees is 1°C higher than in queens and 3°C higher than in drones. In worker bees, the periods of cessation of heart functioning are, on average, 2.2 times those in queens ($P > 0.99$) and 5.5 times those in drones ($P > 0.999$). The periods of resumed electrical oscillations differ in a similar way.

Cast-related differences in the ECG amplitude manifest no distinct relationship to heat tolerance of worker bees, queens, and drones. However, the threshold temperatures causing cessations of heart functioning are different and depend on this tolerance. According to their values, the worker bees are more tolerant to hyperthermia than queens, and the queens are more tolerant than drones. These cast differences are also reflected in the periods of resumption of heart function, whose frequency at sublethal exposure to the thermal factor is the highest in drones and lowest in worker bees.

REFERENCES

Eskov, E.K., Temperature Dependence of Electrical Activity of the Heart Function in Wasps, *Biofizika,* 1998, vol. 43, no. l,pp. 139-142.

Eskov, E.K., Temperature Dependence of Electrical Oscillations of Bumblebee Heart, *Biofizika,* 1999, vol. 44, no. 3, pp. 493-495.

Eskov, E.K. and Dolgov, L.A., Regulation of Temperature in the Nest and Its Importance in the Life of Bumblebee Family, *Zool Zh.,* 1986, vol. 65, no. 10, pp. 1500-1507.

Irisawa, H., Irisawa, A,F., and Kadotani, T., Findings on the Electrograms of the Cicada's Heart *(Ciyptotympana japon-ensis* Kato), *Japan. J. Physiol,* 1956, vol. 6, pp. 150-161.

Tyshchenko, V.R, *Osnovyfiziologii nasekomykh* (Fundamentals of Insect Physiology), Leningrad: Leningrad Gos. Univ., 1976, vol. 1.

In: Bees
Editor: Richard M. Florio

ISBN: 978-1-61324-825-6

Chapter 13

MODELING OF THE TEMPERATURE FIELD DISTRIBUTION IN WINTER CLUSTERS

E. K. Eskov and V. A. Toboev
Russian State Agrarian Correspondence University,
Balashikha, Moscow, Russia

INTRODUCTION

Honeybee as an insect is classed with poikilotherms. However, poikilothermy is typical only of an individual bee outside the family. Bee colonies in moderate and cold climate can winter at ambient temperatures down to -35...-45°C. This is far beyond the individual cold tolerance, which is determined by the freezing temperature of body liquids, -7...-16°C in wintering bees [1]. The ability of the species to survive the long winter is based on a complex of hereditary programmed eto-physiological reactions stimulated by cold. Protection against life-endangering chilling is attained through formation of ordered clusters that ensure accumulation and rational use of heat [2, 3].

The present work undertakes mathematical modeling of thermoregulation mechanisms and temperature field distribution in bee clusters, which is topical for understanding the principles of self-organization in complex biological systems.

EXPERIMENTAL

Construction and study of mathematical models of temperature field distribution is based on investigating the natural processes of heat production and loss taking places in bee clusters. The bees were kept in special hives with hinged combs that could be opened. The temperature field distribution and heat production in the clusters was examined with infrared imagers IRTIS-2000 and ThermaCam S3000.

Analysis of the thermograms reveals alternating periods of relative rest and active heat generation, which result in substantial destabilization of the temperature regime in different zones of bee localization (Fig. 1). This is also contributed to by unordered migration of bees within the cluster. However, the portion of active bees directly adjacent to the lower surface of the cluster in the intercomb space increases with a decrease of ambient temperature.

In the heat core of the model, the temperature ranging 24...32°C is regarded as the preferred one whereat the bees can feed ad libitum and migrate freely. The bees outside the heat core must move toward it or heat themselves at the expense of intensified metabolism. In the bees forming the bottom of the cluster periphery (the most cooled part) the body temperature may drop to 8...13.5°C, which initiates the cold torpor and minimization of metabolic processes.

The premises for modeling the thermoregulation mechanisms were as follows. The bee winter cluster is regarded as a spherically symmetrical structure within which heat transfer takes place by through-body conduction and partly by convection through air gaps between bodies. The main stimulus to which the clustering bees react by changing the volume they occupy is the ambient temperature. The latter and the localization in the cluster determine the behavior of every bee and its part in thermoregulation. The threshold for stimulation of heat production by 'shivering' (microvibration of flight muscles) is about 18°C [4, 5], whereas 13.5 °C corresponds to the onset of superficial cold torpor, which drastically reduces the metabolism. In such a state the bee can survive only for a few days [2].

Thus, in the model the key thermoregulation mechanisms are divided into the movement of bees in the cluster and the heat production by separate bees. Temperature adjustment within certain internal zones of the cluster is considered as a result of activity of every bee tending to stay in the region where the temperature satisfies its physiological needs.

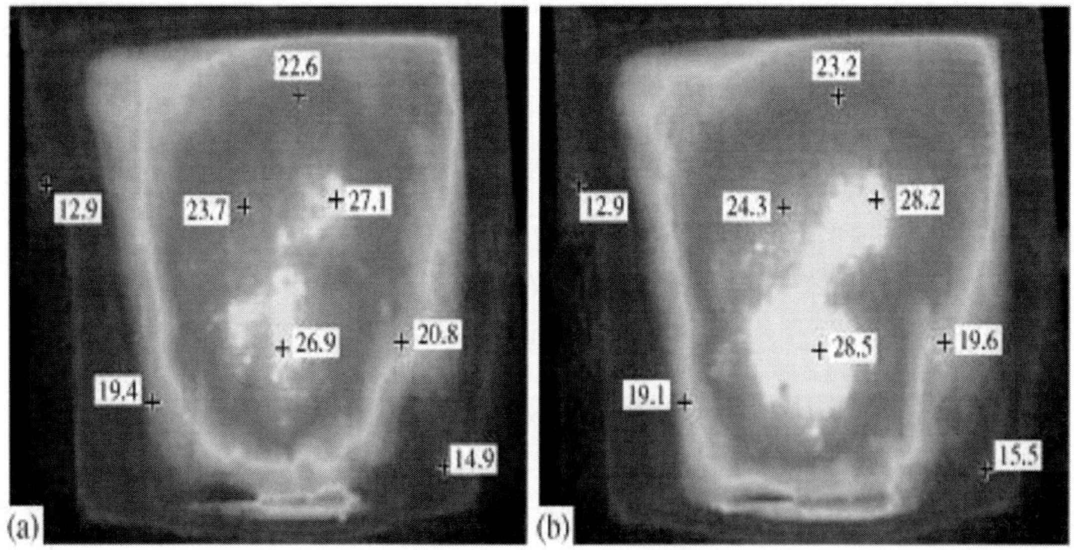

Figure 1. Images of the thermal fields in a bee winter cluster taken 60 s apart.

MODEL DESCRIPTION

Assuming spherical symmetry of the model cluster, the temperature field is unidimensional, i.e., heat spreads only along the sphere radius. The temperature T(r, t) and density p(r, t) of the bees are functions of distance r from the center of the model to the point considered within the cluster and of time t. With these and the above conditions of modeling, the classical heat conductance equation takes the form

$$c\frac{\partial T}{\partial t} = \frac{1}{r^2}\frac{\partial}{\partial r}\left(r^2\lambda(\rho)\frac{\partial T}{\partial r}\right) + \rho f(T) \qquad (1)$$

with boundary conditions

$$\begin{cases} \left.\dfrac{\partial T}{\partial r}\right|_{r=0} = 0, \\ \left.\left(\lambda(\rho)\dfrac{\partial T}{\partial r}\right)\right|_{r=R(t)} = \alpha(T_{\text{amb}} - T(R(t))), \end{cases} \qquad (2)$$

where c is specific heat capacity, λ (p) is the conductivity coefficient, f(T) is a function of individual bee metabolism, a is the heat emission coefficient depending on the properties of the cluster surface, R(t) is the radius of the cluster model at ambient temperature T_{amb}.

The first term in the right-hand part of equation (1) determines the energy received by bees at distance r owing to conduction by adjacent bees. The second sum-

mand characterizes the change in energy released by bees owing to metabolism. The function f(T) takes into account the heat production by individual bees through shivering and in the stationary state (without locomotion), as well as the heat released by bees at the cluster periphery in the state of superficial torpor.

According to Krogh's normal curve, the intensity of metabolism in insects at apparent rest rises with increasing temperature. The minimal heat production by a single bee is 1 to 1.5 mW [6-8]. The temperature factor Q_{10} for metabolism (increase upon heating by 10°C) is 2.4 [1, 9]. When the local temperature falls below 18°C, the bee generates additional heat by shivering. The peripheral bees, having exhausted the resources in their honey stomachs, have to migrate into the heated zones of the cluster.

The experimental data available allow the maximal metabolism to be related to that at rest using a factor of 35 [1]. Hence, the metabolism function of an individual bee can presented as

$$f(T) = \begin{cases} 0.035e^{1.116(T-14)}, & \text{at} \quad 8 \le T < 14°\text{C}; \\ 0.035, & \text{at} \quad 14 \le T < 14°\text{C}; \\ 0.035e^{0.713(T-15)}, & \text{at} \quad 15 \le T < 18°\text{C}; \\ 0.001 \times 2.4^{0.1(T-18)}, & \text{at} \quad 18 \le T < 32°\text{C}. \end{cases} \qquad (3)$$

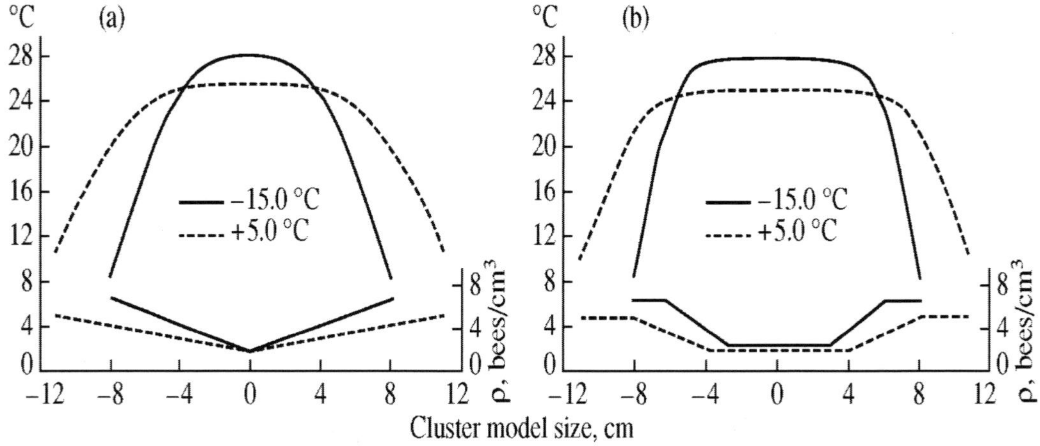

Figure 2. Temperature profiles at (a) linear and (b) step distribution of cluster density and specified ambient temperature; the model parameters were N = 18000; c - 0.8 cal/(g °c); λ = 0.9 x 10-3 W/(cm °C); α = 6 x 10-4 W/(cm2 °C).

Determination of the thermophysical parameters included in (1) is complicated by their dependency on the state of the bees that simultaneously take different parts in thermoregulation. It is impossible to integrate the heat capacity over the cluster volume because of the structural inhomogeneity and unordered bee migration. Therefore it is expedient to consider only the intervals in which the analyzed parameters may change.

The heat flux to the surface of the bee cluster is associated with the body heat capacity and depends on the water content; the heat capacity of the air gaps between the bodies in insignificant. Taking that the water content in the bee body is 65% and does not appreciably change through wintering [3] and that the heat capacity of the hemolymph is close to that of water, being 0.9^0.93 cal/(g °C), with increasing bee density their heat capacity changes roughly linearly from 0.6 to 0.8 cal/(g °c). The specific conductivity also increases with bee density from 7.6 • 10-4 to 3.0 • 10-3 W/(cm °C) [5, 10]; this is commensurate with the conductivity of dry air, 3.4 • 10^{-4} W/(cm °C).

For heterogeneous systems with internal and surface heat sources (such as the bee clusters) the heat transfer coefficient is physically ambiguous, including in the general case the values of specific heat fluxes depending on conductivity, convection, and irradiation. The sum flux of heat from the cluster surface depends on the effective outer surface, i.e. on the area through which heat is removed. This is always smaller than the geometric surface area and depends substantially on the position of the cluster relative to the hive entrance. To add, parameter a strongly depends on the water vapor percentage in the surface layer as well as on the heat loss by evaporation and respiration. Considering only the average heat loss from the cluster surface through convection, this parameter can be taken to range from 6 • 10^{-4} to 3 • 10^{-3} W/(cm^2 °C) [10, 13].

The first equality in boundary conditions (2) stems from the condition that the temperature at the model center is maximal. The second equality is a Newton's law expression with account of heat irradiation.

RESULTS AND DISCUSSION

In equation (1) the bee density ρ(r, t) depends on their motility and number and on ambient temperature. In approximating the dependence of bee density on distance r (constant, linear, stepwise, etc.), the mean values were taken to be 2 bees/cm^3 in the center and 46 bees/cm^3 at the periphery [11].

During wintering the total number of bees N does not change significantly, so

$$N = 4\pi \int_0^{R(T_{amb}, t)} r^2 \rho(r, t)dr \qquad (4)$$

can be regarded as constant. From this relationship, with a given density distribution one can find the cluster radius $R(T_{amb}, t)$ as a function of ambient temperature.

To solve equation (1) with boundary conditions (2), use is made of the numerical methods implemented in software packages MatLab 7.02 and FemLab 3.01; they allow the solutions to be examined at different values of thermophysical parameters in the above-specified ranges. According to the calculations, the most essential parameters affecting the temperature profiles at the same ambient temperature are the heat conductivity and irradiation. The smaller are these coefficients, the broader is the high-temperature zone in the cluster and the steeper is the drop in the temperature curves. Therefore, reduction of the cluster size tells on the temperature rise in the core. With increasing conductivity and irradiation, the model also predicts a temperature rise in the center but at the surface the temperature drops below the preset limit even at maximal bee density (Fig. 2). Thus at $\alpha = 3.0 \cdot 10^{-3}$ W/(cm^2 °C) the model predicts rapid cooling of the surface to subzero temperatures. The same happens upon raising $\lambda(\rho)$ to $1.3 \cdot 10^{-3}$ W/(cm °C). Positive temperature at the cluster surface in this case can be maintained only at the expense of additional heat production by the bees, but heat removal will also increase thereby.

Let us examine the contribution of each term in the right-hand part of (1). The role of the second summand reflecting the individual metabolism becomes manifest at ambient temperatures below -5°C. This is in quantitative agreement with the experimental results that locomotion as such does not suffice for efficient ther-moregulation. These circumstances require active production of heat by shivering [1, 4, 12]. The contribution of this process into thermoregulation is attenuated or maybe completely excluded at 2-9°C because in this temperature range the metabolism intensity is near its minimum as judged by oxygen consumption [1].

At -12°C the production of heat by shivering becomes dominant in maintaining a positive cluster surface temperature. The function chosen here for metabolism (3) allows a conclusion that the main heat-production load at low temperatures is borne by the bees immediately adjacent to those at the air-cooled surface. This permits the bees in superficial torpor to migrate into the warmed zone of the nest. The stimulus for migrating to the heat core harboring the food resources is obviously hunger commencing upon depletion of the honey stomach [1].

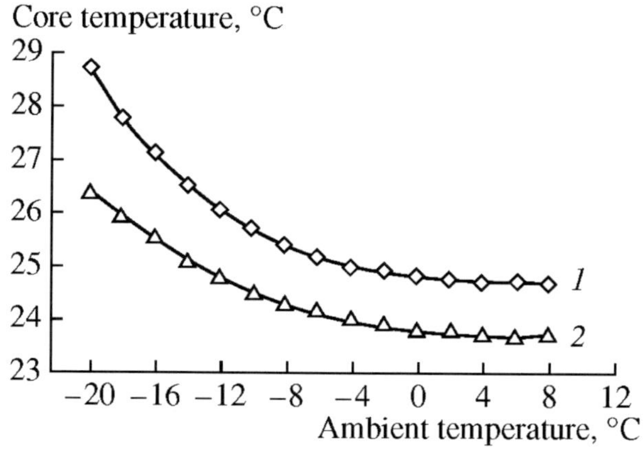

Figure 3. Core temperature in model clusters of (1) 18000 and (2) 12000 bees as a function of ambient temperature; the model parameters were c - 0.8 cal/(g °C); λ = 9.5 • 10-3 W/(cm °C); α= 8.8 • 10-4 W/(cm2 °C).

The model predicts an inverse dependence of the thermal core temperature on ambient temperature (Fig. 3) and an increase in the core temperature with increasing number of bees in the cluster. This is consistent with the data of the literature [1, 5, 10] and of the author's studies of thermal processes in cold-induced bee clusters, but has not been explained heretofore [5, 8, 14].

Conclusion

A bee cluster represents an open spatially demarcated biological system, the thermal processes in which can, with some reserve, be modeled using kinetic equations with preset initial and boundary conditions. However, since a priori such conditions are hard to relate with the real functioning of the biosystem, modeling required an idealized notion of spherical symmetry in the cooled insect cluster. It must be emphasized that the thermophysical parameters of a model based on such an assumption cannot be physically unambiguous. To add, the use of equation (2) for the boundary conditions is valid only if evaporation as a way of heat removal is completely absent.

Despite these limitations, the proposed model quantitatively reproduces the main features characteristic of the dynamics of heat processes in cold-induced insect

clusters. In particular, it reliably predicts the core temperature rise in response to deeper chilling of the periphery and the shape of the temperature profiles, as well as conditions simulating additional heat production and the thermoregulatory function of the bees adjacent to the outer layer. It reacts to ambient temperature variation by changing the cluster shape and density. Consolidation of the cluster associated with intense cooling tells on lower insulation of the surface layer.

The proposed theoretical model does not support the concept that an insect cluster is a superorganism where the dominant role in thermoregulation belongs to the portion of insects localized in the heat core [7, 14]. They have been supposed to react by enhanced activity and heat generation to elevated carbon dioxide resulting from compaction [1, 7, 13]. However, thermal imaging studies did not reveal the influence of bee heating in the core on bee heating at the periphery.

The relatively high body temperature of the outer bees is largely ensured by their heat generation and migration into the warmed zone. Thereby their body temperature significantly exceeds that of maximal over-cooling [3] and does not reach the threshold of chill coma [1]. Probably, migrations favor reduction of excess moisture and carbon dioxide accumulating in the heat core.

The real enhancement of core heating in response to stronger cooling in the model corresponds to a decrease of ambient temperature below the optimal range for a bee cluster (Fig. 3). At a certain relation between the thermal conductivity, heat transfer factor, and cluster radius the temperature in the model center does not exceed 36°C even during severe chilling at -40°C. There can be no core overheating if the heat is produced by all bees to warm the surface layer or if heat production and compaction are performed by the bees in the sub-surface region to maintain the core temperature within a certain range. The increased heat emission from the surface at lowered ambient temperature and the ensuing stimulation of additional heat production by the sub-surface bees allow one to state that the thermoregulation mechanism corresponds to the principle of self-organization. However, the same bees having come in the course of clustering and subsequent migration to different zones perform dissimilar functions in thermoregulation. The greatest load during temperature regulation is carried by the bees at the cluster periphery because they in the greatest measure experience the unfavorable influence of chilling and therefore must react to it more actively than the others.

REFERENCES

[1] E. K. Eskov, *Honeybee Ecology* (Russkoe Slovo, Ryazan, 1995).
[2] E. K. Eskov, *Izv. AN SSSR, No. 4*, 535 (1984).
[3] E. K. Eskov, *Usp Sovr. Biol.* ***123*** (4), 383 (2003).
[4] E. Kammer and B. Heinrich, *J. Exp. Biol.* ***61,*** 219 (1974).
[5] S. Omholt, *J. Theor. Biol.* ***128,*** 219 (1987).
[6] B. Heinrech, *J. Exp. Biol.* ***91,*** 25 (1981).
[7] C. D. Owens, *Agric. Res. Service*, No. 1429, 1 (1971).
[8] J. Jones, M. Myerscough, S. Graham, and B. P. Oldroyd, Science ***305,*** 402 (2004).
[9] B. Heinrech, *Forschr. Zool.*, No. 31, 393 (1985).
[10] J. Watmough and S. Camazine, *J. Theor. Biol.* ***176,*** 391 (1995).
[11] E. K. Eskov and V. A. Toboev, *Pchelovodstvo No. 3*, 14 (2007).
[12] E. E. Soulwick and J. Moritz, *Comp. Biochem. Physiol.* ***40A,*** 935 (1971).
[13] H. Human, S. W. Nicolson, and V. Dieteman, *Naturwissenschaften* ***93,*** 397 (2006).
[14] A. Stabentheiner, H. Pressl, Th. Papst, et al., *J. Exp. Biol.* ***206,*** 353 (2003).

In: Bees
Editor: Richard M. Florio

ISBN: 978-1-61324-825-6

Chapter 14

SOUNDS GENERATED BY BEE COLONIES DURING SOCIOTOMY

E. K. Eskov and V. A. Toboev
Russian State Agrarian Correspondence University,
Balashikha, Moscow, Russia

INTRODUCTION

The colony of honey bees, as well as other species of the genus *Apis,* is a consolidated group of related and mutually dependent organisms which cannot survive and reproduce outside it. In the case of eusocial species, natural selection and other factors of evolution act on colonies rather than on individuals. Reproduction and dispersal of bee colonies proceeds by part (approximately half) of worker bees leaving the parental colony together with a fertilized or unfertilized female (queen). Beekeepers usually call this process "swarming" but this name does not properly reflect the phenomenon and is inconsistent with the strict entomological use of this term. Therefore, the process of fission of a bee colony will be referred to as sociotomy.

Preparation for sociotomy may take from several days to several weeks. During this period many bees keep from replenishing the food reserves, building combs, feeding the brood, etc., which leads to decreasing production of the colony. Since the bees leaving the parental colony tend to settle as far from it as possible (Eskov, 1992, 1995; Camazine et al., 1999; Seeley and Buhrman, 1999), the beekeeper should be able to recognize the colonies about to undergo fission, and to estimate the timing of this process.

There are many publications devoted to diagnostics of the physiological state of bee colonies based on the structure of sounds produced by them (Wenner, 1962, 2002; Frisch, 1967; Eskov, 1969, 1970, 1972, 1979, 1992; Winston, 1987; Michelsen et al., 2003; Schneider et al., 2004; Rybochkin, 2007; Ferrari et al., 2008, etc.). Methods of acoustic diagnostics of preparation for sociotomy were also developed. Specific changes in the acoustic noise spectrum associated with sociot-omy were observed. However, practical use of these results is hindered by their inconsistency. For example, according to Eskov (1979), preparation of bee colonies for sociotomy is marked by increasing noise components within the range 210-240 Hz. A broader range of frequency bands associated with sociotomy (210-240, 300-330, 390-420, 420-450 Hz) was reported by Rybochkin (2007). Other researchers observed an initial increase in relative acoustic energy at about 110 Hz, the peak shifting to 500-600 Hz when the swarm was about to leave (Ferrari et al., 2008). The idea of diagnosing impending sociotomy by changes in temperature and humidity inside the nest (Ferrari et al., 2008) is hardly feasible because these parameters of the hive microclimate depend on the number of bees, ambient temperature and humidity, productivity of the foraging area, and many other ecological factors (Eskov, 1995).

The low reliability of the known methods of acoustic diagnostics of the physiological state of bee colonies can be largely attributed to the fact that the existing methods of analysis fail to reflect adequately the connection between changes in the noise structure and the behavior of bees. In particular, filtration of acoustically significant spectral components of sounds produced by bees during preparation for sociotomy cannot distinguish them from other acoustic processes. During the active phase of colony life, sounds of high intensity are produced by bees ventilating the hive. Before swarm emergence, intensive and irregular communication signals of the queens appear in the noise (Eskov, 1979; Michelsen et al., 1986; Lewis and Schneider, 2008). During daylight hours the spectral structure of colony noise is strongly affected by communication signals of the foragers and guards. Noise is also generated by various mechanical actions (cleaning the cells by the workers, gnawing of the emerging young adults, etc.) as well as by the adult bees brushing against one another. Random external acoustic phenomena of biotic, abiotic or anthropogenic origin add to the noise of the bee hive.

This communication is devoted to the dynamics of acoustic processes marking the physiological changes in the bee colony associated with preparation to so-ciotomy. We tried to detect changes in the colony state by the temporal structure of its noise, based on identification of statistically uniform fragments.

MATERIALS AND METHODS

The work was carried out on 12 bee colonies, 8 of which completed preparation for sociotomy by casting swarms. The acoustic processes in all the colonies were analyzed during the entire spring-and-summer active phase. The nests were examined in the process, and the quantity of brood and the presence of cells with developing females were recorded.

The acoustic processes in the nests were recorded using a Type 4182 electrostatic probe microphone (Bruel & Kjsr, Denmark) with sensitivity 3.16 mV/Pa and a linear amplitude-frequency characteristic within the entire range analyzed (20-1000 Hz). The nonlinear distortion factor did not exceed 0.01% (confidence channel at 1 kHz). In order to reduce the impact of intense sounds produced by bees during daytime hours, the phonograms were recorded between midnight and 3 a.m.

The acoustic noise to be analyzed was represented by a set of univariate series of data recorded at sampling frequency of 44.1 kHz. The signal processing was performed within Matlab and MathCAD software packages and included the following procedures: numerical integration of selected segments of noise, in order to reveal the trends of changes in its temporal structure; revealing the internal (i.e., not imposed by the researcher) diagnostic characters; clustering of the segments with respect to a statistically uniform character; visualization of the clusters by methods allowing the results to be adequately interpreted.

In order to distinguish the statistically uniform fragments of diagnostic value, we considered the sequences of ranked amplitudes of relative fluctuations (SRARF) and trends revealed by the procedure of optimal linear smoothing (POLS) with Gaussian kernel applied to the integrated sequence of original data. The use of sequences of ranked amplitudes allowed us to quantify the relative fluctuations in terms of a "universal" set of reduced (fitting) parameters included in the approximate formula for SRARF, which were needed for quantitative comparison of arbitrary segments of acoustic noise. Quantitative differences between two sections could be revealed by increasing the number of fitting parameters. In case of insignificant differences we determined the confidence interval within which the two sections became "indistinguishable." The generalized mean function (GMF) expressed in terms of higher and fractional moments allowed the possible behavior of the analyzed random noise to be predicted.

The relative fluctuations were determined and ranked in the following order: (1) normalization of selected acoustic segments recorded at sampling frequency of 8 kHz; (2) finding integrated sequences and their smoothing by the POLS method; (3) finding the sequences of ranked amplitudes for the relative fluctuations (Fig. 1).

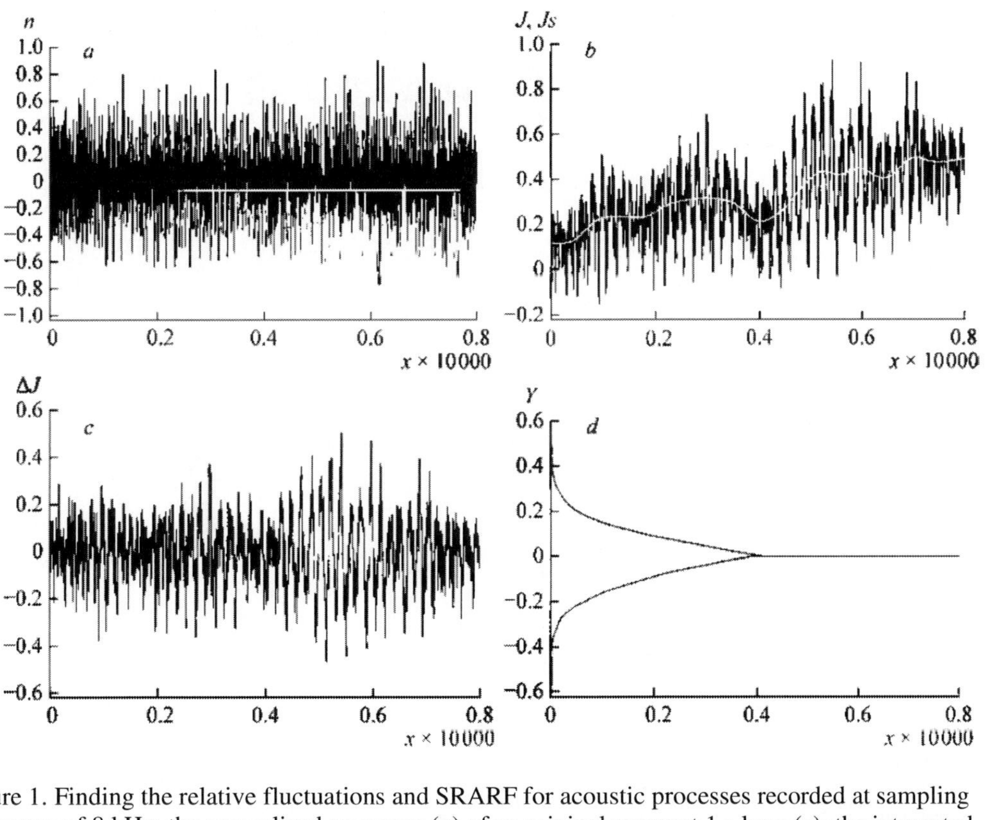

Figure 1. Finding the relative fluctuations and SRARF for acoustic processes recorded at sampling frequency of 8 kHz: the normalized sequence (n) of an original segment 1 s long (a); the integrated sequence (J) and its smoothed values (Js - white curve) (b); the relative fluctuations (c); the sequence of ranked amplitudes for relative fluctuations (Y) (d).

Results and Discussion

Preparation of a bee colony for sociotomy, which usually takes from 2 to 3 weeks depending on the colony state and ecological situation, is accompanied by changes in the sounds produced. In all the eight experimental colonies, considerable changes in the temporal structure of noise were observed 15 days before sociotomy. Analysis of the acoustic processes in terms of SRARF and GMF has shown that with approaching swarm emergence, the statistically uniform fragments in the temporal structure of noise become more prominent; their duration and the optimum interval length τ_{opt} increase. In particular, the mean τ_{opt} value was 48 ms 10 days before swarm emergence and reached 128 ms 1 day before the event. The result of fragmentation of the acoustic noise 3 days before swarm emergence is shown in Fig. 2. At the optimum interval length of 100 ms, the sequence of segments marked with rectangles corresponds to statistically uniform fragments. Quantitative comparison of fragments S2 and S18 by four parameters (a, b, *d*, and A) shows their association with the background noise segments (Fig. 3*a*, 3*b*), confirming the physical basis (synchronism) of the distinguished segments.

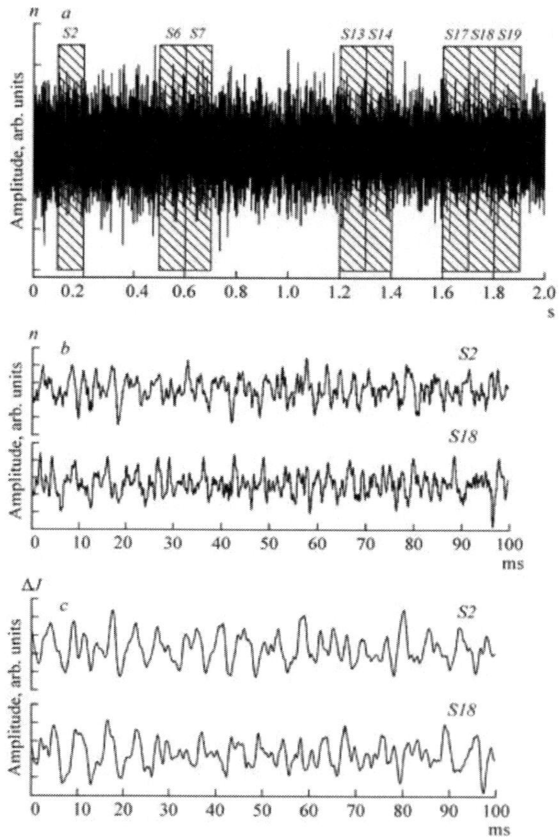

Figure 2. Statistically uniform fragments of noise reflecting the state of bee colonies 3 days before swarm emergence (a), the temporal structure of fragments S2 and S18 (b), and relative fluctuations of the same fragments (c).

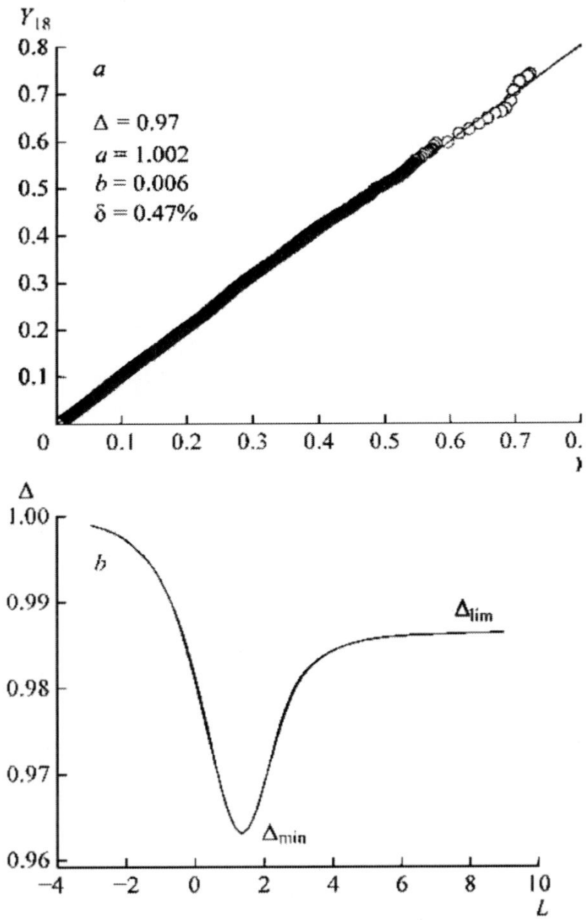

Figure 3. Quantitative comparison of relative fluctuations of fragments S2 and S18 using statistics of fractional moments (a), and the dependence between parameter A characterizing the degree of correlation of the selected segments, and parameter p (L = lnp) (b); τ_{opt} = 100 ms is the optimum interval length of noise fragmentation.

The total average duration of statistically uniform fragments in each 1-s segment was 0.3 s 10 days before swarm emergence and increased to 0.5 s 1 day before sociotomy, probably due to increasing synchronism of the sounds produced by passive bees about to leave the parental colony. This is indicated by the results of analysis of trends and the generalized mean functions. Immediately before swarm emergence, the integrated sequences of the selected noise segments drop to the values of *Js* non-significant relative to the abscissa axis (Fig. 4*a*).

As the day of colony fission approaches, the degree of correlation increases in the dynamics of noise fluctuation, represented by a function *Yn(Y*0) of the reference value *Y*0. A strong correlation of the hive noise (in terms of SRARF) is determined by the linear dependence

$$Y_n(Y_0) \approx a_n Y_0 + b_n,\ n = 1, 2, \ldots, 30, \qquad (1)$$

where $Y0$ is the SRARF of acoustic noise on the day of swarm emergence, whose parameters were obtained by statistical processing of noise of reproducing colonies (sample size $N = 30$), and Y_n is the relative fluctuation of the colony noise n days before swarm emergence ($n = 20$). The coefficients a (the slope ratio of the corresponding line) and b (the initial "cutoff," or distance from the origin of coordinates to the crossing of the line and the ordinate axis), characterizing changes in the colony noise during preparation for sociotomy, were calculated by the least squares method. In the ideal case (with the studied signal completely matching the reference) the former coefficient equals 1, and the latter equals 0.

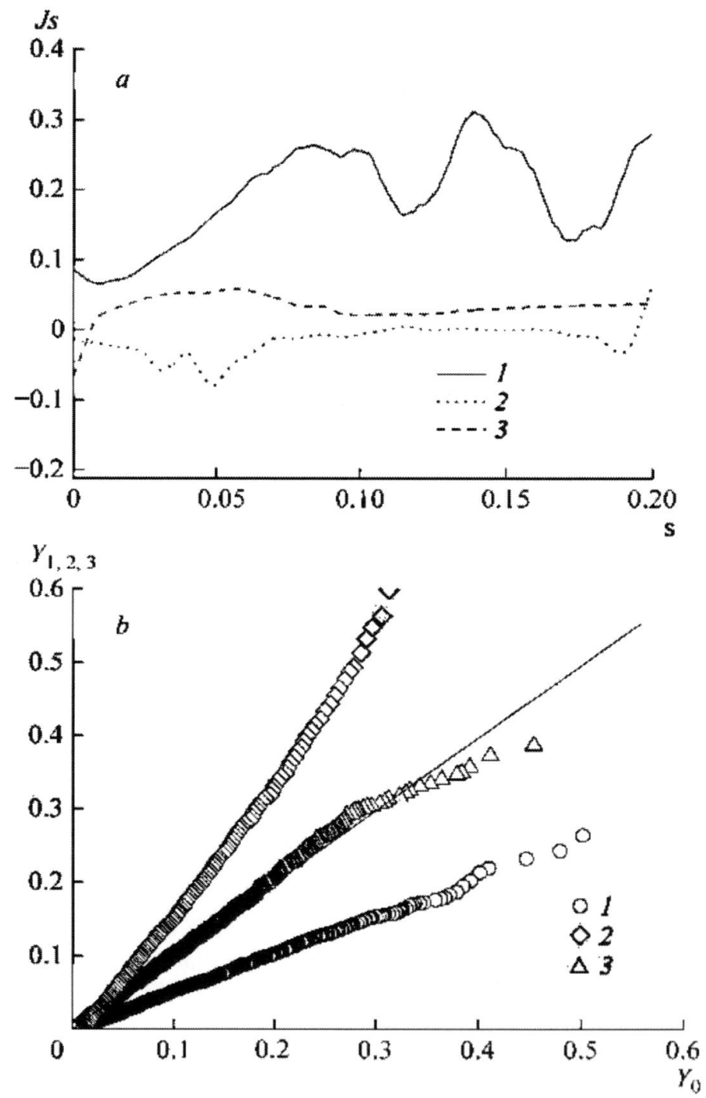

Figure 4. Dynamics of smoothed values (Js) of the integrated sequence for selected segments of acoustic noise (a) and mutual correlations of SRARF (b) in a colony that showed no signs of preparation for sociotomy during the observation season (control) (1), a colony preparing for sociotomy, 10 days before swarm emergence (2), and the same colony, 1 day before swarm emergence (3).

The reliability of analysis can be estimated by the slope ratio of the line and the cutoff value with respect to SRARF related to the reference values of *Y*0 (Fig. 4*b*). Since the functions *Yn*(*Y*0) are different and the GMF parameters for the trends and relative noise fluctuations are also different, they may be used for identification and quantitative assessment of readiness to sociotomy. The initial phases of this process can be easily recognized by a considerable increase in the slope ratio, with $a > 1$ (the linear part of graph 2 in Fig. 4b).

Parameter *a* tends to 1 as the day of swarm emergence approaches. Five days before sociotomy, the mean slope ratio of the lines was 1.12 ± 0.04; 3 days before sociotomy it was 1.04 ± 0.02; and 1 day before the event, 1.01 ± 0.01. The cutoff value *b* (the normalized ordinate value at zero point) did not exceed 0.02 ± 0.01 during 5 days before swarm emergence for all the colonies that underwent sociotomy.

Comparison of functions $Y_{,,}(Y_0)$ allows one more parameter to be considered: the relative fitting error of the line (5), which equals zero in the ideal case. Comparison of relative fluctuations using GMFs in the space of fractional moments (Toboev and Nigmatullin, 2007) reveals the range of correlations (D_{min}, D_{lim}) characterizing the proximity of colony fission. The closer is the day of swarm emergence, the closer is the value of D = 0.5 (Dmin + Dlim) to 1.

The quantitative characteristics of the acoustic process form a cluster of parameters corresponding to different phases of preparation for sociotomy. With the given function $Y_n(Y_0)$, in which Y_0 corresponds to the averaged SRARF values of colonies on the day of fission, the moment of swarm emergence will be determined by the values $a = 1$, $b = 0$, 5 = 0, A = 1.

Based on the four parameters (α, b, δ. and Δ) that determine the statistically uniform clusters, one can build a dynamic model of changes in the acoustic processes associated with preparation for sociotomy. Changes in the parameters of statistically uniform fragments of noise can be visualized and related to the moment of colony fission by means of an elliptical classifier,

$$r(\varphi) = r_0 + \frac{p}{a + (1.2 - \varepsilon)\cos\varphi}, \qquad (2)$$

$$x(\varphi) = r(\varphi)\cos\varphi, \quad y(\varphi) = r(\varphi)\sin\varphi,$$

in which *r*0 characterizes the relative line fitting error, and ε, the class of Pearson's general correlation function. The value of *p* is the slope ratio of the line representing the relative fluctuations in the coordinate system GMFn-GMF0 (or Y_n-Y_0. The value of α, determining the line cutoff on the ordinate axis, is selected in such a way that α = 1 when the line $Y_n(Y_0)$ passes through the origin of coordinates.

The ellipses are plotted based on comparison of GMF and SRARF of the hive noise depending on the colony readiness to swarm emergence (Fig. 5). The position of the ellipse within the shaded area (Fig. 5, curve 2) marks the beginning of preparation for so-ciotomy. A decrease in the ellipse area to the minimum (Fig. 5, curve 3) is associated with accumulation of passive bees ready to leave the parental colony.

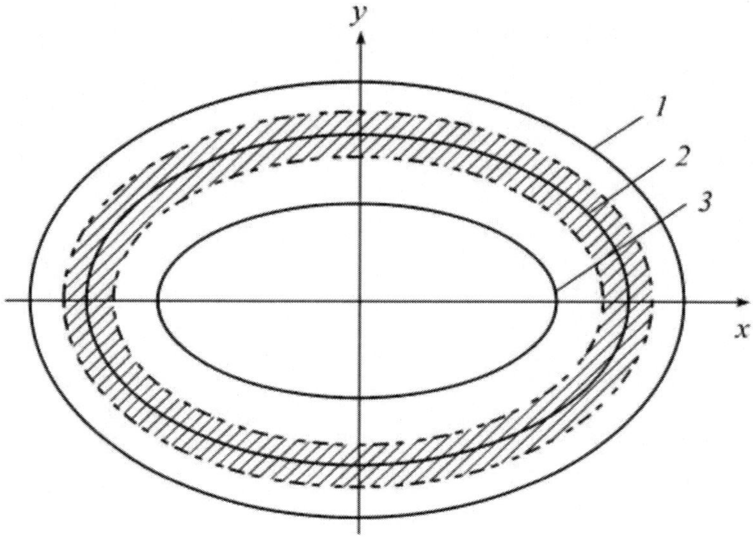

Figure 5. A dynamic model of changes in the acoustic processes associated with preparation for sociotomy in a bee colony (shaded area): 20 days before (1), 15 days before (2), and on the day of swarm emergence (3).

CONCLUSION

An important role in the evolution of sociality in bees belongs to acoustic communications inside the nest, which have reached the highest level of perfection in species of the genus *Apis* and especially in the honey bee *A. mellifera* L. During the period of colony reproduction, the queens of this species detect their rivals by acoustic signals. Complex acoustic communications are used to coordinate the activity of worker bees (Eskov, 1977, 1979). All the acoustic signals produced by the workers and queens have high noise stability, mostly due to active correction of the time-frequency structure of signals depending on the structure of interfering noise (Eskov, 1972b).

Although the acoustic noise produced by a bee colony is a largely random process, its time-frequency structure has a certain degree of orderliness depending on the physiological state of the bees and the ecological situation. In particular, when the ambient temperature rises to the level of nest overheating, the noise component of 75-160 Hz is enhanced and the energy maximum shifts into the high-frequency range. The locomotor activity of bees is reflected by intensive components located within 200-350 Hz. At favorable weather conditions and high productivity of the forag ing area the energy maximum shifts into 300-350 Hz, and at conditions preventing the bees from leaving the hive, into 200-250 Hz (Eskov, 1979).

The ordered time-amplitude-frequency structure of the hive noise, depending on the physiological state of the bees, is a manifestation of autosynchronization of biological systems and appears to play a role in consolidation of the colony. Therefore, preparation for sociotomy, involving a steady increase in the fraction of physiologically young bees, is accompanied by a more frequent generation of statistically uniform fragments in the sequences of ranked amplitudes. Thus, as the bee colony becomes differentiated in the physiological state of its members, it also becomes differentiated with respect to the sounds produced.

The transient acoustic noise generated by the bee colony during preparation for sociotomy and other phases can be described by structured modeling based on selected (most frequent) fragments. Such fragments are correlated with the physiological state of members of the colony and thus reflect its changes. Therefore the statistically uniform noise fragments with a similar distribution function can be used to describe the temporal structure of the noise.

Analysis of the trends of clustering of the parameters of statistically uniform noise fragments may allow one to reveal hidden periodic segments and uncharacteristic (marginal) features. The use of statistics of fractional moments opens wide possibilities of inter-cluster classification of diagnostic characters of noise by the scheme "experiment - cluster sample." Visualization of the acoustic process helps increase the reliability of acoustic analysis and allows the results to be interpreted in an easier way.

References

[1] Camazine, S., Visscher, P.K., Finley, J., and Vetter, R.S., "House-Hunting by Honey Bee Swarms: Collective Decisions and Individual Behaviors," Insectes Soc. *46,* 348-360 (1999).

[2] Eskov, E.K., "The Sound-Producing Apparatus in Bees," Biofizika *14* (1), 158-166 (1969).

[3] Eskov, E.K., "The Background Sound of a Bee Colony," Zool. Zh. *49* (2), 241-248 (1970).

[4] Eskov, E.K., "Structure of the Background Sound of a Bee Colony," Zool. Zh. *51* (7), 1018-1024 (1972a).

[5] Eskov, E.K., "Correction of Acoustic Distance Signal in Bees under the Influence of Noise," Doklady Akad. Nauk SSSR *202* (1), 211-213 (1972b).

[6] Eskov, E.K., "Acoustic Signals in Communication of Social Insects," Uspekhi Sovrem. Biol. *83* (3), 419-431 (1977).

[7] Eskov, E.K., *Acoustic Signals in Social Insects* (Nauka, Moscow, 1979) [in Russian].

[8] Eskov, E.K., *Ethology of the Honeybee* (Kolos, Moscow, 1992) [in Russian].

[9] Eskov, E.K., *Ecology of the Honeybee* (Russkoe Slovo, Ryazan, 1995) [in Russian].

[10] Ferrari, S., Silva, D., Guarino, M., and Berckmans, D., "Monitoring of Swarming Sounds for Early Detection of the Swarming Period," Comp. Electr. Agric. *64,* 72-77 (2008).

[11] Frisch, K. von, *Tanzsprache und Orientierung der Bie-nen* (Springer-Verlag, Berlin, 1965).

[12] Lewis, L.A. and Schneider, S., "'Migration Dances' in Swarming Colonies of the Honey Bee, *Apis mellifera,"* Apidologie *39,* 354-361 (2008).

[13] Michelsen, A., "Signals and Flexibility in the Dance Communication of Honeybees," J. Comp. Physiol. A *189,* 165-174 (2003).

[14] Michelsen, A., Kirchner, W.N., and Lindauer, M., "Sound and Vibrational Signals in the Dance Language of the Honeybee, *Apis mellifera,"* Behav. Ecol. Socio-biol. *18,* 207-218 (1986).

[15] Rybochkin, A.F., "The Background Noise as a Source of Information about a Bee Colony," Pchelovodstvo, no. 3, 15-17 (2007).

[16] Schneider, S.S., Lewis, L.A., and Huang, Z.Y., "The Vibration Signal and Juvenile Hormone Titers in Worker Honeybees, *Apis mellifera,"* Ethology *110,* 977-985 (2004).

[17] Seeley, T.D. and Buhrman, S.C., "Group Decision Making in Swarms of Honey Bees," Behav. Ecol. Sociobiol. *45,* 19-31 (1999).

[18] Toboev, V.A. and Nigmatullin, R.R., "A New Method of Statistical Processing of Temporal Series: a Study of Collective Behavior of Social Insects Based on Their Temperature Regulation Activity," Nelineinyi Mir *5* (4), 183-193 (2007).

[19] Wenner, A., "Sound Production during the Waggle Dance of the Honey-Bee," Anim. Behav. *10,* 79-95 (1962).

[20] Wenner, A., "The Elusive Honey Bee Dance 'Language' Hypothesis," J. Insect Behav. *15* (6), 859-878 (2002).

[21] Winston, M.L., *The Biology of the Honeybee* (Harvard Univ. Press, Cambridge, Massachusetts, 1987).

In: Bees
Editor: Richard M. Florio

ISBN: 978-1-61324-825-6

Chapter 15

BEES ADAPTATION TO WINTERING

E. K. Eskov, M. D. Eskova and V. A. Toboev
Russian State Agrarian Correspondence University,
Balashikha, Moscow, Russia

INTRODUCTION

Although the honey bee, as a representative of the class of insects, belongs to the poikilotherm animals, it can actively regulate the temperature inside the nest, which is usually the most stable in the zones occupied by the brood (Eskov, 1995). However, outside the nest the bee responds to changes in the external temperature as a typical poikilotherm (Eskov, 1992; Eskov and Toboev, 2009; Schmidt-Nielsen, 1982).

The overcoming by a honey bee long wintering in conditions of moderate and cold climate is based on using a large complex of individual and social adaptations. Their acquisition is related to the evolution of sociality. In this direction in the species phylogenesis adaptations providing forage supplies accumulation, their economical utilization during wintering and inner-nest temperature regulation have appeared to be of dominating importance.

The honey bee colonies overwinter in the state of low activity and maintain a relatively high temperature in their nests. Data on the thermal regime in the nests of wintering bees were mostly obtained by various sensors (thermoelectric couples, thermal resistors) installed in the hives (Zhdanova, 1958; Simpson, 1961; Eskov, 1983; Southwick, 1985; Rybochkin and Pustovalov, 2007, etc.). Such methods, however, do not allow one to analyze the entire system of thermal processes taking place in the nest, much less to measure the temperature of individual bees, follow their responses to temperature dynamics, and determine their contribution to regulation of thermal processes.

In this work the role of fat accumulations gained by bees in the period of preparation to wintering and changing of the maximum overcooling temperature (MOT) of the bee's liquid fractions during their wintering are considered. The task of the research was also to study temperature fluctuations dynamics in the nests of wintering bees. However, the influence of external temperature on the thermal processes in the nests of wintering bee colonies was studied by analysis of infrared radiation. In addition, we analyzed the response to cooling in individual bees located in different inter-comb spaces.

Materials and Methods

The study was carried out in honey bee colonies weighing from 1 to 2 kg (1 kg corresponds to about 10 000 worker bees). The colonies were kept in polycarbonate hives with removable front and side walls that remained outdoors during the entire winter and were thus affected by changes in the temperature, illumination, humidity, and other ambient physical factors.

The maximum overcooling temperature was controlled by a microthermosensor introduced into the bee body or into the one in contact with it and which was placed into the refrigerator camera with the temperature – 17.0±0.5 °C. As sensors thermo resistors CT 3-19 (made in Russia) were used, and the surface temperature was measured by sensors with platinum sensitive element (701-102BAB-B00, firm «Honeywell»). The state of the wintering bees adipose body located on the inner side of heart camera tergite was defined by the 5 point scale A. Maurizio (1958).

The infrared radiation of bees was recorded by IRTIS-2000 and ThermaCam SC3000 thermal imagers; the former is characterized by a high temperature resolution (0.05°C), and the latter, by a high spatial resolution (1.1 mrad). These devices allowed us to measure the body temperature of individual bees and determine the distribution of heat flows over the entire surface occupied by bees with high precision. The comb-holding frames of the hive, hinged together on one side like leaves in a book, were drawn apart for thermal imaging. The exposed comb surface was covered with infrared-translucent polyethylene film to protect the bees from direct influence of the environment. A single recording event took no more than 5-6 s. The radiance temperature recorded by the imager was calculated into the actual physical value with a correction factor of 1.01.

Results

The maximum overcooling temperature of bees to the middle of wintering decreases and to the end of the wintering increases. Adipose body reaches its maximum development in the first half of wintering and degrades to its end (see the table). Thus, from the beginning to the end of the winter (from December to February) the adipose body was decreased in average in 1.19 times (*P* is not lower than 0.95), and to the middle of the spring (the second decade of April) – in 3.39 times (P>0.99).

To the higher development of the adipose body relatively low and medium values of freezing temperature in the first half of wintering (in December – January) correspond. However, the decreasing of the freezing temperature from December to the end of January doesn't coincide with statistically significant changes of the adipose body state. The relation between the decreasing of the average MOT values and the adipose body degradation is traced clearly only during the period of the wintering completion (see the table).

Table. Dynamics of MOT and adipose body state of bees from the beginning to the end their wintering

Dates of sample selection	MOT,° C			Adipose body, in points		
	M±m	*Lim*	*Cv,%*	*M±m*	*lim*	*Cv,%*
15 Dec	-8,92± 0,312	-4,1..-15,4	34	4,49± 0,061	3-5	13
30 Dec.	-10,66± 0,091	-2,9..-15,6	24	4,45± 0,039	3-5	12
27 Jan.	-11,09± 0,422	-3,2..-14,6	25	4,51± 0,086	3-5	13
17 Feb.	-8,09± 0,172	-3,5..-13,2	22	3,80 ±0,077	2-5	21
5 March	-8,85± 0,300	-5,6..-11,2	15	2,20± 0,108	1-3	21
12 April	-7,42± 0,246	-5,4…-9,6	16	1,34± 0,099	1-2	37

The bees that were selected simultaneously in different wintering periods didn't reveal any explicit relation between the adipose body development and the freezing temperature. For example, the correlation coefficient between the studied physiological state indexes of the bees taken in December was equal to 0.05±0.04, in January – 0.12±0.08, in February – 0.07± 0.12 and in March – April – 0.02±0.01.

Wintering bees have adapted themselves to spend energy resources economically. This is achieved primarily by means of temporal stopping of queen reproduction function and location of the bee majority outside the thermal centre in the regions of comb spaces with temperatures inhibiting metabolic processes. The state of the bees forming the periphery of their accumulation (except the zone over the thermal centre) that can not allow them to penetrate into the warm part of the nest independently. This penetration becomes possible and occurs after forming of layers by the bees leaving the thermal centre. Migration of satiated and warmed-up bees out of the centre is stimulated by unfavorable physical conditions, for example, by the warmth that is formed under high CO_2 concentration.

To the most important ethological adaptations of bees to wintering one can refer the acquisition by them of an adaptation mechanism of inter-nest thermoregulation. In autumn winter cold period stimulates the bee aggregation around the maximum warming-up zone (heat centre). The phase alternation of a relative rest and active heat generation states essentially destabilizes the temperature in different bee accumulation zones. Their disorderly (chaotic) migration within the accumulation clusters causes local air mixing that generates the heat flow dynamics.

External temperature has the greatest influence on thermal centre warming-up at the beginning of wintering changing its temperature within relatively wide limits. For example, in November-December temperature fluctuations in the thermal centre of the cluster having about 20 thousand bees reached 2.5…2.7 °C, and in February-March – 1.0…1.6 °C. Accordingly the correlation coefficient between the external temperature varying within the limits – 17.1…+11.2 °C and head centre warming-up decreased from -0.81±0.10 to 0.13±0.04. But during the whole wintering its value between the thermal centre temperature and the periphery of bee accumulation settled above was kept at the level of 0.82±0.11 (P>0.95).

The relation between the external and maximum inter-nest temperatures depends on the amount of bees in the colony. In colonies with relatively small amount of bees (≈10 thousand) strong negative correlation between the analyzed temperatures is observed only at the beginning of wintering period in initial phases of adaptation to the autumn temperature falls.

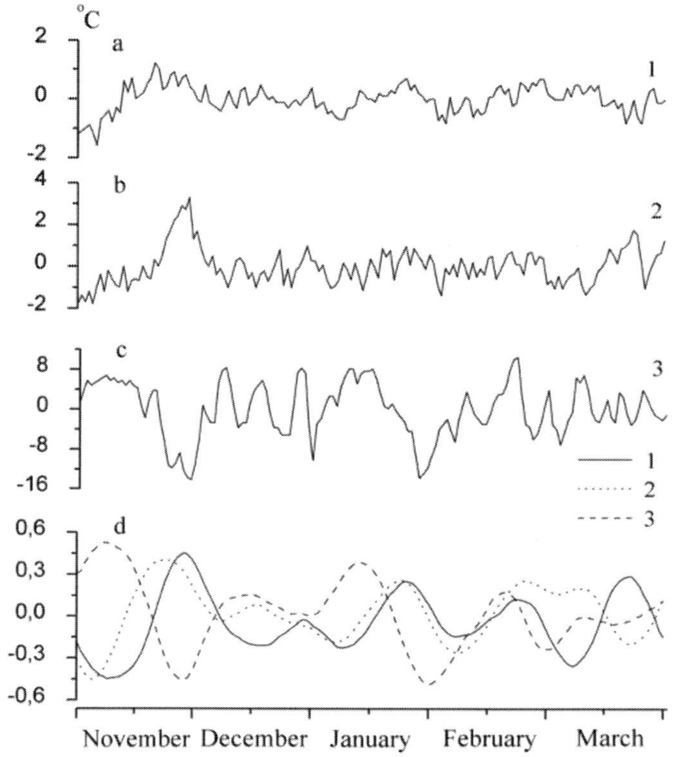

Figure 1. Temperature fluctuations in the heat centre for two clusters having different numbers of bees during five wintering months: (a) – in the cluster having about 10 thousand bees; (b) - in the cluster about 18 thousand bees; (c) – external temperature; d – the normalized values of large-scale temperature fluctuations calculated by the fractional moment method (Toboev and Nigmatullin, 2007).

Starting from December the correlation coefficient value is decreased to – 0.36±0.12. In bee clusters that contained in two times more the number of bees (≈18 thousand) the temperature fluctuations reached in November-December the level of 0.8…2.7 °C, and the correlation coefficient was conserved on the level -0.84±0.09. In the second half of wintering (starting from January) the bee clusters responded weakly by thermo regulating activity changings even to sufficient external temperature fluctuations (see Fig. 1).

Irrespective in number the bees react similarly to the cooling factor by decreasing the accumulation surface (occupied volume) thanks to consolidation that reaches its maximum in the zone of contact with cold air. In this case the heat flow formed in various directions differs due to forming heterogeneity of the cluster accumulation. The total heat flow from its surface depends on the effective outer surface, i.e. on the effective heat area. It is always less than the true geometrical surface, and its value sufficiently depends on the external temperature and location of bees in different inter-comb spaces and the distance from the bee-entrance hole. The latter one provides the basic heat exchange between the inter-hive space and outer surroundings.

The cooling process that stimulates the cold bee aggregation phenomenon in inter-comb spaces influences differently on their density and heat flow distribution. Temperature gradients in the direction from the most warmed-up zone (thermal centre) to upper and low boundaries of bee accumulation differ significantly in their absolute values and variabilities.

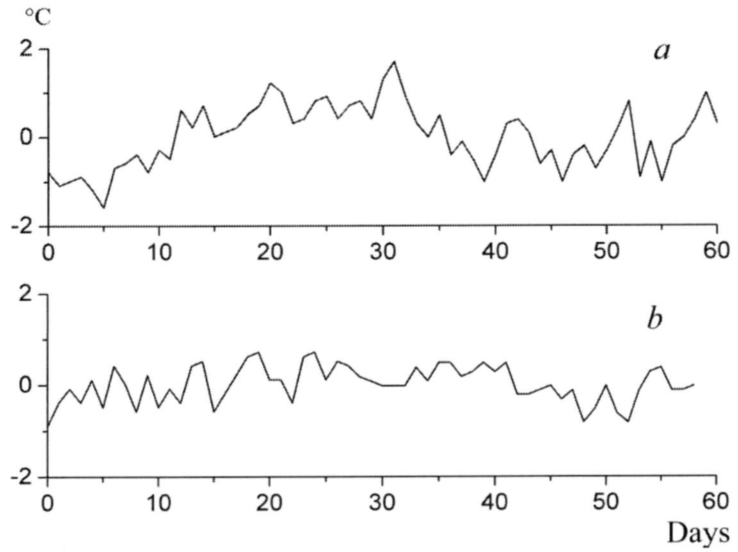

Figure 2. Fluctuations of the daily mean temperatures of the heat center at the beginning (a) and the end (b) of the wintering period.

To the upper side they change insignificantly not exceeding 12 °C during wintering and to the lower side they reach 24…28 °C with constant changing. On the same level under -20…-18 °C the difference is maintained between the external temperature and warming-up of bee bodies forming the lower part of their accumulation.

The drop of temperature during the autumn and winter period stimulates aggregation of bees around the zone of maximum heating (the "heat center"). The alternating phases of rest and active heat production considerably destabilize the temperature in the different zones of the bee clusters. The heat flow pattern is also affected by irregular migrations of bees inside the cluster, which cause local agitation of air.

The external temperature at the beginning of the wintering period has the greatest influence on that of the heat center, modifying it within a relatively broad range (Fig. 2). For example, the fluctuations of temperature in the heat center of a colony comprising about 20 thousand bees reached 2.5….2.7°C in November-December, whereas the corresponding values in February-March were 1…1.6°C. The coefficient of correlation between the external temperature, varying from -17 to +11°C, and that of the heat center correspondingly decreased from -0.81 ± 0.10 to 0.13 ± 0.04. During the entire wintering, the average coefficient of correlation between the temperature of the heat center and the peripheral area of the bee cluster above the center was 0.82 ± 0.11 ($P > 0.95$).

The relation between the external temperature fluctuations and the maximum temperature inside the nest depends on the number of bees in the colony. In relatively small colonies (about 10 thousand ind.), a strong negative correlation between these values could be observed only at the beginning of the wintering period, i.e., during the initial phases of adaptation to the drop of temperature in autumn. Starting from December, the coefficient of correlation decreases to -0.36 ± 0.12. In colonies comprising about twice as many bees (≈ 18 thousand ind.), the fluctuations of temperature in November-December reached 0.8…2.7°C whereas the average coefficient of correlation was -0.84 ± 0.09. During the second half of the wintering period (since January) the colonies showed only a weak response even to considerable fluctuations of external temperature.

Regardless of the colony size, the bees react to cooling by increasing the density of their cluster, so as to reduce its surface area; the maximum density is observed in the zone of direct contact with cold air. The intensity of heat flows in different directions varies due to the non-uniform structure of the cluster. The total heat flux from the cluster surface depends on its effective area, i.e., area through which heat may escape. The effective area is always smaller than the total (geometric) surface area and depends on the external temperature, localization of bees in the inter-comb spaces, and the distance from the hive entrance. The main heat exchange between the inner space of the hive and the environment takes place through the entrance.

At certain values of the external temperature (the optimum zone: from -3.0 to +8.8°C), the rate of heat transfer through 1 cm2 of the cluster surface only weakly depends on the number of bees. In the colonies numbering from 8.2 to 19.5 thousand bees, the heat flow reaches 2.05…3.21 x 10-3 W/cm2. The intensity of the heat flow depends on its direction: within the range specified, the flow towards the upper surface is higher, while that towards the lower surface is lower. When the external temperature is within the optimum zone, no significant changes of temperature in the heat center are observed, since the heat load (production) from each bee remains practically the same (on average 6…8 x 10-4 W).

A drop of external temperature below the optimum values is usually accompanied by warming of the heat center. This process depends on the number of bees in the cluster: smaller clusters have a less stable thermal regime and a greater heat loss, compensated for by

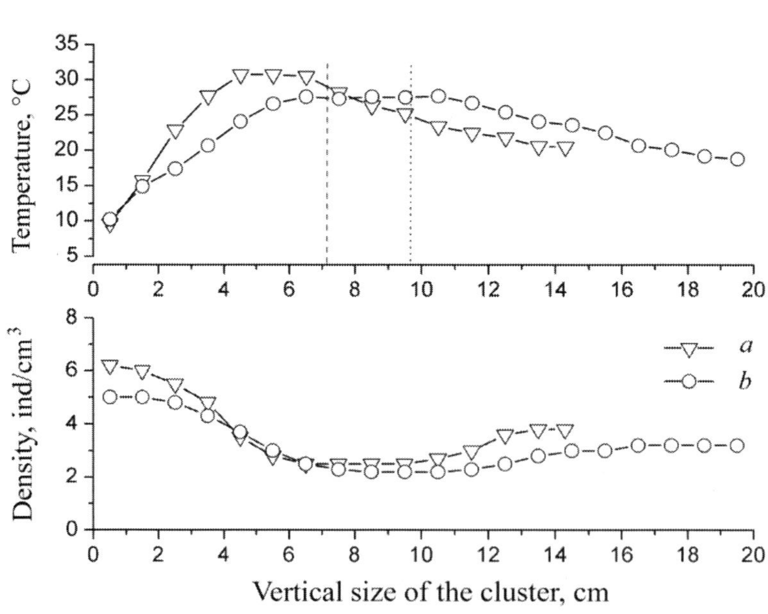

Figure 3. Temperature and density of bee clusters in the central inter-comb space of a hive occupied by a colony weighing 1.8 ± 0.2 kg, at the external temperatures of -15.7°C (a) and +2.7°C (b). The geometric centers of bee clusters are marked with vertical dotted lines; the aggregation densities are determined from the surface temperature patterns.

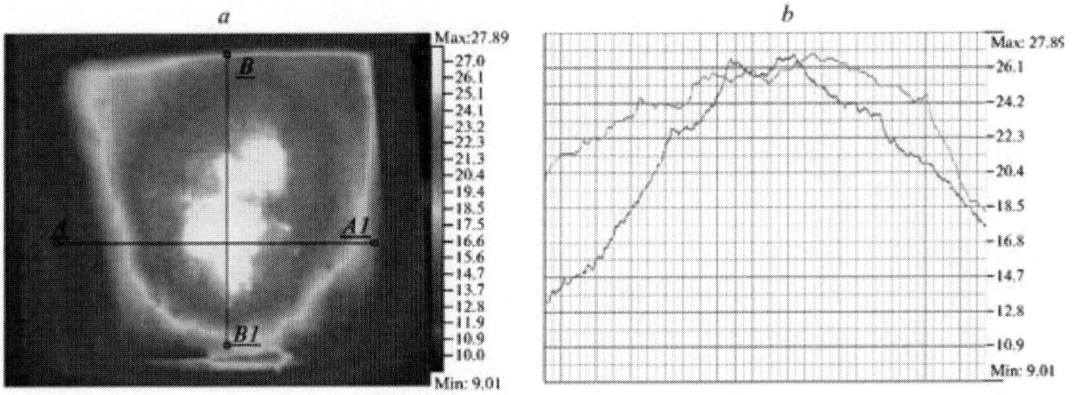

Figure 4. The distribution of heat in the central inter-comb cluster of bees at -2°C: a thermal image (a) and temperature profiles along the.

extra food consumption. In particular, within the optimum temperature zone, the mean basal metabolic rate of a colony comprising about 8 thousand bees is 3.28 W (47 cal/min). As a result of cooling to -15°C during 120 min, the total heat production of the colony increases by 2.6 times, whereas the heat load from a single bee increases from 6 x 10-4 to 1.1 x 10-3 W.

The external temperature and the total number of bees in the nest determine the distribution of bees in the adjacent inter-comb spaces at the beginning of wintering. The highest density of aggregation is observed in the lower part of each cluster, which is determined by the cooling pattern. The density decreases towards the heat center, or the zone of the maximum warming. The density of bees above the heat center is relatively low because this area is heated by the ascending heat flux (Fig. 3).

The temperature gradients directed from the heat center to the upper and lower boundaries of the bee cluster differ considerably in their absolute values and variability. The upward gradients vary insignificantly and do not exceed 12°C during wintering, whereas the downwards ones are variable and may reach 24…28°C. The difference between the external temperature and that of bees in the lower part of the cluster is maintained at the same level at -20…-18°C (Fig. 4).

The temperature of different body parts of bees on the periphery of the cluster depends on their position and the external temperature. At external temperatures varying from -1 to -13°C, the thoracic region is usually the warmest; its mean temperature differs from those of the head and abdominal regions by fractions of a degree centigrade. However, the temperature of the head sometimes reaches or even slightly exceeds that of the thorax. The mean temperature of the abdominal region is always lower than that of the thorax.

A drop of the external temperature stimulates warming of the bees which form the cluster periphery. This process is the most intense in the lower part of the cluster, which normally undergoes the most intensive cooling. For example, as the external temperature drops from -1 to -21°C, the temperature of the bee thorax is increased on average by 3.0°C, that of the head, by 2.6°C, and that of the abdomen, by 2.4°C ($P \geq 0.99$); the corresponding values for the upper part of the cluster are 1.5, 0.6, and 0.9°C, respectively ($P \geq 0.95$) (Fig. 5). The maximum temperature of bees in the lower part of the nest is 27.9°C, and the minimum, -7.5°C at the external temperature -13°C.

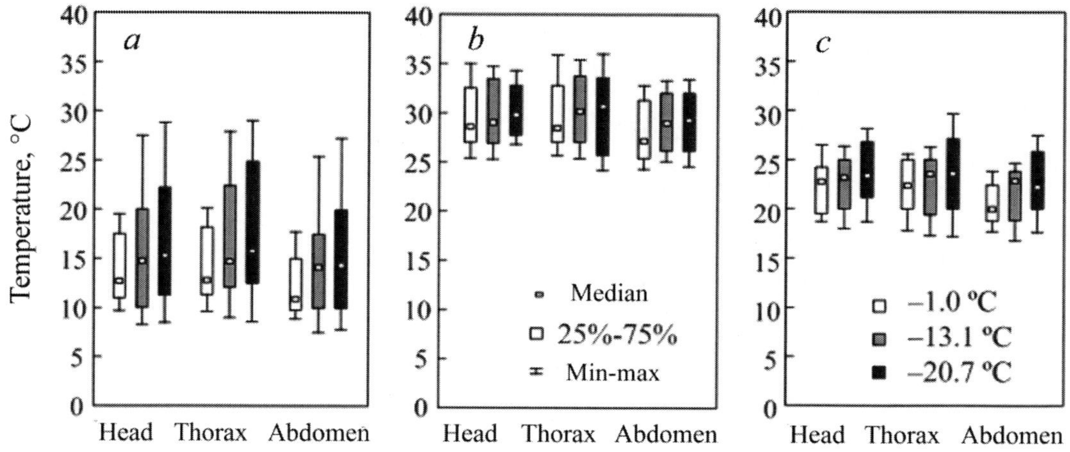

Figure 5. Temperature of different body parts of bees (°C) located in the periphery and in the center of inter-comb bee clusters, at different external temperatures: the lower surface (a), center (b), and upper surface (c) of the cluster.

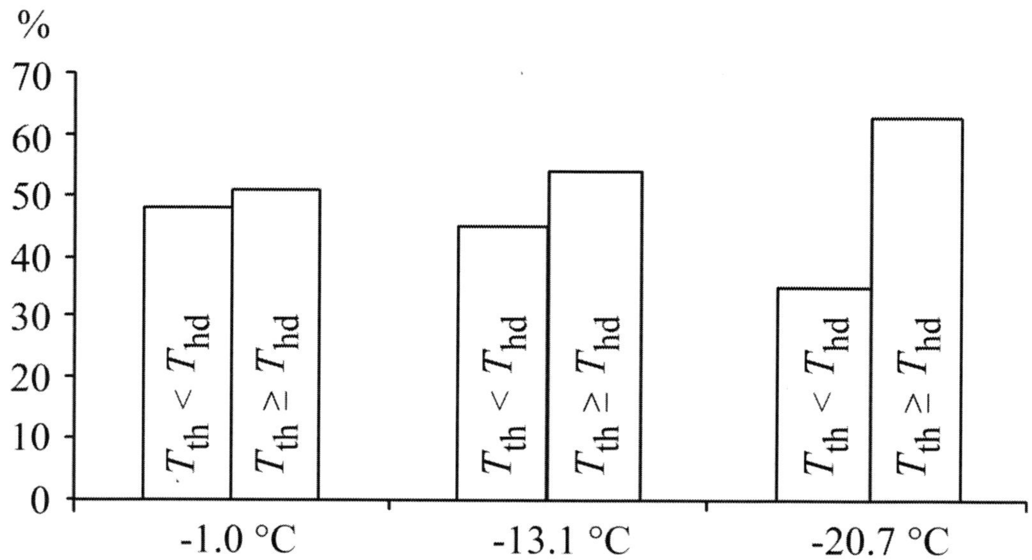

Figure 6. The fraction (%) of bees with different temperatures of the thorax (Tth) and head (Thd) on the lower surface of the cluster, at different external temperatures.

As the external temperature decreases, the fraction of bees in which the thorax is warmer than the head increases (Fig. 6). The maximum difference between the temperatures of the head and thorax increases as the external temperature drops. This difference reaches 5.7°C at -1°C, 8.5°C at -13°C, and 10.3°C at -21°C, whereas the fraction of bees in which the thorax is warmer than the head by no less than 2°C is 25.4, 36.0, and 47.2%, respectively. At the same time, as the external temperature drops, the fraction of bees

with body temperatures below the chill coma threshold also decreases: the average values are 52.4 ± 4.9% at -1°C, 45.7 ± 4.1% at -13°C, and 32.1 ± 3.3% at -21°C.

The body temperature of bees located close to the heat center varies within a smaller range. In particular, at the external temperature of -20°C, the temperature of the head varied from 26.8 to 34.3°C in different bees (n = 158), that of the thorax, from 24.2 to 36.0, and that of the abdomen, from 24.6 to 33.4°C, whereas the mean values were 29.8, 30.7, and 29.3°C, respectively (Fig. 4).

DISCUSSION

A honey bee has not acquired adapting itself to cold climate or, possibly, has lost with development of sociality the ability to long and deep vital activity interruption that is typical for many species of individual insects. Obviously, it is biologically not expedient for a honey bee presenting of six species of a present "Apis" genus bees to diapauses in unfavorable periods of life because they themselves and their forage supplies are needed in protection from many animals. Therefore, bees were forced to develop some adaptation mechanisms giving at least some part of their cluster a possibility to be always in an active state that is necessary and sufficient for their safety and protection of forage supplies. This fact involves the necessity to maintain the temperature on the level making possible rapid activation of locomotions (local motions). However, inter-nest temperature regulation by bees is not connected with the presence of some coordinating centre in their accumulations. Depending on the colony state and external conditions every bee is included into a hierarchy that implies a systematically-organized order; it is involved dynamically in inter-communication processes that are determined by levels of structuring. From this conclusion it follows that self-organization in the life of the bee colony is accompanied by the changing of the structural symmetry and increasing/decreasing of the number of coordinating external stimulators.

The bees respond to cooling by increasing the cluster density (stimulated by cold air), which decreases the surface area of the cluster. The total heat flow from the cluster surface depends on its effective area, i.e., area through which heat escapes. The effective area is always smaller than the geometric surface area, which in turn depends on the external temperature. The direction and intensity of heat flows are affected by non-uniform density and shape of bee clusters.

The index of heat circulation, i.e., the ratio of the external and internal temperature gradients, is mostly determined by the intensity of heat flow from the inner zones towards the periphery, which changes under the influence of fluctuations of external temperature. A drop of external temperature has the greatest effect on the index of heat circulation near the lower cluster surface, and the smallest effect on that near the upper surface. An increase in the circulation index towards the lower surface reduces its heat insulation properties. Therefore the bees forming the lower part of the cluster are subject to the strongest influence of the temperature factor, which stimulates them to respond effectively to temperature fluctuations. A decrease in temperature stimulates active heat production by the bees which prevents them from entering chill coma. In the chilled state the bees remain viable for several days only, but may be reactivated when heated (Eskov, 1995), usually by contact with warm bees migrating from the heat center into the cluster periphery.

Heat transfer within the cluster occurs through the bodies of bees and partly by convection of air through the spaces between the individuals, whereas the surface heat loss results from free convection and thermal radiation. In view of this, the heat flux directed from the heat center towards the surface of the bee cluster can be described by the equation:

$$dq_{int} = h_{int} \cdot f \quad \text{grad}\, T \cdot \vec{n} \;, T_{int}^4 - T_{sf}^4, T_{int} - T_{sf} \;, \tag{1}$$

where f is a function taking into account all the ways of heat transfer inside the cluster (its arguments describe heat transfer by conductivity, radiation, and convection, respectively), T_{int} and T_{sf} are the temperatures in some point inside the cluster and on its elementary surface (dS), and **n** is the external surface normal. The value $I_{int} = 1/h_{int}$ may be determined as dynamic resistance to heat flux, considering the non-stationary nature of thermal processes inside the bee cluster. This value describes the heat insulation properties of bees forming the periphery of the cluster. The greater the dynamic resistance Iint, the less heat will pass through the elementary surface dS at the gradient of 1°C.

If we assume that specific heat fluxes resulting from conductivity, convection, and radiation have a linear relationship with the difference T_{int}-T_{sf}, then the value of I_{int} can be estimated by the temperature differences and gradients calculated from the thermal images. Considering these assumptions, the dynamic heat insulation towards the cluster periphery will be determined by the formula:

$$I_{int} = a \cdot \frac{I_a}{r}, \tag{2}$$

In this formula, r is the index of heat circulation determined by the ratio:

$$r = \frac{T_{sf} - T_{ext}}{T_{int} - T_{sf}}, \tag{3}$$

where I_{int} is the temperature of air around the bee cluster, I_a is the heat insulation capacity of air for the same elementary surface dS.

In the absence of active evaporation from the surface of the bee cluster, the ratio of the internal and external heat fluxes (a) may be considered a constant close to 1. If we assume that the heat insulation properties of air do not change, i.e., that I_a is a constant, the change of r will characterize the heat insulation capacity of bees located on the cluster periphery. Therefore, an increase in the heat circulation index leads to a less efficient insulation and, correspondingly, to a more extensive loss of heat from the cluster surface (Fig. 6).

Although the bees are genetically programmed to respond to fluctuations of external temperature, their aggregation under the influence of cooling is not related to activity of any special coordinating center. The bees in the cluster act as simple independent units with a small number of effective degrees of freedom; the states of these units vary depending on their position and functional capacities. The most active bees contribute to warming of the cluster but spend much energy in the process. The minimal energy expenditures are observed

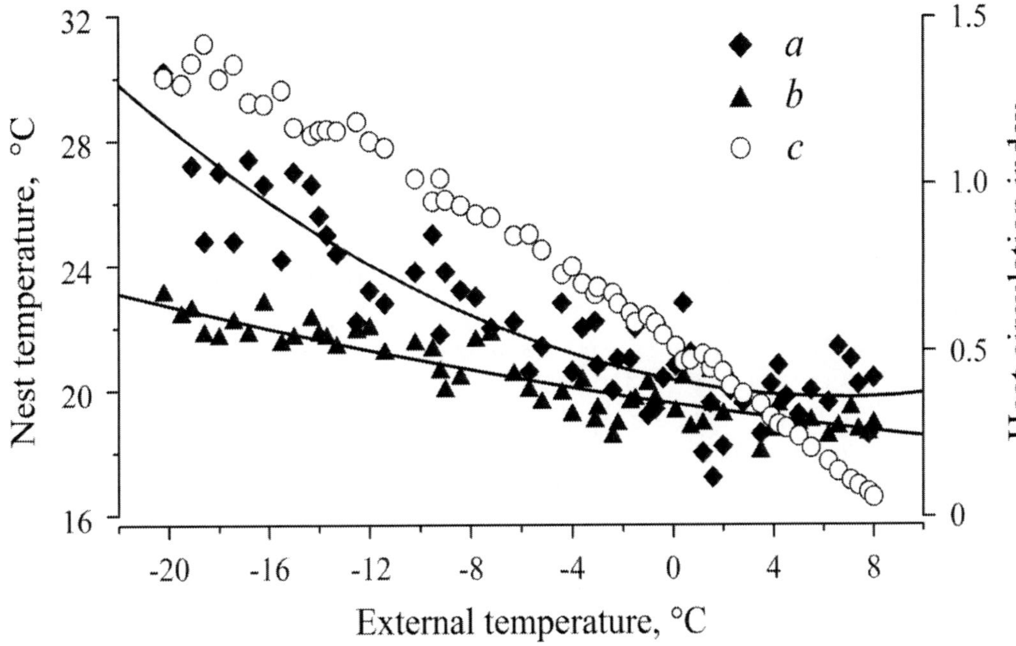

Figure 7. Dependence of temperature in the central (a) and upper (b) parts of the nest and the heat circulation index (c) on the external temperature.

in some passive bees located on the periphery of the parts of cluster subject to the most intensive cooling. These bees show relatively low body temperatures and are likely to be in a weak chill coma, suppressing their metabolism.

Conservation of energy resources of the wintering colony is facilitated not only by efficient mechanisms of heat loss regulation, but also by low metabolic activity in some of the bees. The same purpose is served by interruption of oviposition by the queen, at least from the beginning to the middle of the wintering period (Eskov, 2003). Using these adaptive mechanisms, the bee colony uses a relatively small amount of food during wintering (about 20 times less than during the spring and summer period). Food consumption is reduced to the minimum in the wintering bees at the external temperatures of -3.0...+8.8°C. Under such conditions, the bees maintain a favorable thermal regime by aggregating into regular spatio-temporal structures which stabilize the temperature of the heat center.

Changes in the variance of temperatures from the beginning to the end of wintering may be interpreted as an indication of changes in the adaptive potentials of the colony. A high variance at the beginning of the wintering period corresponds to a greater number of possible states and a higher probability of change, resulting in a relatively high adaptive potential. This potential is reduced as the fluctuations of temperatures inside the nest decrease by the end of wintering. This means that the wintering bee colony behaves as an open system and shows a trend to autonomism, manifesting itself in an increased independence from random temperature fluctuations.

REFERENCES

[1] Eskov, E.K., *The Microclimate of a Bee Hive* (Rossel'k-hozizdat, Moscow, 1983) [in Russian].

[2] Eskov, E.K., *Ethology of the Honeybee* (Kolos, Moscow, 1992) [in Russian].

[3] Eskov, E.K., *Ecology of the Honeybee* (Russkoe Slovo, Ryazan, 1995) [in Russian].

[4] Eskov, E.K., "Individual and Social Wintering Adaptations in Bees," Uspekhi Sovrem. Biol. ***123*** (4), 383-390 (2003).

[5] Eskov, E.K. and Toboev, V.L., "Heating of Wintering Bee Bodies Related to External Air Temperature," Zool. Zh. ***88*** (1), 125-126 (2009) [Entomol. Rev. ***89*** (1), 111-112 *(2009)]*.

[6] Maurizio A. Feeding bees with pollen and vital processes of a honey bee// New developments in bee-keeping. M.: State publishing of agricultural literature. 1958.

[7] Rybochkin, A.F. and Pustovalov, S.N., "Determining the Position of Bee Clusters by the Thermal Pattern," in Beekeeping in Cold and Temperate Climates. Proc. of the 2nd Int. Conf., Pskov, 17-18 March, 2007 *(Moscow, 2007), pp. 51-54.*

[8] Toboev V.A., Nigmatullin R.R. New method of the statistical treatment of temporal series: the investigation of the collective behavior of social insects based on their thermal regulation activity// *Non-linear world*. 2007. V.5. № 4. pp. 183-193.

[9] Schmidt-Nielsen, K., Animal Physiology: Adaptation and Environment (Cambridge Univ. Press, Cambridge, 1979; Mir, Moscow, 1982) [in Russian].

[10] Simpson, J., "Nest Climate Regulation in Honeybee Colonies," Science 133, 1327-1333 (1961).

[11] Southwick, E.E., "Allometric Relations, Metabolism and Heat Conductance in Clusters of Honey Bees at Cool Temperatures," J. Comp. Physiol. B **156**, 143-149 (1985).

[12] Zhdanova, T.S., "The Thermal Regime of a Wintering Bee Colony," Pchelovodstvo, No. 10, 36-40 (1958).

INDEX

A

B

C

F

G

H

I

J

K

L

M

N

O

P

Q

R

S

T

U

V

W

Y

Z